T0135760

Blowups, slicings and permutation groups in combinatorial topology

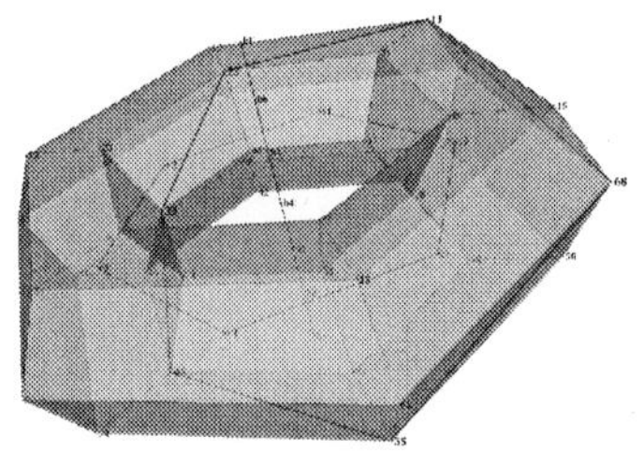

Von der Fakultät Mathematik und Physik der Universität Stuttgart zur Erlangung der Würde eines Doktors der Naturwissenschaften (Dr. rer. nat.) genehmigte Abhandlung

Vorgelegt von

Dipl. Math. Jonathan Spreer

aus Stuttgart – Bad Cannstatt

Hauptberichter: Prof. Dr. Wolfgang Kühnel

1. Mitberichter: Prof. Dr. Wolfgang Kimmerle
2. Mitberichter: Prof. Edward Swartz, Ph.D. (Cornell University)
3. Mitberichter: Priv.-Doz. Dr. Frank Lutz (TU Berlin)

Tag der mündlichen Prüfung: 29. Juni 2011

Institut für Geometrie und Topologie der Universität Stuttgart

2011

Bibliografische Information der Deutschen Nationalbibliothek

Die Deutsche Nationalbibliothek verzeichnet diese Publikation in der Deutschen Nationalbibliografie; detaillierte bibliografische Daten sind im Internet über http://dnb.d-nb.de abrufbar.

ISBN 978-3-8325-2983-3

Logos Verlag Berlin GmbH
Comeniushof, Gubener Str. 47,
10243 Berlin
Tel.: +49 (0)30 42 85 10 90
Fax: +49 (0)30 42 85 10 92
INTERNET: http://www.logos-verlag.de

Acknowledgements

First of all, I would like to thank my supervisor *Prof. Dr. Wolfgang Kühnel* for his constant support and for providing me with perfect conditions to pursue the research for my dissertation. Also, I want to thank him for countless enlightening discussions which many times led to new results – and sometimes away from wrong conclusions.

My thanks also go to *Prof. Dr. Wolfgang Kimmerle* for numerous fruitful conversations about the theory of permutation groups and representation theory, and for pointing me out to some interesting algebraic problems. Additionally, I would like to thank *Priv.-Doz. Dr. Frank Lutz* for many helpful explanations about computational topology and for writing some of the most valuable lines of GAP-code in the world. Furthermore, my thanks go to *Prof. Edward Swartz, Ph.D.* for many useful remarks and comments on a preliminary version of this work.

I want to thank my colleague and companion *Dr. Felix Effenberger* for working next door during my time at the University of Stuttgart. The projects we worked on together meant a welcome change of life to me in my otherwise slightly solitary mathematical life.

Also, I want to thank all members of the Institute of Geometry and Topology, especially *Prof. Dr. Michael Eisermann*, *Prof. Dr. Markus Stroppel* and *Prof. Dr. Eberhard Teufel* for plenty of interesting talks and discussions during our weekly workshop at the institute.

Finally, my thanks are going to my family and friends for their never-ending appreciation and encouragement which had a significant impact on the positive outcome of this work.

This thesis was funded in main parts by the German Research Foundation (DFG) under the grant Ku 1203/5-2 and Ku 1203/5-3.

Contents

List of Figures

List of Tables

Notation and Symbols

$\lvert M\rvert$	cardinality of a set M
$\mathbb{N} = \{1, 2, 3, \ldots\}$	set of natural numbers
$\mathbb{N}_0 = \{0, 1, 2, 3, \ldots\}$	set of natural numbers including 0
$\mathbb{Z}$	set of integer numbers (also as ring)
$\mathbb{Q}, \mathbb{R}, \mathbb{C}$	sets of rational, real and complex numbers (also as fields)
$\mathbb{F}$	arbitrary field
$\mathbb{F}_q$	finite field with q elements
E^n	n-dimensional Euclidean space
E_n	unity matrix of dimension $n \times n$
$\mathrm{Tr}(A)$	trace of a quadratic matrix A
$\gcd(k_1, \ldots, k_m)$	greatest common divisor of the integers $k_1, \ldots, k_m$
$a \mid d$	a divides b
$a \equiv b(p)$	a is congruent to b modulo p
Δ^d	d-simplex
$\langle v_1, \ldots, v_{d+1}\rangle$	d-simplex with vertices $v_1, \ldots, v_{d+1}$
$\langle a_1, \ldots, a_n\rangle$	closed path of length n passing through the vertices $a_1, \ldots, a_n$
P	convex polytope P
C	simplicial complex C
M	(combinatorial) manifold / pseudomanifold
S	triangulated surface / "pinch-point" surface

S_g	surface of genus g
β^d	d-dimensional cross polytope
C^d	d-cube
$C_d(n)$	cyclic d-polytope with n vertices
S^d	(combinatorial) d-dimensional sphere
$\mathbb{T}^d$	(combinatorial) d-dimensional torus
$\mathbb{R}P^d$	d-dimensional real projective space
$\mathbb{C}P^d$	d-dimensional complex projective space (of real dimension $2d$)
$\mathbb{K}^2$	(combinatorial) Klein bottle
K^d	(combinatorial) Kummer variety
$K3$	$K3$ surface
$L(p,q)$	Lens space defined by torus knot (p,q)
Δ	facet of a simplicial complex
δ	face of a simplicial complex
e	edge of a simplicial complex
v	vertex of a simplicial complex
∂C	boundary of a simplicial complex C
$\dim(C)$	dimension of a simplicial complex C
$f(C), f$	f-vector of a simplicial complex C
$f_k(C), f_k$	kth entry of the f-vector of a simplicial complex C
$h(C), h$	h-vector of a simplicial complex C
$\mathrm{g}(C), \mathrm{g}$	g-vector of a simplicial complex C
$g(S), g(M)$	genus of a surface S, Heegaard genus of a 3-manifold M
$\chi(C), \chi$	Euler characteristic of a simplicial complex C
$\mathrm{Aut}(C)$	automorphism group of a simplicial complex C
$\lvert C \rvert$	support of a simplicial complex C
$V(C)$	vertex set of a simplicial complex C
$\mathrm{skel}_k(C)$	k-skeleton of the simplicial complex C
$\mathrm{lk}_C(\delta)$	link of the face δ in a simplicial complex C

$\mathrm{st}_C(\delta)$	star of the face δ in a simplicial complex C
$\mathrm{span}_C(X)$	span of the vertex set $X \subseteq V(C)$ in a simplicial complex C
$M_1 \# M_2$	connected sum of two combinatorial manifolds M_1, M_2
$M^{\# k}$	k-fold connected sum of a combinatorial manifold M
$C_1 \times C_2$	cartesian product of two simplicial complexes C_1, C_2
$C_1 \ltimes C_2$	twisted product of two simplicial complexes C_1, C_2
$C_1 \star C_2$	join of two simplicial complexes C_1, C_2
$C_1 \cong C_2$	combinatorial isomorphy of two simplicial complexes C_1, C_2
$V_1 \dot{\cup} V_2$	disjoint union of two sets V_1, V_2
(C, o_σ)	ordered simplicial complex
∂_k	kth simplicial boundary operator
$H_*(C;G)$	homology groups of the simplicial complex C with coefficients in G
$H_*(C)$	(integral) homology groups of a simplicial complex C with coefficients in $\mathbb{Z}$
d^k	kth simplicial coboundary operator
$H^*(C;G)$	cohomology groups of a simplicial complex C with coefficients in G
$H^*(C)$	(integral) cohomology groups of the simplicial complex C with coefficients in $\mathbb{Z}$
$\beta_i(C;G)$	ith Betti number of a simplicial complex C with respect to the group of coefficients G
$\beta_i(C)$	ith Betti number of a simplicial complex C with respect to the group of coefficients $\mathbb{Z}$
$\tau_i(C;G)$	torsion part of the ith homology group of a simplicial complex C with respect to the group of coefficients $\mathbb{Z}$
$a \smile b$	cup-product of two cocylces a and b
Q_M	intersection form of a manifold M
$\mathrm{MCG}(M)$	mapping class group of a manifold M

$S_{(V_1,V_2)}$	slicing defined by the vertex partition $V(C) = V_1 \dot\cup V_2$ of a simplicial complex C
T	a triangle of a slicing
Q	a quadrilateral of a slicing
n	number of vertices of a slicing (or a simplicial complex)
e	number of edges of a slicing
t	number of triangles of a slicing
q	number of quadrilaterals of a slicing
$f(S_{(V_1,V_2)})$	f-vector of a slicing, $f(S_{(V_1,V_2)}) := (n,e,t,q)$
$f : M \to \mathbb{R}$	rsl-function, not to confuse with the f-vector of a simplicial complex / slicing
$\mu_i(f;\mathbb{F})$	number of critial points of index i of the rsl function f with respect to the field $\mathbb{F}$
$\kappa_{(\gamma,\delta)}(C)$	bistellar move defined by face γ and non-face δ of a simplicial complex C
G	group
$G_1 \cong G_2$	isomorphy of two groups G_1, G_2
$G_1 \times G_2$	direct product of the groups G_1, G_2
$G_1 \rtimes G_2$	semidirect product of the groups G_1, G_2
$\mathrm{stab}_G\langle\alpha\rangle$	pointwise stabilizer of the set α under the action of a group G
$\mathrm{stab}_G(\alpha)$	stabilizer of α (as a set) under the action of a group G
$(a_1,\dots,a_n)$	cycle of length n (an element of a permutation group)
σ_n	symmetric group of order $n!$
$\mathbb{Z}_n$	cyclic group of order n
D_n	dihedral group of order $2n$ (please note the difference to the GAP-notation)
$\binom{n}{k}$	binomial coefficient "n choose k"
$\{q_1,\dots,q_{d-1}\}$	Schläfli symbol of a regular d-polytope
$\partial D = (a_0 : \dots : a_m)$	difference cycle (interpreted as a set of m-simplices)

$(c_1, \ldots, c_{m+1})_k$ — orbit of length k of the set $\{c_1, \ldots, c_{m+1}\}$ under the action of a suitable permutation group (interpreted as a set of m-simplices)

See also Table 5.4 in Chapter 5 for further details about the GAP-notation for groups that will be used in certain cases.

Zusammenfassung

Die vorliegende Dissertation beschäftigt sich mit Fragestellungen der kombinatorischen Topologie. Dieses Teilgebiet der Topologie vereinigt die Methoden der geometrischen Topologie, der Kombinatorik, der algebraischen Topologie und der Polytoptheorie.

Während die Beschreibung von topologischen Mannigfaltigkeiten mittels Triangulierungen (also durch eine Zerlegung der Mannigfaltigkeit in Simplizes) zu den ersten Methoden gehörte, um topologische Invarianten zu berechnen (auch heute noch wird dieser Zugang zur Topologie in vielen Lehrbüchern gewählt, siehe zum Beispiel [8]), erwies sich diese Methode bald als zu umständlich im Falle größerer Objekte. Neue Zerlegungsmethoden wurden entwickelt, in der zwar die Verklebeoperation der einzelnen geometrischen Primitive der Komplexe nicht mehr intuitiv möglich war, dieser Verlust jedoch mit einer signifikanten Einsparung von Komplexität einherging (siehe zum Beispiel [141] von Whitehead über CW-Komplexe).

Doch was für die Berechnung von topologischen Invarianten von Hand einen Gewinn darstellte, erwies sich als ungeeignet für den Computer. Und als dann ab den 1980er Jahren computergestützte Methoden an Bedeutung gewannen, spielten folglich auch die Simpizialkomplexe wieder eine wichtigere Rolle – die kombinatorische Topologie begann sich rasant zu entwickeln (siehe [81, 83, 92, 15, 50, 28, 49, 122, 34, 91, 41, 40, 132, 95, 39, 128] für eine Reihe von Programmen für Simplizialkomplexe und Publikationen, welche Computerbeweise über Simplizialkomplexe enthalten).

In diesem Zusammenhang ist es von besonderer Bedeutung, dass die Lösung des Königsberger-Brücken-Problems [44] einerseits als die wohl erste Arbeit auf dem Gebiet der Topologie, andererseits als die Begründung der Graphentheorie (und damit eines Teilgebiets der Kombinatorik) gilt. In gewisser Weise greift damit die kombinatorische Topologie den kanonischen Zugang zur Topologie mit den modernen Methoden der heutigen Mathematik neu auf.

Ziel dieser Arbeit ist es, den Werkzeugkasten der kombinatorischen Topologie um einige gängige Konzepte verwandter Disziplinen zu erweitern. Im Besonderen handelt es sich hierbei um i) algebraische Aufblasungen von Singularitäten in Pseudomannigfaltigkeiten, ein Konzept der algebraischen Geometrie (vgl. [123]), ii) Normalflächen in topologischen 3-Mannigfaltigkeiten, ein Hilfsmittel aus der Theorie der 3-Mannigfaltigkeiten (siehe [71, 55, 137]) und iii) die Analyse und Erzeugung von Simplizialkomplexen über das Studium von Permutationsgruppen (hier wurde bereits schon im Kontext der kombinatorischen Topologie gearbeitet, siehe [83, 67, 92, 37, 91]). Konkret sollen also mathematische Konzepte und Theorien für topologische Mannigfaltigkeiten so erweitert werden, dass sie auf *kombinatorische Mannigfaltigkeiten*, also Simplizialkomplexe, deren *Eckenlinks* (der Rand einer simplizialen Umgebung einer Ecke) Sphären mit stückweise linearer Standardstruktur sind, angewendet werden können.

In Kapitel 2 wird die kombinatorische Umsetzung des Konzepts einer algebraischen Aufblasung zusammen mit der ersten kombinatorischen Triangulierung einer $K3$-Fläche vorgestellt, bei der die kanonische stückweise lineare Struktur der Mannigfaltigkeit durch den Konstruktionsprozess sichergestellt ist. Das Verfahren eignet sich zudem zur Auflösung weiterer Arten von Singularitäten von kombinatorischen 4-Pseudomannigfaltigkeiten.

Bei algebraischen Aufblasungen von *gewöhnlichen Doppelpunkten* einer algebraischen Varietät (also Punkte, deren Umgebungen Bällen nach Antipodenidentifizierungen gleichen) entfernt man eine Umgebung des Punktes und ersetzt diese durch die Menge aller Richtungen durch diesen Punkt. In gewisser Weise wird also die singuläre Umgebung des Punktes durch einen entsprechend berandeten projektiven Raum ersetzt – und somit die Singularität aufgelöst. So kann aus der singulären 4-dimensionalen Kummervarietät K^4 (siehe [75]) eine nichtsinguläre $K3$-Fläche konstruiert werden (siehe [28, 123]).

In Kapitel 3 wird eine kombinatorische Version der Normalflächentheorie entwickelt. Diese wurde ursprünglich von Kneser [71] und Haken [55] entworfen, um das Homöomorphieproblem für 3-Mannigfaltigkeiten zu lösen. Normalflächen von 3-Mannigfaltigkeiten sind eingebettete Flächen, welche eine gegebene Zellzerlegung der umgebenden 3-Mannigfaltigkeit respektieren. Für kombinatorische 3-Mannigfaltigkeiten bedeutet dies, dass die Fläche keine Ecke der Triangulierung trifft und jedes Tetraeder von der Fläche in einem Dreieck oder

Viereck geschnitten wird.

Im Zentrum der neuen kombinatorischen Theorie stehen die sogenannten *diskreten Normalflächen*. Diese sind Normalflächen, die i) nur auf kombinatorischen 3-Mannigfaltigkeiten definiert sind und ii) pro Tetraeder nur ein Dreieck oder Viereck enthalten dürfen. Von besonderem Interesse ist der Fall, in dem die diskrete Normalfläche gleichzeitig das Urbild einer polyedrischen Morsefunktion auf der Triangulierung ist. In diesem Fall wird auch von *Schnitten* der Mannigfaltigkeit gesprochen. Für die neu entwickelte Theorie wird zudem eine Reihe von Anwendungen vorgestellt. So lassen sich mittels diskreter Normalflächen triangulierte 3-Mannigfaltigkeiten visualisieren, kombinatorische Aussagen über den Zusammenhang des Geschlechts der diskreten Normalflächen und der Anzahl ihrer Vierecke treffen, die Einbettungen schwach nachbarschaftlicher Karten als normale Untermengen von triangulierten 3-Mannigfaltigkeiten klassifizieren und zahlreiche neue Serien von hochsymmetrischen 3-dimensionalen Sphärenbündeln konstruieren.

In Kapitel 4 und 5 wird die vorhandene Theorie über Permutationsgruppen als Automorphismengruppen von Simplizialkomplexen erweitert und angewendet auf *transitive* oder *mehrfach transitive* kombinatorische Mannigfaltigkeiten. Dabei heißt eine kombinatorische Mannigfaltigkeit (mehrfach) transitiv, wenn ihre Automorphismengruppe (mehrfach) transitiv ist.

Die Ergebnisse der neuerlichen Betrachtung hochsymmetrischer Komplexe enthalten unter anderem i) die Partition der Dreiecke des d-dimensionalen Kreuzpolytops in triangulierte Tori, Sphären und Klein'sche Flaschen mit transitiver Symmetrie (Kapitel 4.3), ii) die Konstruktion von mehreren unendlichen Serien von kombinatorischen 3-Mannigfaltigkeiten mit zyklischer Symmetrie (Kapitel 4.5) und iii) die Beschreibung einer unendlichen Serie von kombinatorischen 4-Pseudomannigfaltigkeiten mit 2-fach transitiver Symmetrie (Kapitel 5.8).

Viele der vorgestellten Resultate bedienen sich der Hilfe von Algorithmen. Alle hierfür entwickelten Programme und konstruierten Serien von Simplizialkomplexen wurden in das hierfür entwickelte Softwarepaket simpcomp [40, 41, 42], welches auf dem Computeralgebrasystem GAP aufbaut, integriert und so für jedermann zugänglich gemacht. Die Software wurde in Zusammenarbeit mit Felix Effenberger [38] entwickelt. Sie stellt eine Vielzahl ausführlich dokumentierter Funktionen zur Verfügung und ist in ihren Grundzügen in Anhang A beschrieben.

Preface

Topology is one of the major areas of research in pure mathematics. However, unlike geometry or analysis, the term is largely unknown to non-mathematicians. Topology deals with classes of objects which can be transformed into each other by continuous deformations. Such classes are called *topological objects*. The main difference to the related field of geometry is the focus on qualitative rather than quantitative properties. The exact definition is as follows:

Topology (from the Greek topos, "place", and logos, "study") *The branch of pure mathematics that deals only with the properties of a figure X that hold for every figure into which X can be transformed with a one-to-one correspondence that is continuous in both directions.*[1]

There are many different aspects of topology. Set-theoretic topology discusses the term proximity generalizing the concept of metric spaces, algebraic topology makes use of algebraic structures to define invariants of topological objects, geometric topology focuses on the study of topological properties of geometric objects, etc.

This work deals with an even more specialized field of topology that lies in the intersection of geometric topology, combinatorics, algebraic topology and polytope theory: Namely, we study topological manifolds where the manifold is given as a union of a finite number of geometric primitives, which form a so-called abstract simplicial complex (see Figure 0.0.1). In this way, the topological information is encoded in a form that can be understood by computers. Therefore, problems in the field are often approached with the help of algorithms. We will call this area of topology *combinatorial topology* or *discrete geometric topology.* If one wants to emphasize the algorithmic aspect, the terms *discrete computational topology* or *experimental topology* can be used.

[1] George A. Miller, WordNet - About Us, Princeton University, http://wordnet.princeton.edu, 2009

The solution to the Königsberg Bridge problem [44] by Leonhard Euler in 1736 is considered to be the first important work in the field of topology (see Figure 0.0.2 for an illustration of the problem by Euler in [44]).

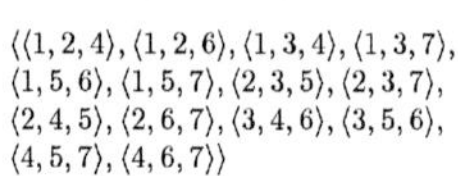

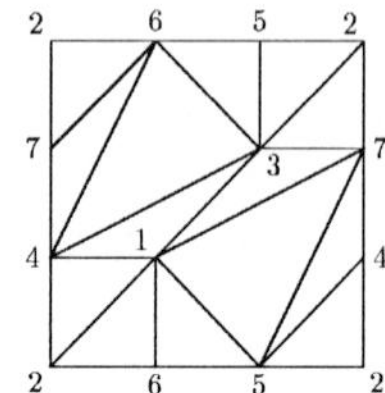

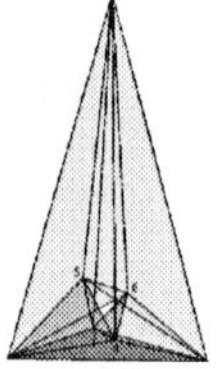

Figure 0.0.1: The 7-vertex Möbius torus as an abstract simplicial complex (on the left), a surface with identified boundaries (in the middle), an embedding into Euclidian 3-space (on the right: Császár's torus, see [31], figure taken from [94]).

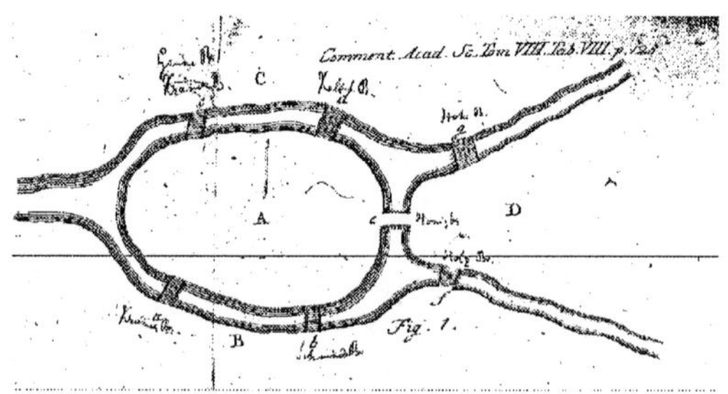

Figure 0.0.2: The seven bridges of Königsberg. Figure taken from Euler's original publication in 1736 [44].

It is remarkable that this work also stands for the first contribution to graph theory and thus combinatorics. Hence, combinatorial methods have been associated to topological questions from the very beginning. Although the interest in combinatorial descriptions of manifolds declined with the upcoming of more efficient cell-decompositions (regarding the number of cells) of manifolds, the beginning of the digital era caused a regain of interest in simplicial constructions due to the compatibility of simplicial complexes to any kind of algorithmic computation.

Thus, combinatorial topology can be regarded as a fundamental as well as a modern branch of topology. However, in contrast to most other areas of research in pure mathematics that were developed recently, combinatorial topology is built on elementary observations and does not require a lot of advanced results and theories to get started. In my opinion it is this characteristic that reveals the beauty of the field.

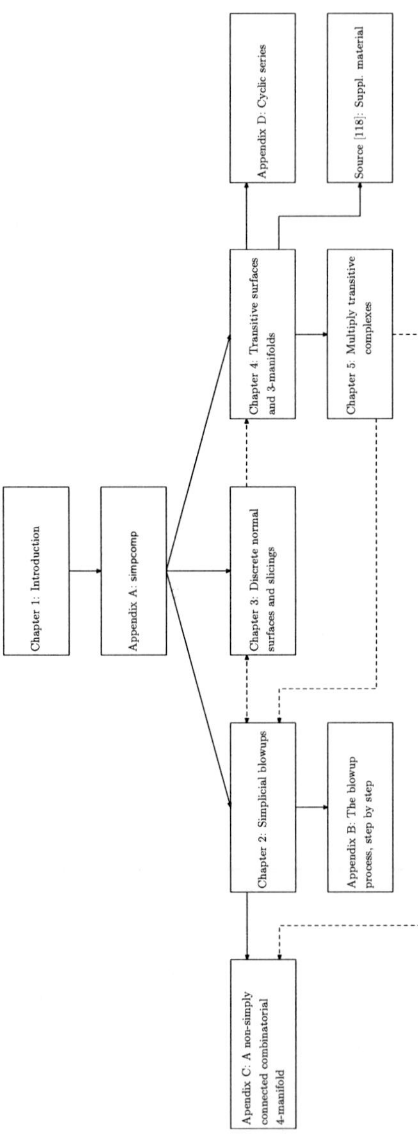

Figure 0.0.3: Logical dependence of the chapters of this work. For a more detailed outline of the results as well as the content of this work see Section 1.8.

CHAPTER 1

Introduction and results

1.1 Polytopes and simplicial complexes

Simplicial complexes are sets of simplices which in turn are convex polytopes. Hence, it is natural to treat a simplicial complex as an object in polytope theory. For an introduction to the field of convex polytopes we refer to the books of Coxeter [30], Grünbaum [53] and Ziegler [142] where the terminology of the latter will be used in the following.

The *(geometric) d-simplex* Δ^d, i. e. the convex hull of $d+1$ points in $\mathbb{R}^d$ in general position, is the smallest of all convex d-polytopes (with respect to the number of vertices). It is one of the three infinite series of regular polytopes, the other two being the *d-dimensional cube* C^d and its dual, the *d-dimensional cross polytope* β^d. The d-simplex Δ^d can be distinguished from all other convex d-polytopes by one remarkable property: any $k+1$ points, $0 \leq k \leq d$, of Δ^d span a k-dimensional face of Δ^d, which is a k-simplex. It follows that any $(d+1)$-tuple of distinct vertices can be interpreted as an *(abstract) d-simplex* which, a priori, does not need any embedding into an ambient space. It is because of this property that in the following simplices will be the only convex polytopes of further interest.

A finite set of simplices is called *simplicial complex* if the intersection of any two simplices is either empty or a common face of both. Note that any intersecting set of the vertices of two simplices is either empty or defines such a face. In this way a gluing of any two simplices of a simplicial complex is given by their common (abstract) vertices. Thus, every finite set of tuples of vertices (abstract simplices) defines a simplicial complex. Such a complex has always a geometric realization by identifying its n labels with the vertices of a geometric $(n-1)$-simplex. We call such a set an *abstract simplicial complex* (see Figure 0.0.1 for the

7-vertex Möbius torus as an abstract simplicial complex, a geometric realization in the plane and an embedding into 3-space (Császár's torus)).

An abstract simplicial complex C is already determined by its maximal simplices, i. e. all simplices that do not occur as a proper face of another simplex. The maximum over all dimensions of all simplices is called the *dimension* $\dim(C)$ of the simplicial complex. The empty complex $\varnothing$ is of dimension -1. If all maximal simplices are of the same dimension, the simplicial complex is called *pure*. In this case, the maximal simplices are referred to as *facets*. The 0-dimensional simplices are called *vertices*, the set of all vertices of C is usually denoted by V, the 1-dimensional faces are called *edges*. The set of all k-dimensional faces $\mathrm{skel}_k(C)$ is called the *k-skeleton* of C. The graded set of all k-skeletons $0 \leq k \leq \dim(C)$ is referred to as *face lattice*. It carries a natural partial ordering given by inclusion – the so-called *face poset*. The number of k-dimensional faces of a simplicial complex C is denoted by $f_k(C) = |\mathrm{skel}_k(C)|$. The vector

$$f(C) = (n := f_0(C), f_1 := f_1(C), \ldots, f_d := f_d(C))$$

is said to be the *face vector* or the *f-vector* of C. Whenever necessary, we additionally set $f_{-1}(C) := 1$ and $f_{d+1}(C) := 1$. This is motivated by the face poset of C.

If $v_1, \ldots, v_{k+1}$ are vertices that span a face of C we will denote it by $\langle v_1, \ldots, v_{k+1} \rangle$. If δ is a face of C we will write $\delta \in C$. If v is one of its vertices we write $v \in \delta$. For two disjoint simplices γ and δ we will write $\gamma \star \delta$ for the simplex spanned by all vertices of γ and δ. For simplicial complexes B and C we define the *join* to be the following complex

$$B \star C := \{ \gamma \star \delta \mid \gamma \in B, \delta \in C \}.$$

Two abstract simplicial complexes are said to be *combinatorially isomorphic* if there exists a bijective map on the sets of vertices transforming one complex into the other. The set of permutations on the set of vertices, which does not change the complex as a whole, forms a group, the so-called *automorphism group* of the complex. An abstract simplicial complex is called *k-transitive* if its automorphism group is k-transitive. If $k > 1$ it is also referred to as *multiply transitive*, if $k = 1$ it is often just called *transitive*. If G is a permutation group acting on a set of vertices V and $\langle v_0, \ldots, v_d \rangle$ is a d-simplex, $v_i \in V$, then by $G\langle v_0, \ldots, v_d \rangle$ we mean the G-orbit given by the action of G on subsets of V of order $d+1$ interpreted as a

d-dimensional abstract simplicial complex. If G is given by the context we will sometimes also write $(v_0, \ldots, v_d)_m := G\langle v_0, \ldots, v_d \rangle$, where m indicates the number of elements in the orbit.

The underlying set $|C|$ of a simplicial complex C is referred to as the *support* of C. Now let M be a topological manifold. If there exist a homeomorphism $h : |C| \to M$, the complex C is called a *triangulated manifold* or, more specific, a *triangulation of* M.

The subcomplex

$$\operatorname{star}_C(v) := \{ \delta \mid v \in \delta \in C \}$$

is called the *star* of a vertex v in C. The complex

$$\operatorname{lk}_C(v) := \{ \delta \setminus v \mid v \in \delta \in C \}$$

is called the *link* of v in C. By $\delta \setminus v$ we mean the face spanned by all vertices $w \neq v$ of δ. Analogously, we can define the link or the star of a k-face. An n-dimensional simplicial complex M is said to be the standard *piecewise linear sphere* (or *PL sphere* for short) if its underlying set is PL homeomorphic to the d-sphere with standard PL structure. M is a *combinatorial manifold* if all vertex links are $(d-1)$-dimensional standard PL spheres. M is called a *combinatorial pseudomanifold* if the link of every vertex is a combinatorial manifold.

Remark 1.1.1. Note that for $d \leq 4$ the definition of a triangulated manifold and a combinatorial manifold are equivalent. However, in dimension $d \geq 5$ not all triangulated manifolds have a PL structure whereas this is always true for combinatorial manifolds. For $d \geq 5$ there are triangulated spheres with $d + 13$ vertices that are not combinatorial manifolds. See [92] for a construction of the so-called Edward's sphere which is a triangulated 5-sphere with no PL structure (cf. Section 1.5 and Appendix A.4.7). Since in combinatorial topology the objects of interest are always piecewise linear, the notion of a combinatorial manifold is necessary.

1.2 Simplicial (co-)homology

A very powerful family of invariants in combinatorial topology is given by the concept of simplicial homology and cohomology. To define the (co-)homology groups of a simplicial complex we first need some basic terminology.

Definition 1.2.1 (Ordered simplicial complex). An *ordered simplicial complex* (C, o_σ) is a simplicial complex C on the set of vertices $V = \{v_1, \dots, v_n\}$ together with an ordering $o_\sigma : v_{\sigma(1)} < \dots < v_{\sigma(n)}$ on V, $\sigma \in \sigma_n$. Its faces are called *ordered simplices* or *oriented simplices* where two oriented simplices $\delta \in (C, o_\sigma)$, $\gamma \in (C, o_\tau)$ are considered equivalent whenever they share the same vertices and σ and τ differ at most by an even permutation. Thus, for any face of a simplicial complex there exist exactly two orientations which will be denoted by $\pm\delta \in (C, o_\sigma)$ for a fixed ordering o_σ.

In what follows we will just write C for an ordered simplicial complex (C, o_σ) whenever no particular ordering is needed.

Definition 1.2.2 (Simplicial chain complex). Let $C := (C, o_\sigma)$ be a d-dimensional ordered simplicial complex, $\text{skel}_k(C)$ its ordered k-skeleton and G an Abelian group. A *k-chain of C with coefficients in G* is the formal linear combination

$$\sum_{\delta \in \text{skel}_k(C)} \lambda_\delta \, \delta,$$

where $\lambda_\delta \in G$ and G is called the *group of coefficients*. The set of all k-chains together with componentwise addition of the coefficients forms an Abelian group, called the *group of k-chains of C with coefficients in G*, written $C_k(C; G)$. It is by definition trivial for all $k < 0$ and $k > d$. The sequence of groups $C_\star(C; G) := (C_k(C; G))_{k \in \mathbb{Z}}$ is called the *chain complex of C with coefficients in G*.

Definition 1.2.3 (Simplicial boundary operator). Let $C_\star(C; G)$ be the chain complex of an ordered simplicial complex C with coefficients in G. The homomorphism

$$\partial_k : C_k(C; G) \to C_{k-1}(C; G); \quad \langle v_0, \dots, v_k \rangle \mapsto \sum_{i=0}^{k} (-1)^i \langle v_0, \dots, \hat{v}_i, \dots, v_k \rangle$$

is called *kth simplicial boundary operator of C*. Here $\langle v_0, \dots, \hat{v}_i, \dots, v_k \rangle$ denotes the $(k-1)$-simplex $\langle v_0, \dots, v_{i-1}, v_{i+1}, \dots, v_k \rangle$.

Definition 1.2.4 (Cycles and boundaries). Let $C_\star(C; G)$ be the chain complex of an ordered simplicial complex C with coefficients in G and ∂_k, $k \in \mathbb{Z}$, its simplicial boundary operators. Then

$$Z_k(C; G) := \ker(\partial_k) = \{c \in C_k(C; G) \mid \partial_k(c) = 0\}$$

is called the group of *k-cycles of C* and

$$B_k(C;G) := \text{im}(\partial_{k+1}) = \{\partial_{k+1}(d) \mid d \in C_{k+1}(C;G)\}$$

is called the group of *k-boundaries of C*. Both $Z_k(C;G)$ and $B_k(C;G)$ are subgroups of the free Abelian group $C_k(C;G)$ and thus again free.

A basic calculation shows that $\partial_k \circ \partial_{k+1} = 0$ holds, what implies $B_k(C;G) \subset Z_k(C;G)$. This motivates the following definition.

Definition 1.2.5. Let C be an ordered simplicial complex and G an Abelian group, then

$$H_k(C;G) := Z_k(C;G)/B_k(C;G)$$

is called the *kth homology group of C with coefficients in G*.

In most applications we will set $G = \mathbb{Z}$ and just write $H_k(C) := H_k(C;\mathbb{Z})$. $H_k(C)$ is referred to as the *kth integral homology group of C*. By construction, $H_k(C)$ is a finitely generated $\mathbb{Z}$-module and thus can be written in the form

$$H_k(C;G) = \mathbb{Z}^{\beta_k} \oplus \mathbb{Z}_{t_1} \oplus \ldots \oplus \mathbb{Z}_{t_m}$$

where $t_i \leq t_{i+1}$, $1 \leq i \leq m-1$, $\beta_k \in \mathbb{N}_0$.

It is well known that the simplicial homology groups $H_\star = (H_k(C;G))_{k\in\mathbb{Z}}$ form a topological invariant of C which, in particular, is independent of the ordering chosen for C. In fact, two simplicial complexes already have the same homology groups whenever their supports are homotopy equivalent.

The rank $\beta_k(C;G)$ of the free part of $H_k(C;G)$ is called the *kth Betti number of C with respect to G*. Using the Betti numbers we are able to define another well known topological invariant.

Definition 1.2.6 (Euler characteristic)**.** Let C be an ordered d-dimensional simplicial complex and G an Abelian group, then the alternating sum

$$\chi(C) := \sum_{i=0}^{d} (-1)^i \beta_i(C;G)$$

is called the *Euler characteristic of* C. By virtue of the universal coefficient theorem its value is not only independent of the ordering of C but also of the choice of G.

Clearly, the Betti numbers of a simplicial complex encode more information than the Euler characteristic. Thus, at this point the definition of the Euler characteristic does not seem to be useful at all. However, due to the following theorem, there is a way to compute the Euler characteristic of a simplicial complex without knowing its simplicial homology groups.

Theorem 1.2.7 (Euler-Poincaré formula). *Let* C *be a* d*-dimensional simplicial complex and* G *an Abelian group, then the following equality holds:*

$$\sum_{i=0}^{d}(-1)^i\beta_i(C;G) = \chi(C) = \sum_{i=0}^{d}(-1)^i f_i(C).$$

Since in many cases the f-vector of a simplicial complex is computed much faster than the homology groups, the Euler characteristic of a simplicial complex turns out to be a very powerful tool to distinguish combinatorial manifolds.

A d-dimensional pure simplicial complex C is said to fulfill the *weak pseudomanifold property* if any $(d-1)$-dimensional face occurs in at most two facets. For such complexes we can define the following.

Definition 1.2.8 (Boundary and orientability). Let C be an ordered d-dimensional pure simplicial complex that fulfills the weak pseudomanifold property and let $\Delta := \sum_{\delta \in C} \delta \in C_d(C;\mathbb{Z}_2)$ be the formal sum of all facets of C with coefficients in $\mathbb{Z}_2$. Then the set of $(d-1)$-faces of C with non-zero coefficients in $\partial_d(\Delta) \in C_{d-1}(C;\mathbb{Z}_2)$ is called the *boundary* ∂C *of* C. A weak pseudomanifold is called *closed* if $\partial_d(\Delta) = 0$.

If for a closed complex C there exists a d-chain $F := \sum_{\delta \in C} \epsilon_\delta \delta \in C_d(C;\mathbb{Z})$, $\epsilon_\delta \in \{\pm 1\}$, such that $\partial_d(F) = 0$, then C is said to be *orientable*. The set of all ϵ_δ is called an *orientation* of C and F is referred to as *fundamental cycle* which is, by definition, an element of the dth homology group of C.

Simplicial cohomology

The notion of simplicial cohomology groups of a simplicial complex C is closely related to the simplicial homology groups by duality. The elements of the cohomology groups are, roughly speaking, 1-forms (so-called *cochains*) on the chain complex of a simplicial complex. While cohomology theory supports more algebraic structure it has the disadvantage of being less geometrically intuitive.

Similar to homology theory we look at the *cochain complex* of an ordered simplicial complex C which is the set of all k-cochains $C^k(C;G) := \mathrm{Hom}(C_k(C;G);G)$. The *kth simplicial coboundary operator* $d^k : C^k(C;G) \to C^{k+1}(C;G)$ can be described best as the transposed of the simplicial boundary operator in matrix representation. Similar as in the case of the simplicial boundary operator we have $d^k \circ d^{k-1} = 0$ and $B^k(C;G) := \mathrm{im}(d^{k-1}) \subset \ker(d^k) =: Z^k(C;G)$ for the *coboundaries* and *cocycles* of C. The *kth cohomology group* of C is defined by

$$H^k(C;G) := Z^k(C;G)/B^k(C;G).$$

Since homology and cohomology are dual concepts it is not surprising that they often look very similar as the following statement will show.

Theorem 1.2.9 (Poincaré duality). *Let M be an orientable connected combinatorial d-dimensional manifold and G a group, $k \in \mathbb{Z}$. Then*

$$H_k(M;G) \cong H^{d-k}(M;G).$$

If M is not orientable then the statement just holds for $G \cong \mathbb{Z}_2$.

Using Theorem 1.2.9 it can be shown that for closed connected orientable triangulated d-manifolds M we have in fact

$$\beta_k(M;G) = \beta_{d-k}(M;G) \tag{1.2.1}$$

and $\tau_k(M;G) \cong \tau_{d-k-1}(M;G)$ for all groups G where β_k denotes the kth Betti number and τ_k is the torsion part of the kth homology group. Again, the assumption that M has to be orientable can be dropped if $G \cong \mathbb{Z}_2$. Thus, we have the following corollary.

Corollary 1.2.10. *Let M be an odd dimensional combinatorial manifold, then $\chi(M) = 0$.*

Proof. Just combine the Euler-Poincaré formula with equation (1.2.1) in the case $G \cong \mathbb{Z}_2$ and apply it to every connected component of M separately. □

Corollary 1.2.10 illustrates why a lot of combinatorial approaches to proofs depending on the Euler characteristic fail in odd dimensions.

The concept of cohomology will be a useful tool when examining simply connected combinatorial 4-manifolds and thus will be discussed in further detail in Section 1.5.

1.3 Lower and upper bound theorems

To illustrate how combinatoric arguments are in involved in combinatorial topology, let us first look at the following basic example.

Example 1.3.1. Let S be a *triangulated surface* (i. e. a 2-dimensional combinatorial manifold) with n vertices, e edges and t triangles. By the classification theorem for surfaces, its topological type is determined up to orientability by its Euler characteristic

$$\chi(S) = n - e + t.$$

Moreover, since S is a combinatorial manifold, every edge is contained in exactly two triangles, i. e. $3t = 2e$. It follows that

$$3(n - \chi(S)) = e \le \binom{n}{2},$$

which leads to the following lower bound on the number of vertices needed to triangulate S:

$$n \ge \frac{1}{2}(7 + \sqrt{49 - 24\chi(S)}). \tag{1.3.1}$$

Inequality (1.3.1) is known as Heawood's inequality as it was first conjectured in 1890 by Heawood (see [58]). It is closely related to the famous map color problem which was solved much later by Ringel and Youngs in [116].

Starting from the Heawood inequality, there is a whole family of theorems and conjectures giving upper and lower bounds to the number of faces of simplicial polytopes and other combinatorial manifolds. A lot of these statements are easier to prove if a combinatorial manifold M is of even dimension. This is due to the fact that the Euler characteristic of even dimensional manifolds contains essential topological information whereas $\chi(M) = 0$ always holds in the odd dimensional case (cf. Corollary 1.2.10).

In order to present further results efficiently, we need some more definitions: Clearly, the entries of the f-vector of a complex fulfill the upper bound $f_k \le \binom{n}{k+1}$, $0 \le k \le d$, for any d-dimensional complex on n vertices with equality for all k if and only if C is the d-simplex. If a simplicial complex C fulfills $f_{k-1} = \binom{n}{k}$, the complex is said to be *k-neighborly*. In particular, any simplicial complex is at least 1-neighborly by definition.

For any $0 \le k \le d+1$ we define

$$\sum_{i=0}^{d+1} h_i x^{d+1-i} = \sum_{i=0}^{d+1} f_{i-1}(x-1)^{d+1-i}$$

or equivalently

$$h_k = \sum_{i=-1}^{k-1} (-1)^{k-i-1} \binom{d-i}{k-i-1} f_i$$

and we call $h(C) = (h_0, \dots, h_{d+1})$ the *h-vector* of a d-dimensional simplicial complex C. In particular, we have $h_{d+1} = (-1)^{d+1}(1 - \chi(C))$. In addition, for any $0 \le j \le \frac{d}{2}$ we set $\mathbf{g}_j := h_{j+1} - h_j$ and refer to $\mathbf{g}(C) = (\mathbf{g}_0, \dots, \mathbf{g}_{\lfloor \frac{d}{2} \rfloor})$ as the *g-vector* of the complex.

As shown in Theorem 1.2.7, the alternating sum over the entries of the f-vector equals the Euler characteristic of C. However, if C is not an arbitrary simplicial complex, further restrictions to the f-vector do apply. For example, if a pure d-dimensional simplicial complex C fulfills the weak pseudomanifold property, then $(d+1)f_d = 2f_{d-1}$ holds. If, in addition, C is a combinatorial manifold, the link of any k-face, $0 \le k \le d$, has to be a combinatorial $(d-k-1)$-sphere with Euler characteristic $1 + (-1)^{d-k-1}$. These conditions can be combined to a linear system of $\lfloor \frac{d+2}{2} \rfloor$ equations of the form

$$\sum_{i=0}^{d} (-1)^i f_i - \chi(C) = 0 \tag{1.3.2}$$

$$\sum_{i=2j-1}^{d} (-1)^i \binom{i+1}{2j-1} f_i = 0 \qquad \text{for } 1 \le j \le \frac{d}{2} \text{ if } d \text{ is even} \tag{1.3.3}$$

$$\sum_{i=2j}^{d} (-1)^i \binom{i+1}{2j} f_i = 0 \qquad \text{for } 1 \le j \le \frac{d-1}{2} \text{ if } d \text{ is odd.} \tag{1.3.4}$$

The equations (1.3.2), (1.3.3), (1.3.4) are called *Dehn-Sommerville equations* for combinatorial d-manifolds. In particular, the f-vector of a combinatorial d-manifold is completely determined by only $\lfloor \frac{d+2}{2} \rfloor$ entries. In terms of the h-vector, the above equations transform into

$$h_j - h_{d+1-j} = (-1)^{d+1-j} \binom{d+1}{j} (\chi(M) - 2) \tag{1.3.5}$$

for $0 \leq j \leq \frac{d}{2}$ and d even, and

$$h_j - h_{d+1-j} = 0 \tag{1.3.6}$$

for $0 \leq j \leq \frac{d-1}{2}$ and d odd.

The Dehn-Sommerville equations were first stated for simplicial polytopes. The only adjustment that has to be made for the general case, is to replace $1 + (-1)^d$ by $\chi(C)$ (see [69, 117, 53, 103]). For a short proof of (1.3.2), (1.3.3) and (1.3.4) see [77].

Theorem 1.3.2 (Generalized Lower Bound Theorem). *Let P be a simplicial d-polytope. Then for $0 \leq j \leq \frac{d-1}{2}$ the following inequalities hold:*

$$\mathrm{g}_j \geq 0.$$

In terms of the f-vector this is equivalent to

$$\sum_{i=-1}^{j} (-1)^{j-i} \binom{d-i}{j-i} f_i \geq 0.$$

The generalized lower bound theorem was first conjectured 1971 by McMullen and Walkup in [104] and proved by Billera and Lee (the sufficiency part, [13], 1981) and by Stanley (the necessity part, [131], 1985). In 1992, McMullen himself gave a more geometric and less technical proof in [102].

Remark 1.3.3. Most upper and lower bound theorems were first stated for polytopes. Note here that the boundary of a d-polytope topologically is a $(d-1)$-dimensional sphere, more precisely a so-called *polytopal sphere.* In the course of time, most of the theorems presented in this section could be shown to also hold (in a modified form) for arbitrary triangulated spheres and, in even greater generality, for arbitrary combinatorial (or Eulerian) manifolds. As the focus in this work is on manifolds, for ease of notation we will usually fix the dimension of the manifold to be d instead of $d-1$, the latter being the historically motivated indexing of the dimensions in terms of polytope theory.

Theorem 1.3.4 (Lower Bound Conjecture). *Let M be a combinatorial d-manifold, $d \geq 3$ and let $\mathrm{g}_1 = h_2 - h_1 = f_1 - (d+1) f_0 + \binom{d+2}{2}$. Then we have*

$$\mathrm{g}_1 \geq \binom{d+2}{2} \beta_1(M; \mathbb{Q}), \tag{1.3.7}$$

or equivalently

$$f_1 \geq (d+1) f_0 + \binom{d+2}{2} (\beta_1(M;\mathbb{Q}) - 1)$$

where $\beta_1(M;\mathbb{Q})$ is the first Betti number of M with respect to $\mathbb{Q}$.

Theorem 1.3.4 was first conjectured by Kalai in 1987 (see [64]). He also gave some examples of combinatorial manifolds for which equality is attained in Inequality (1.3.7). The theorem was proved most recently by Novik and Swartz in [113] using an algebraic approach.

In the following, a d-dimensional simplicial complex is called an *Eulerian manifold* if the link of any k-face has the Euler characteristic of the $(d-k-1)$-sphere.

Theorem 1.3.5 (Upper Bound Theorem). *Let M be an Eulerian manifold with the Euler characteristic of a d-sphere and $C_{d+1}(n)$ the cyclic $(d+1)$-polytope, each with n vertices. Then*

$$f_i(M) \leq f_i(C_{d+1}(n))$$

holds for all $0 \leq i \leq d$.

Theorem 1.3.5 was first proposed for convex polytopes by Motzkin at the November meeting of the AMS in Evanston 1957 (see [110, p. 35]). In 1964, Klee conjectured Theorem 1.3.5 for Eulerian manifolds and proved it for sufficiently large values of n in [70]. In 1970, McMullen was able to give a complete proof of Motzkin's conjecture for polytopes (see [101]) and 5 years later, Stanley proved it for triangulated spheres (see [130]). In 1998, the theorem was proved by Novik for an even larger class of homology manifolds in [112]. Further generalisations were settled by Novik and Swartz in [113, Theorem 3.4 and 4.3].

The following theorem about the minimal number of vertices needed to combinatorially triangulate a $2k$-manifold was first conjectured by Kühnel and became to be known as Kühnel's conjecture.

Theorem 1.3.6 (Conjecture B in [77]). *Let M be a combinatorial $2k$-manifold with n vertices. Then*

$$\binom{n-k-2}{k+1} \geq (-1)^k \binom{2k+1}{k+1} (\chi(M) - 2). \tag{1.3.8}$$

It is closely related to Theorem 1.3.5 and was also proved by Novik and Swartz in [113]. Inequality 1.3.8 is also known as *generalized Heawood inequality*.

Another related statement about the minimal number of vertices of a combinatorial manifold M in the case that M is not a sphere was established by Brehm and Kühnel.

Theorem 1.3.7 (Theorem A in [20], Brehm-Kühnel-Bound)**.** *Let M be a d-dimensional combinatorial manifold with n vertices. Then the following implications hold:*

i) $n < 3\left\lceil \frac{d}{2} \right\rceil + 3 \Rightarrow$ *M is a combinatorial d-sphere.*

ii) $n = 3\left\lceil \frac{d}{2} \right\rceil + 3 \Rightarrow$ *Either M is a combinatorial d-sphere or $d \in \{2, 4, 8, 16\}$ and M is a "manifold like a projective plane" (see [36]).*

Remark 1.3.8. There are several examples of combinatorial manifolds different to the d-sphere for which $n = 3\left\lceil \frac{d}{2} \right\rceil + 3$ holds. Namely there exist

- a triangulation of the real projective plane $\mathbb{R}P^2$ with 6-vertices ($d = 2$),
- a triangulation of the complex projective plane $\mathbb{C}P^2$ with 9-vertices ($d = 4$, see [80]) and
- six combinatorially distinct triangulations of 8-manifolds with the homology of $\mathbb{H}P^2$ with 15-vertices (see [21] and [95]).

In the case $d = 16$, no example of a combinatorial manifold with 27 vertices is known. However, a number of combinatorial arguments suggest that there might exist a triangulation of a "manifold like the Cayley-plane" $\mathbf{Ca}P^2$ satisfying these properties (see [77, Section 4C]).

In the case $d = 3$, Lutz, Sulanke and Swartz recently established a number of lower bounds for f-vectors of combinatorial manifolds of specific topological types (see [97]).

Some of the results mentioned above were adapted to the case of centrally symmetric polytopes by Sparla in [124]. Note that a simplicial complex (polytope) is said to be *centrally symmetric* if its automorphism group contains a fixed point free involution, i. e. a permutation on the set of vertices of order 2 which does not fix any vertex and which does not contain any 2-cycle spanning an edge of the simplicial complex (boundary complex of the polytope).

1.4 3-manifolds and normal surfaces

Throughout this section every 3-manifold is assumed to be compact, closed and connected if not stated otherwise.

Manifolds of dimension 3 have been studied extensively during the last century. The main problem in the field has been the attempt to classify all 3-manifolds. Although the problem is far from being solved as of today, a lot of progress has been made and a solution seems not completely hopeless. For a more thorough introduction to the field of 3-manifolds and knot theory see the books by Saveliev [120], Lickorish [90], Thurston [136] and the introduction of [51] by Gompf and Stipsicz where further references are given.

A first step towards a classification was the attempt to decompose arbitrary 3-manifolds into "simpler" pieces: If a 3-manifold is not homeomorphic to the 3-sphere it is referred to as *non-trivial.* A 3-manifold M is called *irreducible* if any embedded 2-sphere $S \subset M$ bounds a 3-ball, it is called *prime* if it cannot be represented as the *connected sum* $M = M_1 \# M_2$ (i. e. the manifold obtained by removing a 3-ball from both M_1 and M_2 and gluing the resulting boundaries together along an orientation reversing homeomorphism) of two non-trivial 3-manifolds. Every irreducible 3-manifold is prime. If, on the other hand, a 3-manifold M is prime then either M is irreducible or a bundle over the circle.

In 1929, Kneser proved the existence of a decomposition of any 3-manifold into a connected sum of prime manifolds. For the proof he developed a theory dealing with certain properly embedded surfaces in 3-manifolds which he called *normal surfaces* (see [71, p. 256]). However, it was not until 1962 that Milnor was able to prove the following theorem.

Theorem 1.4.1 (Prime decomposition theorem for 3-manifolds, Theorem 1 in [107]). *Every non-trivial compact 3-manifold M is isomorphic to a connected sum $P_1 \# \ldots \# P_k$ of prime manifolds. The summands P_i are uniquely determined up to order and isomorphism.*

Any compact, connected, orientable 3-manifold is triangulable. This was first proved by Moise in 1952 in [109] and by Bing in 1957 in [14]. In addition, every compact topological 3-manifold admits a smooth or a PL structure. Thus, any 3-manifold can be realized as a combinatorial manifold (cf. Remark 1.1.1).

The fact that every 3-manifold M admits a triangulation makes it easy to show that every 3-manifold can be decomposed into two *handlebodies* X and Y (i. e. thickened graphs or, equivalently, a 3-ball with a certain number of solid cylinders glued to its boundary in a handle-like manner) with common boundary (for a prove see [90, Lemma 12.12]). Such a decomposition $M = X \cup Y$ is called a *Heegaard splitting* [90, Definition 12.11]. A handlebody is completely determined by its boundary which is an oriented surface. Thus, every handlebody X can uniquely be described by the genus of its boundary ∂X. This number is called the *genus* of X, written $g(X)$. If $M = X \cup Y$ like above, then $g(X) = g(Y)$ and this number is called the *genus of the Heegaard splitting*. Since every 3-manifold admits a Heegaard splitting of finite genus, there is a splitting of minimal genus in the sense that any surface of lower genus embedded into M cannot bound two handlebodies. Such a splitting is called a *minimal Heegaard splitting*, its genus is an invariant of M which is referred to as *Heegaard genus $g(M)$ of M*.

Hence, every 3-manifold is determined by its Heegaard genus $g(M)$ and an automorphism on the orientable surface of genus $g(M)$. Since handlebodies are easy to understand, the classification problem of 3-manifolds is thus reduced to the group of isotopy classes of automorphisms on an orientable surface S_g of genus g which is called the *mapping class group* $\mathrm{MCG}(S_g)$. This emphasizes the central role of embedded surfaces, in particular normal surfaces, in the theory of 3-manifolds.

As already mentioned, the concept of (classical) normal surfaces is due to Kneser [71]. A surface S, properly embedded into a 3-manifold M, is said to be *normal* if it respects a given cell decomposition of M in the following sense: The surface does not contain any vertex of the manifold nor intersect with any 2-cell in a circle, a point or an arc starting and ending in a point of the same edge (see Figure 1.4.1 for the simplicial case).

The precise definition of the term *normal surface* is due to Haken [55]. Haken developed an algebraic theory of normal surfaces to advance the research on the homeomorphism problem of 3-manifolds (for any pair of 3-manifolds (M_1, M_2) decide after a finite number of operations whether $M_1 \cong M_2$ or not, cf. [56]). In the theory of (hyperbolic) 3-manifolds, normal surfaces are often examined using special kinds of cell decompositions: If $\tilde{\Delta}$ is a set of tetrahedra together with a set of gluing instructions Φ on the set of triangles of $\tilde{\Delta}$ such that each triangle is identified with at most one other triangle, then $P = \tilde{\Delta}/\Phi$ is called a *pseudotriangulation*. Pseudotriangulations where every vertex link is a surface of genus 1 are called *ideal triangulations*. Ideal triangulations are a very efficient way to describe the exterior of a link in Euclidian 3-space where every vertex is identified with one connected component of the link (i. e. every vertex corresponds to a knot in 3-space). Ideal triangulations are understandable to computers (see the software packages `SnapPea` [139, 140] by Weeks or `Regina` [26] by Burton). More generally, they can be thought of as bounded 3-manifolds with some solid tori removed which can be closed in many different ways using Dehn-surgery. Most of the resulting 3-manifolds can be endowed with a hyperbolic structure. Although in most cases we cannot expect an ideal triangulation to be combinatorial there are some examples.

One of them is examined in Section 5.4.

In this work we will exclusively consider normal surfaces embedded into combinatorial manifolds that can be seen as a special class of pseudotriangulations of 3-manifolds.

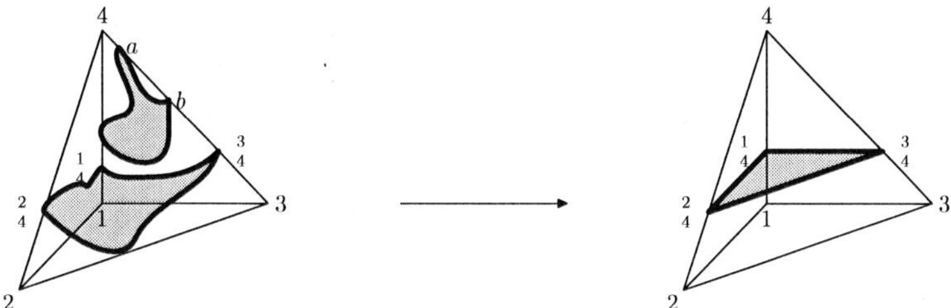

Figure 1.4.1: Intersection of an embedded surface with a tetrahedron of the ambient combinatorial 3-manifold and the corresponding normal subset.

Definition 1.4.2 (Polyhedral manifold, polyhedral map, cf. [25]). A *polyhedral complex* C is a finite family of convex polytopes such that *(i)* for every polytope $P \in C$ all of its faces $F \in P$ are contained in C and *(ii)* the intersection $P_1 \cap P_2$ of any two polytopes $P_1, P_2 \in C$ is either empty or a common face of P_1 and P_2.

A *polyhedral manifold* is a polyhedral complex M such that any simplicial subdivision of M is a combinatorial manifold. If M is a surface we will call it a *polyhedral map*. If, in addition, M entirely consists of m-gons, we call it a *polyhedral m-gon map*.

Definition 1.4.3 (Discrete normal surface). Let M be a combinatorial 3-manifold or 3-pseudomanifold, $\Delta \in M$ one of its tetrahedra and P the intersection of Δ with a plane that does not include any vertex of Δ. Then P is called a *normal subset* of Δ. Up to an isotopy that respects the face lattice of Δ, P is equal to one of the model subsets P_i, $1 \leq i \leq 7$, shown in Figure 1.4.2.

A polyhedral map $S \subset M$ that entirely consists of facets P_i such that every tetrahedron contains at most one facet is called a *discrete normal surface* of M.

Remark 1.4.4. In classical normal surface theory a tetrahedron of a combinatorial manifold can contain several facets of a normal surface of possibly different types. Hence, we can assign a vector $v \in \mathbb{N}_0^7$ to each tetrahedron counting the number of parallel cuts of each type. The set of all such vectors of all tetrahedra of a combinatorial manifold is called the *set of*

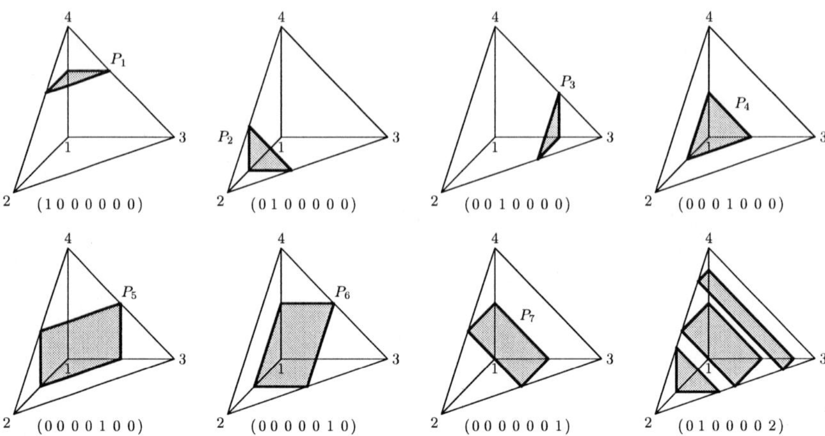

Figure 1.4.2: Possible cuts of a tetrahedron by a normal surface and associated normal coordinates. Note that the rightmost picture of the bottom row cannot be part of a discrete normal surface.

normal coordinates of a normal surface (cf. Figure 1.4.2). Since all cuts of a tetrahedron have to be disjoint, not all combinations of cuts are valid. In particular, each tetrahedron may contain only one type of quadrilaterals at most. In addition, every tetrahedron has to contain cuts compatible with the ones of adjacent tetrahedra as a normal surface has to be a closed polyhedral map. For a combinatorial 3-manifold M we have $3f_2$ such linear *compatibility equations*, where f_2 denotes the number of triangles of M (3 restrictions per triangle). A set of normal coordinates with compatible entries is called *admissible*. It is an interesting fact that a normal surface without vertex linking connected components is already determined by its quadrilaterals (see [137, Theorem 2.4]). This leads to a more compressed type of normal coordinates with a vector $v \in \mathbb{N}_0^3$ for each tetrahedron.

The description of normal surfaces in terms of normal coordinates gives rise to the concept of the *geometric sum*: The union of two normal surfaces, defined by the componentwise sum Σ of their normal coordinates, is well defined if and only if Σ is admissible (see [137] for further details). This implies that the theory of normal surfaces has a much less algebraic structure than homology theory, although Haken himself emphasized a close connection between the two theories in [55].

1.5 Simply connected 4-manifolds

The theory of 4-manifolds has been developing intensively over the last decades. This is most likely due to several phenomena which do not occur in dimensions other than 4. First of all, in dimension 4 the class of topological manifolds differs from the class of smooth manifolds, whereas this is not the case in lower dimensions. Secondly, 4-dimensional manifolds seem to be even more complex than manifolds of dimension 5 and higher. In any dimension $d \neq 4$, there are at most a finite number of smooth structures for any topological manifold. In fact, by a theorem of Stallings (see [129]) we know that for any $d > 4$, the Euclidian space $\mathbb{R}^d$ admits a unique smooth structure. This is false in dimension 4: First examples of exotic smooth structures of $\mathbb{R}^4$ were discovered by Freedman and Kirby and presented in [68]. Later Taubes proved in [134] that there even exist uncountably many exotic $\mathbb{R}^4$.

It seems that while 4-space provides enough "space" for complicated structures it is still too "narrow" for certain simplifying constructions (for example the so-called Whitney trick).

On the other hand, it has been known since the 1950s that there exist exotic smooth structures on S^d for some $d > 4$ (see for example [105, 66]). Since S^d is the one-point compactification of $\mathbb{R}^d$, the exoticism of these smooth structures has to be concentrated in one point as $\mathbb{R}^d$ only admits the smooth standard structure for $d \neq 4$. Remarkably, no exotic smooth structures of S^4 are known although $S^4 \setminus \{0\} \cong \mathbb{R}^4$ admits uncountably many exotic smooth structures. The conjecture that no exotic smooth structure of S^4 exist is referred to as the *smooth Poincaré conjecture in dimension* 4 which is an important open problem in topology.

However, a landmark in the field of 4-manifold theory was the classification of simply connected 4-manifolds due to Freedman (see [47]). It was his classification theorem that marked the beginning of a rapid development. In the following, we will take a closer look at Freedman's work from the viewpoint of combinatorial topology.

First of all we need to take a closer look on the cohomology groups of an ordered simplicial complex. To simplify matters we will restrict ourselves in the following to the *integral cohomology groups*, i. e. cohomology groups with integer coefficients. A standard choice for a base of the cochain complex $C^k(C;\mathbb{Z})$ of a simplicial complex is the one induced by the

k-skeleton of C. For all $\alpha \in \mathrm{skel}_k(C)$ consider the k-cochain

$$\alpha_\star : C_k(C;\mathbb{Z}) \to \mathbb{Z}; \sum_{\delta \in \mathrm{skel}_k(C)} \lambda_\delta \delta \mapsto \lambda_\alpha.$$

Thus, the base elements are in one-to-one correspondence to the simplices of $\mathrm{skel}_k(C)$. The main advantage of looking at 1-forms on the chain complex (cochains) rather than linear combinations of simplices (chains) is the existing of a multiplication which endows $H^\star(C;\mathbb{Z})$ with the structure of a ring.

Definition 1.5.1 (Cup product, Paragraph 5.49 in [111]). Let M be an ordered combinatorial d-manifold with integral cohomology $H^\star(M;\mathbb{Z})$, $a \in H^p(M;G)$ a p-cochain and $b \in H^q(M;G)$ a q-cochain. The mapping

$$\smile : H^p(C;\mathbb{Z}) \times H^q(C;\mathbb{Z}) \to H^{p+q}(C;\mathbb{Z})$$

defined by $a \smile b(\langle v_0,\dots,v_{p+q}\rangle) := a(\langle v_0,\dots,v_p\rangle) \cdot b(\langle v_p,\dots,v_{p+q}\rangle)$ for all ordered faces $\langle v_0,\dots,v_{p+q}\rangle$ of C is bilinear, associative and called the *simplicial cup product of a and b*.

The cup product and the coboundary operator interact in the following manner: Let a be a p-cochain, b a q-cochain and d^k the kth coboundary operator, then

$$d^{p+q}(a \smile b) = d^p a \smile b + (-1)^p (a \smile d^q b).$$

Moreover, we have $a \smile b = (-1)^{pq}(b \smile a)$. To learn more about singular and simplicial cup products as well as some example computations see [111, Paragraph 5.48 and 5.49].

In the case of a connected and simply connected, oriented 4-manifold M we have $H^\star(M,\mathbb{Z}) = (\mathbb{Z},0,\mathbb{Z}^{\beta_2},0,\mathbb{Z})$ by Poincaré duality. Following the terminology of surfaces we will call β_2 the *genus* of M. For combinatorial 4-manifolds M, Inequality (1.3.6) (case $k = 2$) is sharp if and only if the manifold is 3-neighborly and thus simply connected (in a simplicial complex where every triple of vertices spans a triangle, every closed path is contractible). Hence, only simply connected 4-manifolds can attain this boundary and in fact there are such examples (cf. Chapter 5). Moreover, in this case the manifold has the minimum number of vertices possible and it is tight (cf. Section 1.6). This is one reason why simply connected combinatorial 4-manifolds are of great interest in combinatorial topology. For connected and simply connected oriented combinatorial 4-manifolds we can define the following.

Definition 1.5.2 (Intersection form). Let M be a connected and simply connected, oriented, combinatorial 4-manifold with second cohomology group $H^2(M,\mathbb{Z}) \cong \mathbb{Z}^{\beta_2}$ and let $\{a_i \in H^2(M;\mathbb{Z}) \mid 1 \leq i \leq \beta_2(M;\mathbb{Z})\}$ be a base of $H^2(M;\mathbb{Z})$. Then the cup product

$$Q_M : H^2(M;\mathbb{Z}) \times H^2(M;\mathbb{Z}) \to H^4(M;\mathbb{Z}) \cong \mathbb{Z},$$

represented by the quadratic matrix $N_{Q_M} := (a_i \smile a_j)_{1 \leq i,j \leq \beta_2(M,\mathbb{Z})}$ is called *intersection form* of M.

The intersection form Q_M is a *unimodular*, i. e. $\det(N_{Q_M}) = \pm 1$, symmetric bilinear form (see [51, Introduction]). Up to *equivalence*, i. e. similarity transformations of the corresponding matrix N_{Q_M}, it is an invariant of M. The difference of the number of positive and negative eigenvalues of the corresponding matrix is called the *signature of the intersection form* $\sigma(Q_M)$. If for all $a \in H^2(M;\mathbb{Z})$ the integer $Q_M(a,a)$ is even, Q_M is called *even*, otherwise it is referred to as *odd*. If Q is even, it follows from [51, Lemma 1.2.20] that $\sigma(Q_M)$ is divisible by 8. β_2 is called the *rank of the intersection form* written $\mathrm{rk}(Q_M)$. If $\mathrm{rk}(Q_M) = \sigma(Q_M)$ then Q_M is called *definite*, otherwise it is called *indefinite*. We now can state Freedman's famous classification theorem.

Theorem 1.5.3 (Freedman, Theorem 1.5 in [47]). *For every unimodular symmetric bilinear form Q there exists a simply connected, closed, topological 4-manifold M such that $Q = Q_M$. If Q is even, this manifold is unique (up to homeomorphism). If Q is odd, there are exactly two different homeomorphism types of manifolds with the given intersection form. At most one of these homeomorphism types carries a PL structure. Consequently, simply connected PL 4-manifolds are determined up to homeomorphism by their intersection forms.*

Theorem 1.5.3 states that the class of simply connected 4-manifolds is in one-to-one (one-to-two) correspondence to even (odd) unimodular symmetric bilinear forms up to equivalence. In the case of an indefinite symmetric biliniear form Q Milnor was able to prove the following.

Theorem 1.5.4 (Milnor [106], Theorem 1.2.14 [51]). *If two indefinite unimodular forms Q_1, Q_2 have the same rank, signature and parity, then they are equivalent.*

More precisely, we have the following theorem.

Theorem 1.5.5 (Theorem 1.2.21 in [51]). *Suppose that Q is an indefinite, unimodular form of rank n. If Q is odd, then it is isomorphic to $k\langle 1\rangle \oplus (n-k)\langle -1\rangle$ (a diagonal matrix with k 1-entries and $n-k$ (-1)-entries on the diagonal); if Q is even then it is isomorphic to $\frac{\sigma(Q)}{8}E_8 \oplus \frac{n-|\sigma(Q)|}{2}H$, where*

$$H := \begin{pmatrix} 0 & 1 \\ 1 & 0 \end{pmatrix}$$

and

$$E_8 := \begin{pmatrix} 2 & -1 & 0 & 0 & 0 & 0 & 0 & 0 \\ -1 & 2 & -1 & 0 & 0 & 0 & 0 & 0 \\ 0 & -1 & 2 & -1 & 0 & 0 & 0 & 0 \\ 0 & 0 & -1 & 2 & -1 & 0 & 0 & 0 \\ 0 & 0 & 0 & -1 & 2 & -1 & 0 & -1 \\ 0 & 0 & 0 & 0 & -1 & 2 & -1 & 0 \\ 0 & 0 & 0 & 0 & 0 & -1 & 2 & 0 \\ 0 & 0 & 0 & 0 & -1 & 0 & 0 & 2 \end{pmatrix}.$$

In the case of definite symmetric bilinear forms Q the situation is more difficult. However, for a given rank there are only finitely many definite unimodular forms (see [108]) although this number can be very large (e.g.: there are more than 10^{50} definite forms of rank 40, cf. [51, p. 14]).

So far, the classification theorem is purely topological. However, since combinatorial 4-manifolds always allow a piecewise linear or, equivalently, a smooth structure, it would be interesting to know which simply connected topological 4-manifolds admit PL structures. There are a number of results partially answering this question.

Theorem 1.5.6 (Rohlin's theorem, [118]). *Let Q_M be the even intersection form of a simply connected closed oriented PL 4-manifold, then the signature of Q_M is divisible by 16.*

As a consequence, we see that the 4-manifold corresponding to E_8 cannot support a PL structure. In order to classify the topological types of simply connected PL 4-manifolds, the following observation is even more interesting.

Theorem 1.5.7 (Donaldson, Theorem 1 in [32]). *Let M be a compact simply connected oriented PL 4-manifold with positive (negative) definite intersection form Q_M of rank n. Then Q_M is equivalent to the standard diagonal form $n\langle 1\rangle$ $(n\langle -1\rangle)$.*

Altogether, Theorem 1.5.7 combined with Theorem 1.5.5 give an upper bound for the number of distinct simply connected topological 4-manifolds admitting a PL structure.

1.6 Discrete Morse theory and slicings

When analyzing smooth manifolds in differential topology, Morse theory can be used as a tool to learn more about the topology, the homology or possible cell decompositions of the manifold. In the field of combinatorial topology Kühnel developed what one might call a polyhedral Morse theory (compare [76, 77]).

Definition 1.6.1 (Rsl-function, [77]). Let M be a d-dimensional combinatorial manifold. A function $f : M \to \mathbb{R}$ is called *regular simplexwise linear (rsl)* if $f(v) \neq f(w)$ for any two vertices $v \neq w$ of M and f is linear when restricted to any simplex of M.

A point $x \in M$ is said to be *critical* for an rsl-function $f : M \to \mathbb{R}$ if

$$H_\star(M_x, M_x \smallsetminus \{x\}, \mathbb{F}) \neq 0$$

where $M_x := \{y \in M \mid f(y) \leq f(x)\}$ and $\mathbb{F}$ is a field. Here, $H_\star$ denotes an appropriate homology theory.

It follows that no point of M can be critical except possibly the vertices. More precisely we call a vertex v *critical of index i and multiplicity m* if $\beta_i(M_v, M_v \smallsetminus \{v\}, \mathbb{F}) = m$. The vector

$$(\beta_0(M_v, M_v \smallsetminus \{v\}, \mathbb{F}), \dots, \beta_d(M_v, M_v \smallsetminus \{v\}, \mathbb{F}))$$

is called *multiplicity vector of v with respect to $\mathbb{F}$*. The sum $\mu_i(f, \mathbb{F}) := \sum_{v \in M} \beta_i(M_v, M_v \smallsetminus \{v\}, \mathbb{F})$ is called the *number of critical points of f of index i with respect to $\mathbb{F}$*. The sum over all $\mu_i(f, \mathbb{F})$ is called the *number of critical points with respect to $\mathbb{F}$*. Note that in polyhedral Morse theory it is not always possible to isolate critical points. In particular, higher order multiplicities cannot always be avoided.

Proposition 1.6.2 (Morse relations, [87]). *Let M be a combinatorial d-manifold, $f : M \to \mathbb{R}$ an rsl-function and $\mathbb{F}$ a field. Then the following statements hold:*

i) $\beta_i(M, \mathbb{F}) \leq \mu_i(f, \mathbb{F})$,

ii) $\sum_{i=0}^{d}(-1)^i\mu_i(f,\mathbb{F}) = \chi(M) = \sum_{i=0}^{d}(-1)^i\beta_i(M,\mathbb{F})$.

If there exist a field $\mathbb{F}$ such that $\mu_i(f,\mathbb{F}) = \beta_i(M,\mathbb{F})$ for all $0 \le i \le d$, then f is called *perfect* or *tight*. Geometrically spoken this means that M, seen from the direction described by f, is "as convex as possible". More generally we define

Definition 1.6.3 (Tightness, [77]). A compact connected subset $M \subset E^d$ is called *tight* with respect to a field $\mathbb{F}$ if for every open or closed half space $h \subset E^d$ the induced homomorphism

$$H_\star(h \cap M,\mathbb{F}) \to H_\star(M,\mathbb{F}).$$

is injective.

A combinatorial d-manifold with n vertices is said to be tight if its underlying set embedded into the $(n-1)$-simplex $M \subset \Delta^{n-1} \subset E^{n-1}$ is tight. This is the case if and only if all rsl-functions $f : M \to \mathbb{R}$ are perfect (note that if no rsl-function of M has more than the minimum number of critical points required by the topology of M, no non-trivial element of $H_\star(h \cap M,\mathbb{F})$, h arbitrary, can vanish in $H_\star(M,\mathbb{F})$). In other words one could say that a manifold is tight if and only if it contains no such things like bumps or dents.

Definition 1.6.4 (Slicing). Let M be a combinatorial pseudomanifold of dimension d and $f : M \to \mathbb{R}$ an rsl-function. We call the pre-image $f^{-1}(x)$ a *slicing* of M whenever $x \neq f(v)$ for any vertex $v \in M$.

By construction, a slicing is a polyhedral $(d-1)$-manifold and for any ordered pair $x < y$ we have $f^{-1}(x) \cong f^{-1}(y)$ whenever $f^{-1}([x,y])$ contains no vertex of M. In particular, a slicing S of a combinatorial 3-manifold M is a discrete normal surface (cf. Section 1.6): It follows from the simplexwise linearity of f that the intersection of the pre-image with any tetrahedron of M either forms a triangle, a quadrilateral, or it is empty. In addition, if two facets of S lie in adjacent tetrahedra, they either are disjoint or glued together along the intersection line of the pre-image and the common triangle.

Remark 1.6.5. Any partition $V = V_1 \dot\cup V_2$ of the set of vertices of M already determines a slicing: Just define an rsl-function $f : M \to \mathbb{R}$ with $f(v) < f(w)$ for all $v \in V_1$ and $w \in V_2$ and look at a pre-image $f^{-1}(x_0)$ such that $f(v) < x_0 < f(w)$ for all $v \in V_1$ and $w \in V_2$. In the following we will write $S_{(V_1,V_2)} := f^{-1}(x_0)$ for the slicing defined by the vertex partition $V = V_1 \dot\cup V_2$.

Every vertex of a slicing is given as an intersection point of the corresponding pre-image with an edge $\langle u, w \rangle$ of the combinatorial manifold. Since there is at most one such intersection point per edge, we usually label this vertex of the slicing according to the vertices of the corresponding edge, that is $\binom{u}{w}$, with $u \in V_1$ and $w \in V_2$.

By construction, every slicing decomposes the ambient combinatorial manifold M into at least two pieces (an upper part M^+ and a lower part M^-). This is not the case for discrete normal surfaces in general. However, in what follows we will focus on discrete normal surfaces that are slicings and we will apply the above notation for discrete normal surfaces whenever this is possible.

An algorithmic description of the computation of slicings is as follows: Let M be a combinatorial 3-manifold and $V = V_1 \dot{\cup} V_2$ a vertex partition.

```
slicing := ∅
loop over all tetrahedra Δ ∈ M:
  if Δ ⊂ V1 or Δ ⊂ V2 then
    do nothing
  else
    slicing := slicing ∪ (Δ ∩ V1) × (Δ ∩ V2)
  end if
end loop
```

Since every combinatorial pseudomanifold M has a finite number of vertices, there exist only a finite number of slicings of any fixed M. Hence, if f is chosen carefully, the induced slicings admit a useful visualization of M. This has been done already in a number of publications: see [82] for a visualization of a 15-vertex version of the 3-torus or [77, Corollary 5.4] for the visualization of a tight embedding of a Klein bottle into E^8 and various further examples. For examples of slicings in this work see Section 3.7 for some 3-dimensional slicings of the Casella-Kühnel triangulation of the $K3$ surface and Chapter 3 for a number of highly symmetric examples in dimension 2.

Furthermore, there is a method to use slicings to construct highly symmetric combinatorial 3-manifolds which will be presented in Section 4.5.

1.7 Bistellar moves

Since two combinatorial manifolds are already considered distinct as soon as they are not combinatorially isomorphic, a topological PL manifold is represented by a whole class of combinatorial manifolds. Thus, a frequent question when working with combinatorial manifolds is whether two such objects are PL homeomorphic or not. One possibility to approach this problem, i. e. to find combinatorially distinct members of the class of a PL manifold, is a heuristic algorithm using the concept of bistellar moves.

Definition 1.7.1 (Bistellar moves [114]). Let M be a combinatorial d-manifold (d-pseudomanifold), $\gamma = \langle v_0, \ldots, v_k \rangle$ a k-face and $\delta = \langle w_0, \ldots, w_{d-k} \rangle$ a $(d-k+1)$-tuple of vertices of M that does not span a $(d-k)$-face in M, $0 \leq k \leq d$, such that $\{v_0, \ldots, v_k\} \cap \{w_0, \ldots, w_{d-k}\} = \emptyset$ and $\{v_0, \ldots, v_k, w_0, \ldots w_{k-d}\}$ spans exactly $d-k+1$ facets. Then the operation

$$\kappa_{(\gamma,\delta)}(M) = M \setminus (\gamma \star \partial\delta) \cup (\partial\gamma \star \delta)$$

is called a *bistellar* $(d-k)$*-move on* M.

In other words: If there exists a bouquet $D \subset M$ of $d-k+1$ facets on a subset of vertices $W \subset V$ of order $d+2$ with a common k-face γ and the complement δ of the vertices of γ in W does not span a $(d-k)$-face in M we can remove D and replace it by a bouquet of $k+1$ facets $E \subset M$ with vertex set W with a common face spanned by δ. By construction we have $\partial D = \partial E$ and the altered complex is again a combinatorial d-manifold (d-pseudomanifold). See Figure 1.7.1 for a bistellar 1-move of a 2-dimensional complex, see Figure 1.7.2 for all possible types of bistellar moves in dimension 3.

A bistellar 0-move is a *stellar subdivision*, i. e. the subdivision of a facet δ into $d+1$ new facets by introducing a new vertex at the center of δ (cf. Figure 1.7.2 on the left). In particular, the vertex set of a combinatorial manifold (pseudomanifold) is not invariant under bistellar moves. For any bistellar $(d-k)$-move $\kappa_{(\gamma,\delta)}$ we have an inverse bistellar k-move $\kappa^{-1}_{(\gamma,\delta)} = \kappa_{(\delta,\gamma)}$ such that $\kappa_{(\delta,\gamma)}(\kappa_{(\gamma,\delta)}(M)) = M$. If for two combinatorial manifolds M and N there exists a sequence of bistellar moves that transforms one into the other, M and N are called *bistellarly equivalent*. So far, bistellar moves are local operations on combinatorial manifolds that change its combinatorial type. However, the strength of the concept in combinatorial topology is a consequence of the following theorem.

$$M := \langle\langle 1,2,3\rangle, \langle 1,2,5\rangle, \langle 1,3,4\rangle, \langle 1,4,8\rangle, \langle 1,5,8\rangle, \langle 2,3,6\rangle, \langle 2,5,6\rangle, \langle 3,4,7\rangle, \langle 3,6,7\rangle, \langle 4,7,8\rangle\rangle;$$
$$\gamma := \langle\langle 1,3\rangle\rangle; \quad \delta := \langle\langle 2,4\rangle\rangle;$$

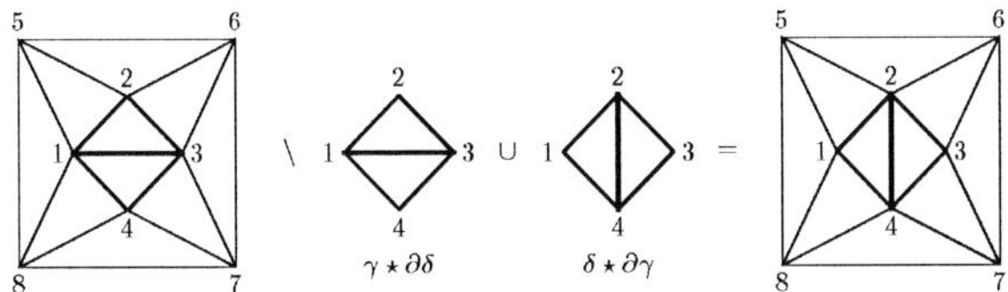

$$\kappa_{(\gamma,\delta)}(M) = \langle\langle 1,2,4\rangle, \langle 1,2,5\rangle, \langle 2,3,4\rangle, \langle 1,4,8\rangle, \langle 1,5,8\rangle, \langle 2,3,6\rangle, \langle 2,5,6\rangle, \langle 3,4,7\rangle, \langle 3,6,7\rangle, \langle 4,7,8\rangle\rangle;$$

Figure 1.7.1: Bistellar 1-move in dimension $d = 2$ with $W = \{1,2,3,4\}$.

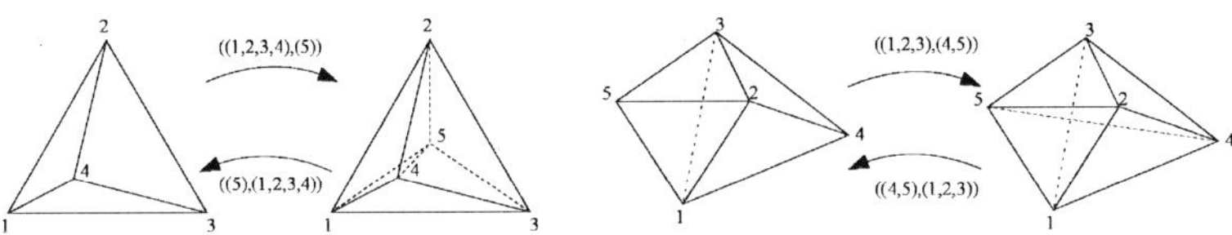

Figure 1.7.2: Bistellar moves in dimension $d = 3$ with $W = \{1,2,3,4,5\}$. On the left side a bistellar 0-, and its inverse, a bistellar 3-move, on the right side a bistellar 1-, and its inverse, a bistellar 2-move.

Theorem 1.7.2 (Pachner [114]). *Two combinatorial manifolds (pseudomanifolds) M and N are PL homeomorphic if and only if they are bistellarly equivalent.*

Unfortunately, Pachner's proof is not constructive, i. e. it does not guarantee that the search for a connecting sequence of bistellar moves between M and N terminates. Hence, using bistellar moves, we cannot prove that M and N are not PL homeomorphic. However, there is a very effective simulated annealing approach that is able to give a positive answer in a lot of cases where in fact $N \cong_{PL} M$. The heuristic was first implemented by Björner and Lutz in [15] and is available in the original version as the GAP (see [49]) programs BISTELLAR or BISTELLAR_EQUIVALENT on [91], but also as built-in function in the GAP-package simpcomp (see Appendix A.3 or [40, 41, 42]), maintained by Effenberger and the author. The heuristic can be used for several tasks:

- Decide, whether two combinatorial manifolds are PL homeomorphic.

- For a given triangulation of a PL manifold, try to find a smaller one with less vertices.
- Check, if an abstract simplicial complex is a combinatorial manifold by reducing all vertex links to the boundary of the d-simplex.

In many cases, the heuristic reduces a given triangulation but does not reach a minimal triangulation after a reasonable amount of moves. Thus, we usually cannot expect the algorithm to terminate. However, in some cases the program usually stops after a small number of moves:

- Whenever $d = 1$ (in this case the algorithm is deterministic).
- Whenever a complex is PL homeomorphic to the boundary of the d-simplex.
- In the case of some 3-manifolds, namely $S^2 \underline{\times} S^1$, $S^2 \times S^1$ or $\mathbb{RP}^3$.

Because of that, bistellar moves have proved to be an essential tool for the work with simplicial complexes.

1.8 Results

This dissertation translates elements and concepts of algebraic geometry, such as blowups (cf. Section 2.4), Morse theory as well as group theory into the field of combinatorial topology in order to establish new tools to study combinatorial manifolds. Most of the methods developed are integrated into the GAP-package simpcomp [40, 41, 42], making them available for future research. The results of this work are presented in five chapters organized as follows:

- Chapter 2 presents a combinatorial version of an algebraic blowup. The concept is used to resolve the singularities of a triangulated Kummer variety. In particular, this chapter contains the first construction of a combinatorial triangulation of a $K3$ surface, where the standard PL structure is assured by the construction process. Moreover, the resolution of isolated singularities of combinatorial 4-pseudomanifolds is discussed in a more general setting and the resolution of a complex from the series of 2-transitive 4-pseudomanifolds from Chapter 5 is presented.

- In Chapter 3, a combinatorial theory of (discrete) normal surfaces in terms of slicings is developed. A linear connection between the genus and the number of quadrilaterals of a 2-dimensional slicing is established. In addition, 2-dimensional slicings that are weakly neighborly polyhedral maps are classified and observations in higher dimensions are made, yielding new insights into the topology of the $K3$ surface and the 4-dimensional Kummer variety and supplementing the observations made in Chapter 2.

- Chapter 4 focuses on simply transitive triangulated 2- and 3-manifolds. In dimension 2, a focus is given on centrally symmetric surfaces: It is shown that the 2-skeleton of β^k is a partition of such surfaces of genus ≤ 1. The partition is explicitly constructed and expressed in terms of a 3-parameter family of such surfaces. In dimension 3, the theory of discrete normal surfaces from Chapter 3 is used to find numerous new infinite series of transitive combinatorial 3-dimensional sphere bundles over the circle. An algorithm to classify infinite series of combinatorial 3-manifolds is provided. Moreover, three infinite series of neighborly combinatorial 3-manifolds are presented.

- Chapter 5 focuses on doubly transitive combinatorial manifolds and pseudomanifolds: It is specified by a simple criterion, when a sharply 2-transitive permutation group generates a 2-transitive "pinch point" surface. In addition, a geometric interpretation of the only 3-transitive 3-pseudomanifold is illustrated, and an infinite series of 2-transitive 4-pseudomanifolds, where the vertex links seem to coincide with two of the infinite series of transitive sphere bundles described in Chapter 4, is constructed.

- Finally, in Appendix A the `GAP`-package `simpcomp` [40, 41, 42] is presented. It contains all algorithms that resulted from the work of the previous chapters and the dissertation of Effenberger [38].

CHAPTER 2

Combinatorial blowups

The 4-dimensional abstract Kummer variety K^4 with 16 nodes leads to the $K3$ surface by resolving the 16 singularities[1]. In this chapter we present a simplicial realization of this minimal resolution. Starting with a minimal 16-vertex triangulation of K^4 we resolve its 16 isolated singularities – step by step – by simplicial blowups. As a result we obtain a 17-vertex triangulation of the $K3$ surface with the standard PL structure. A key step is the construction of a triangulated version of the mapping cylinder of the Hopf map from the real projective 3-space onto the 2-sphere with the minimum number of vertices.

2.1 Introduction

The problem of finding a combinatorial version of an abstract d-pseudomanifold is not trivial in general. Especially, if some additional properties such as vertex minimality is required. It is well known that there are *products* and *connected sums* in the class of combinatorial manifolds in order to solve this problem. Products require a simplicial subdivision of prisms but that is available (cf. Appendix A.4.3 for an example construction using the GAP-package simpcomp [40, 41, 42]). In algebraic geometry there is a third operation on a certain type of pseudomanifolds, namely the *resolution of singularities*. A fourth operation would be a combinatorial version of *Dehn twists*. If these could be applied to simply connected combinatorial 4-manifolds we could make progress towards a solution of some interesting problems:

[1] The findings of this chapter coincide with [128] and are joint work with Kühnel. The work was partially supported by the German Research Foundation (DFG) under the grant Ku 1203/5-2.

Problem 1. *Find a pair of orientable PL d-manifolds* (M_1, M_2) *such that*

(i) M_1 *and* M_2 *are not homeomorphic,*

(ii) *there are combinatorial triangulations of* M_1 *and* M_2 *with* n *vertices but not with* $n-1$ *vertices,*

(iii) *the* f*-vector of such an* n*-vertex triangulation is unique for both* M_1 *and* M_2*.*

Problem 2. *Find two combinatorial triangulations of the same topological* 4*-manifold such that the underlying PL manifolds are not PL homeomorphic. It is known that some compact topological* 4*-manifolds admit exotic PL structures. Furthermore, any combinatorial triangulation induces a unique PL structure and thus a unique smooth structure.*

Problem 3. *For any given abstract compact PL d-manifold find the minimum number* n *of vertices for a combinatorial triangulation of it, and find out which topological invariants are related to this number.*

For pseudomanifolds admitting some combinatorial triangulation we have the same problem.

Concerning Problem 1 there are pairs of non-orientable and orientable surfaces with the same minimum number of vertices, e. g. the two surfaces with $\chi = -10$ admit triangulations with the f-vector $(12, 66, 44)$ but no smaller triangulations. For another example, see Table 5.2 in Chapter 5 for triangulations of $(\mathbb{T}^2)^{\#50}$ and $(\mathbb{K}^2)^{\#50}$, both with the minimum number of 28 vertices and f-vector $(28, 378, 252)$. Moreover, the existence of pairs of non-homeomorphic lens spaces with the same minimum number of vertices is known due to Brehm and Swiatkowski [24]. However, so far no such pair of specific combinatorial manifolds was constructed. Concerning Problem 2, it is well known that in a topological classification of simply connected 4-manifolds the relevant pieces are $\mathbb{C}P^2$ (with two orientations), $S^2 \times S^2$ and the $K3$ surface (with two orientations). However, for topological 4-manifolds it can happen that there are possibly many distinct PL structures (cf. Section 1.5). There is a method to construct *exotic* (i. e. non-standard) PL structures on 4-manifolds using *Akbulut corks*: Akbulut and Yasui investigated bounded submanifolds of a 4-manifold M. These so-called corks can be cut out and glued back into the original manifold, thus changing the PL type of M (see [1] and [2]). However, applying Akbulut corks to combinatorial manifolds requires more experiments. For $\mathbb{C}P^2$ and $S^2 \times S^2$ we have standard triangulations. For the

$K3$ surface we have one optimal triangulation with the minimum number of 16 vertices [28] but so far the PL type has not been identified. Presumably it is the standard structure of the classical $K3$ surface. Concerning Problem 3, the concept of bistellar moves (cf. Section 1.7) together with some theoretical lower bounds (for example the generalized Heawood inequality (1.3.8)) can be used to find combinatorial triangulations with the minimum number of vertices starting with a larger triangulation. However, in many cases this approach seems to fail (cf. Section 1.7).

In this chapter we describe a purely combinatorial version of resolving ordinary nodes or double points in real dimension 4. In particular, we describe this procedure for the $K3$ surface as a resolution of the Kummer variety with 16 nodes. "Purely combinatorial" here means that we are dealing with simplicial complexes (or subdivisions of such) with a relatively small number of vertices such that topological properties or modifications can be recognized or carried out by an efficient computer algorithm. The construction itself is fairly general. We are going to illustrate it for the example of the $K3$ surface as a desingularization of what we call a *Kummer variety*, following [123]. In particular, we describe a straightforward and "canonical" procedure how a specific triangulation of the $K3$ surface with a small number of vertices and with the classical PL structure can be obtained. As we will see in Section 2.5, this procedure also gives some insights to Problem 1. In principle, such a procedure seems to be possible in any even dimension.

For all this, computer algorithms are employed and implemented in the GAP-package simpcomp (see Appendix A.4.4 or [40, 41, 42]). Here, a key operation is the concept of bistellar moves. Since computer algorithms based on finding bistellar move sequences are not deterministic, the character of the procedure is rather experimental and needs a lot of computer calculation. Nonetheless, bistellar moves turned out to be a suitable tool to resolve singularities of combinatorial 4-pseudomanifolds.

2.2 The Kummer variety and the $K3$ surface

An abstract d-dimensional Kummer variety $K^d = \mathbb{T}^d/_{x\sim -x}$ can be interpreted as the d-dimensional torus modulo involution [123]. It is a d-dimensional *flat orbifold* in the sense that the neighborhood of any point of K^d is a quotient of Euclidean d-space by an orthogonal group. Topologically, K^d can be seen as a pseudomanifold with 2^d isolated singularities which are the fixed points of the involution. A typical neighborhood of a singularity is a cone over a real projective $(d-1)$-space, where the apex represents the singularity. Thus, any combinatorial triangulation of K^d needs at least 2^d vertices as a kind of *absolute vertices* [46]. More specifically, a series of minimal triangulations of K^d for any $d \geq 3$ has been given in [75]. These combinatorial pseudomanifolds are 2-neighborly and highly symmetric with a transitive automorphism group of order $(d+1)! \cdot 2^d$. Moreover, they contain a specific combinatorial real projective space $\mathbb{R}P^{d-1}$ with $2^d - 1$ vertices as each vertex link. This vertex link happens to coincide with a 2-fold non-branched quotient of the vertex link of a series of combinatorial d-tori with $2^{d+1} - 1$ vertices [84], which presumably all have the minimum possible number of vertices among all combinatorial d-tori (in the subclass of lattice triangulations this is in fact the minimum number of vertices to triangulate a d-torus, see [23]).

In particular, we have a minimal 2-neighborly 16-vertex triangulation of the 4-dimensional Kummer variety, which will be denoted by $(K^4)_{16}$. A few of its properties are the following: The f-vector is given by $f = (16, 120, 400, 480, 192)$, the Euler characteristic is $\chi(K^4) = 8$, and the integral homology groups are

$$H_*(K^4) = (\mathbb{Z}, 0, \mathbb{Z}^6 \oplus (\mathbb{Z}_2)^5, 0, \mathbb{Z}). \tag{2.2.1}$$

Its intersection form is even, of rank 6 and has signature 0. We use an integer vertex labeling ranging from 1 to 16. The automorphism group of order $5! \cdot 2^4 = 1920$ is generated by two permutations as follows:

$$\langle (1,7,12)(2,8,11)(3,10,16)(4,9,15), (1,9,10,14,16,8,7,3)(2,13,12,6,15,4,5,11) \rangle.$$

The complex coincides with the orbit $(1,2,4,8,16)_{192}$ of length 192 (cf. [75], where the labeling is chosen to be $0,1,2,\dots,15$ instead of $1,2,3,\dots,16$).

The $K3$ surface on the other hand is a prime (i. e. indecomposable by non-trivial connected sums, see [33]), compact, oriented, connected and simply connected 4-manifold, admitting a smooth or PL structure. By Freedman's classification theorem (Theorem 1.5.3 from the introduction, see [47]) it is, up to homeomorphism, uniquely determined by its intersection form. The Euler characteristic is $\chi(K3) = 24$, the integral homology groups are

$$H_*(K3) = (\mathbb{Z}, 0, \mathbb{Z}^{22}, 0, \mathbb{Z}) \tag{2.2.2}$$

and the intersection form is even, of rank 22 and has signature 16. In a suitable basis it is represented by the unimodular matrix

$$\mathbb{E}_8 \oplus \mathbb{E}_8 \oplus 3H \tag{2.2.3}$$

(see Section 1.5 for the definition of $\mathbb{E}_8$ and H). This distinguishes the $K3$ surface from a topological point of view. Also from the combinatorial point of view, this 4-manifold is fairly special, since the data $n = 16, \chi = 24$ coincides with the case of equality in Inequality (1.3.8). Inequality (1.3.8) is also related to Problem 3. As already mentioned, equality in (1.3.8) can occur only for 3-neighborly triangulations. Consequently, the f-vector has to start with $\left(n, \binom{n}{2}, \binom{n}{3}\right)$ in this case. In other words: Any combinatorial triangulation of the $K3$ surface has at least 16 vertices (the same number as required for the Kummer variety K^4), and one with precisely 16 vertices must necessarily be 3-neighborly (or *super-neighborly*). Such a 3-neighborly, vertex minimal 16-vertex triangulation of a PL manifold homeomorphic to the $K3$ surface $(K3)_{16}$ was found by Casella and Kühnel in [28]. The f-vector is $f = (16, 120, 560, 720, 288)$. Observe the 3-neighborliness $f_2 = 560 = \binom{16}{3}$. Its automorphism group is isomorphic to the affine linear group $\mathrm{AGL}(1, 16)$ and is generated by two permutations as follows:

$$\langle(1,3,8,4,9,16,15,2,14,12,6,7,13,5,10),(1,11,16)(2,10,14)(3,12,13)(4,9,15)(5,7,8)\rangle$$

The group is of order $16 \cdot 15 = 240$ and acts (sharply) 2-transitively on the set of vertices $\{1, \ldots, 16\}$ of $(K3)_{16}$. The triangulation $(K3)_{16}$ itself is defined as the union of the orbits $(1,2,3,8,12)_{240}$ and $(1,2,5,8,14)_{48}$ under this permutation group (see [28] where the labeling is chosen to be $0, 1, 2, \ldots, 15$ instead of $1, 2, 3, \ldots, 16$).

2.3 The Hopf σ-process

By the *Hopf σ-process* we mean the blowup process of a point and, simultaneously, the resolution of nodes or ordinary double points of a complex algebraic variety. This was described by Hopf in [61], compare [60] and [57]. From the topological point of view, the process consists of cutting out some subspace of the variety and gluing in some other space. In complex algebraic geometry, one point of a manifold is replaced by all complex projective lines $\mathbb{C}P^1 \cong S^2$ through that point. This is often called *blowing up* of the point. In general, the process can be applied to non-singular 4-manifolds and yields a transformation of a manifold M to $M\#(+\mathbb{C}P^2)$ or $M\#(-\mathbb{C}P^2)$, depending on the choice of an orientation. The same construction is possible for nodes or ordinary double points (a special type of singularity). The ambiguity of the orientation is the same for the blowup process of a node. Similarly it has been used in arbitrary even dimension by Spanier [123] as a so-called *dilatation process*. In the particular case of the 4-dimensional Kummer variety with 16 nodes, a result of Hironaka [59] states that the singularities of a 4-dimensional Kummer variety K^4 can be resolved into a smooth manifold birationally equivalent to K^4. It is also well known that the minimal resolution of the 4-dimensional Kummer variety is a $K3$ surface. This raises the question, whether it is possible to carry out the Hopf σ-process in the combinatorial category. In this case one would have to cut out a certain neighborhood A of each of the singularities and glue in an appropriate simplicial complex B.

The spaces A which have to be cut out are the following: The Kummer variety K^4 is the quotient of a 4-dimensional torus $\mathbb{T}^4 = \mathbb{R}^4/_{\mathbb{Z}^4}$ by the central involution $\sigma : x \mapsto -x$ with precisely 16 fixed points x_i, $1 \leq i \leq 16$. Let X_i be a suitable neighborhood of x_i, then σ acts on $X = \mathbb{T}^4 \backslash \bigcup X_i$ without fixed points. The involution σ acts as the antipodal map on each connected component of ∂X. Therefore, the quotient of ∂X_i is a projective space $\mathbb{R}P^3$ of dimension 3 for each $1 \leq i \leq 16$, and the quotient of X_i itself is a cone over $\mathbb{R}P^3$ and will be denoted by A_i. Thus, the quotient $\widetilde{X} = X/_\sigma$ is a manifold having 16 disjoint copies of $\mathbb{R}P^3$ as its boundary, and the quotient $K^4 = \mathbb{T}^4/_\sigma$ contains the disjoint subsets $A_1, \ldots, A_{16}$ as neighborhoods of the 16 singularities.

The spaces B which have to be glued in are the following: The Hopf map $h : S^3 \to \mathbb{C}P^1$ induces a map $\tilde{h} : \mathbb{R}P^3 \to \mathbb{C}P^1$ since it identifies antipodal pairs of points. We consider the cylinder $C = \mathbb{R}P^3 \times [0,1]$ with the identification along the bottom of the cylinder by an equivalence relation $\sim$ defined by $(x,0) \sim (\tilde{h}(x),0)$. The quotient $C/_\sim$ is a manifold with boundary $\mathbb{R}P^3$ – the space B we are looking for. If we identify the boundary of $\widetilde{X}$ with the union of the boundaries of 16 copies $B_1, \ldots, B_{16}$ of B, we get a closed manifold S. Alternatively, each B_i can be seen as a copy of $(\mathbb{C}P^2 \setminus B^4)/_{\tilde{\sigma}}$, where the involution $\tilde{\sigma} : \mathbb{C}P^2 \to \mathbb{C}P^2$ is defined by $\tilde{\sigma}[z_0 : z_1 : z_2] = [-z_0 : z_1 : z_2]$ with a fixed point set consisting of the point $[1:0:0]$ at the center of the ball B^4 and the polar projective line $z_0 = 0$. Spanier [123] proved that S is in fact a $K3$ surface. Our main result is a simplicial realization of this construction. In principle, one can expect that such a combinatorial construction is possible, but there are a number of technical difficulties to overcome. One of the problems is to make the procedure efficient and to keep the number of vertices sufficiently small at each intermediate step.

2.4 An alternative construction of a combinatorial $K3$ surface

Our goal is to construct a simplicial version of the $K3$ surface out of $(K^4)_{16}$ by a combinatorial version of Spanier's dilatation process. More precisely, we present a way to cut out a certain simplicial version of A_i and to glue in a simplicial version of B_i. We prefer a description of B_i as the mapping cylinder of the Hopf map $\tilde{h}$, defined on $\mathbb{R}P^3 = \partial B^4/_\sigma$. In the combinatorial setting this is possible if the corresponding boundaries are combinatorially isomorphic, i. e. if they are equal up to a relabeling of the vertices. However, in general the boundaries are PL homeomorphic but not combinatorially isomorphic. This is the main difficulty here. Therefore, we need an efficient procedure to change the combinatorial type of the manifold while preserving its PL homeomorphism type. One possibility of such a procedure is the well established concept of bistellar moves. For an introduction see Section 1.7 or [114, 92]. However, in order to solve this problem we first have to construct a simplicial version of the mapping cylinder B_i.

2.4.1 A triangulated mapping cylinder of the Hopf map $\tilde{h} : \mathbb{R}P^3 \to \mathbb{C}P^1$ with the minimum number of vertices

From the topology of the complex projective plane it is fairly clear that one can construct a triangulation of $\mathbb{C}P^2$ from a triangulated version of the Hopf map $h : S^3 \to S^2$. Conversely, every triangulation of $\mathbb{C}P^2$ implicitly contains a triangulation of the Hopf map (possibly with a collapsing of certain simplices) by considering a neighborhood of a triangulated $\mathbb{C}P^1$ inside the triangulation.

Theorem 2.4.1 (Madahar and Sarkaria [99]). *There is a simplicial version of the Hopf map $h : S^3 \to S^2$ with the minimum number of 12 vertices for S^3, which are mapped in triplets onto the 4-vertex S^2. From this simplicial Hopf map one can reconstruct the unique 9-vertex triangulation of $\mathbb{C}P^2$ which was known before, see [77].*

Roughly, the procedure for the construction of a triangulated $\mathbb{C}P^2$ is the following:

1. Find a simplicial subdivision of the mapping cylinder of the Hopf map which is a triangulated $\mathbb{C}P^2$ minus an open 4-ball.

2. Close it up on top by a suitable simplicial 4-ball.

3. Finally reduce the number of vertices by bistellar moves as far as possible.

For our purpose we can follow an analogous procedure here:

1. Find a simplicial version of the Hopf map $\tilde{h} : \mathbb{R}P^3 \to S^2$.

2. Find a simplicial subdivision of the mapping cylinder $\widetilde{C}$, which is nothing but a triangulated complex projective plane with one hole modulo the involution $\tilde{\sigma}$. There is one boundary component which is homeomorphic to $\mathbb{R}P^3$.

3. Finally reduce the number of vertices by bistellar moves as far as possible. It is well known that any combinatorial triangulation of $\mathbb{R}P^3$ has at least 11 vertices [138]. Therefore, 11 is the minimal number of vertices also for the space we are looking for.

Theorem 2.4.2. *There is an 11-vertex triangulation of the mapping cylinder of the Hopf map $\tilde{h} : \mathbb{R}P^3 \to S^2$ such that all vertices and edges are contained in the boundary. This is the minimum possible number of vertices since it is the minimum for the boundary.*

Proof. On the boundary of $\mathbb{C}P^2 \setminus B^4$ the involution σ coincides with $\tilde{\sigma}$ and leads to a twofold quotient map $S^3 \to \mathbb{R}P^3$. From this it is clear that a triangulated version of the Hopf map from $\mathbb{R}P^3$ onto S^2 requires a simplicial Hopf map $h : S^3 \to S^2$ which is centrally symmetric on S^3 and thus is invariant under σ. Therefore, we need to construct a centrally symmetric triangulation of S^3 first. This should allow a simplicial fibration by Hopf fibres.

For the construction we start with two regular hexagons (2-polytopes) P_1, P_2 in the plane and take the product polytope $P := P_1 \times P_2$ (cf. [142, p. 10]). The vertices will be denoted by a_{ij}, where i, j are ranging from 1 to 6. The facets of P are $6+6$ hexagonal prisms, where one of them has vertices $a_{11}, \ldots, a_{16}$ on the top and $a_{21}, \ldots, a_{26}$ on the bottom. The subcomplex $\partial P_1 \times \partial P_2 \subset \partial P$ is the standard 6×6-grid torus as a subcomplex decomposing ∂P into two solid tori, one on each side of the torus. One of the squares has vertices $a_{11}, a_{12}, a_{21}, a_{22}$, see Figure 2.4.2 where the labeling is simply ij instead of a_{ij}. For a simplicial version we need to subdivide the prisms. In a first step we subdivide each square in the torus by the main diagonal, as indicated in Figure 2.4.2. Next we introduce one extra vertex b_i at the center of the six prisms on one side and another vertex c_i at the center of the six prisms on the other side, $i = 1, \ldots, 6$. That is to say, $b_1, \ldots, b_6$ represent the core of one solid torus and $c_1, \ldots, c_6$ the core of the other. Furthermore, we introduce the cones from each b_i and c_i to the $2 \cdot 6 = 12$ triangles of each corresponding prism. Finally, the remaining holes are closed by copies of the join of the edge between two adjacent center vertices and the edge of a hexagon. Typical tetrahedra of this type are $\langle b_1 b_2 a_{11} a_{12} \rangle$ and $\langle c_1 c_2 a_{11} a_{21} \rangle$. This procedure is carried out for each of the two solid tori, see Figure 2.4.1.

Thus, we get a centrally symmetric triangulation S^3_{cs} of the 3-sphere with 48 vertices and $2 \cdot (6 \cdot (12 + 6)) = 216$ tetrahedra. It has an automorphism group G of order 144.

On this triangulation of S^3 we define the simplicial Hopf map $h_{cs} : S^3_{cs} \to S^2$ by the following identifications:

$$\begin{aligned}
\{a_{ij} \mid j - i \equiv 1 \ (6)\} &\mapsto a_1 \\
\{a_{ij} \mid j - i \equiv 2 \ (6)\} &\mapsto a_2 \\
\{a_{ij} \mid j - i \equiv 3 \ (6)\} &\mapsto a_3 \\
\{a_{ij} \mid j - i \equiv 4 \ (6)\} &\mapsto a_4 \\
\{a_{ij} \mid j - i \equiv 5 \ (6)\} &\mapsto a_5 \\
\{a_{ij} \mid j - i = 0\} &\mapsto a_6 \\
\{b_i\} &\mapsto b \\
\{c_i\} &\mapsto c
\end{aligned}$$

The image is a simplicial 2-sphere with 8 vertices, namely a double pyramid from b and c over the hexagon $\langle a_1, a_2, a_3, a_4, a_5, a_6 \rangle$ where the image of the torus in Figure 2.4.1 under h forms the hexagon, and the solid torus on each side gets mapped to one of the hexagonal pyramids. Note that the antipodal map

$$\sigma : \ a_{ij} \mapsto a_{i+3,j+3}, \quad b_i \mapsto b_{i+3}, \quad c_i \mapsto c_{i+3} \tag{2.4.1}$$

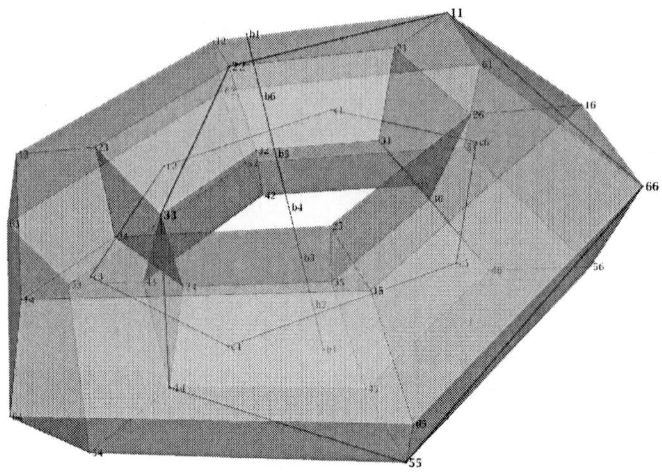

Figure 2.4.1: A solid torus as one half of S^3_{cs} with two Hopf fibres.

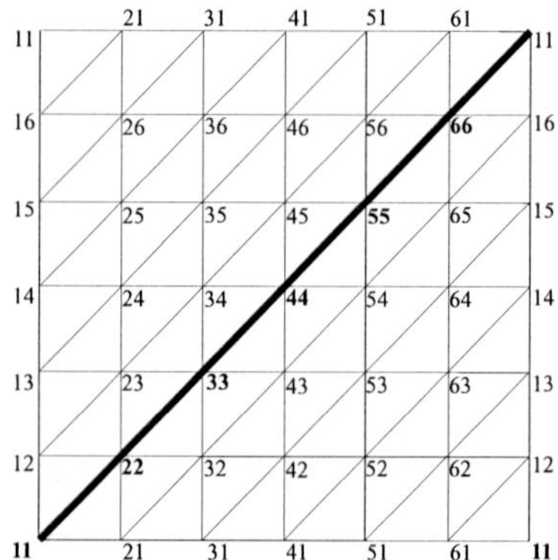

Figure 2.4.2: Combinatorial 6 × 6-grid torus with Hopf fibres.

(all indices taken modulo 6) is compatible with the simplicial Hopf map. By construction, the quotient $\mathbf{P} = S^3_{cs}/_\sigma$ is a 24-vertex triangulation of $\mathbb{R}P^3$. The automorphism group is a normal subgroup of G of index 2. It follows that $\mathbf{P}$ again allows a simplicial version of the Hopf map $\tilde{h} : \mathbf{P} \to S^2$ onto the same 8-vertex triangulation of S^2.

A suitable simplicial decomposition C of the cylinder $\mathbf{P} \times [0,1]$ is compatible with the projection map

$$\begin{array}{rcl} \tilde{h} \times \{0\} : \mathbf{P} \times \{0\} & \to & S^2 \times \{0\} \\ (x,0) & \mapsto & (\tilde{h}(x),0) \end{array}$$

on the bottom of C and leads to a triangulated mapping cylinder $C/_\sim$.

For the purpose of a better handling of the blowup process we computed a minimal version of $\mathbf{C} \cong_{PL} C/_\sim$ by bistellar moves. In this reduced version the boundary is isomorphic to a vertex minimal triangulation of $\mathbb{R}P^3$ with the f-vector $(11,51,80,40)$. Moreover, the boundary $\partial\mathbf{C}$ and $\mathrm{lk}_{(K^4)_{16}}(v)$ are bistellarly equivalent and hence PL homeomorphic for any vertex v, a property which will be needed for the construction of a triangulated $K3$ surface. On 11 vertices labeled by $1,2,\ldots,11$ this complex is the following:

$$\begin{aligned} \mathbf{C} = \big(& \langle 1,3,5,6,11\rangle, \langle 2,3,5,6,11\rangle, \langle 2,4,5,6,11\rangle, \langle 2,3,6,9,11\rangle, \langle 3,6,7,9,11\rangle, \\ & \langle 1,3,6,7,11\rangle, \langle 6,7,8,10,11\rangle, \langle 1,6,7,10,11\rangle, \langle 4,6,8,9,11\rangle, \langle 6,7,8,9,11\rangle, \\ & \langle 2,4,6,9,11\rangle, \langle 1,2,3,5,8\rangle, \langle 1,2,3,5,11\rangle, \langle 1,5,7,8,9\rangle, \langle 1,2,5,7,8\rangle, \\ & \langle 1,4,5,9,11\rangle, \langle 4,5,7,9,11\rangle, \langle 1,4,5,7,9\rangle, \langle 1,2,4,5,11\rangle, \langle 1,2,4,5,7\rangle, \\ & \langle 3,4,7,8,11\rangle, \langle 1,3,4,7,8\rangle, \langle 1,2,3,8,11\rangle, \langle 1,3,7,8,11\rangle, \langle 1,2,8,10,11\rangle, \\ & \langle 1,7,8,10,11\rangle, \langle 1,2,7,8,10\rangle, \langle 4,7,8,9,11\rangle, \langle 1,4,7,8,9\rangle, \langle 1,2,4,9,11\rangle \big). \end{aligned}$$

The f-vector of $\mathbf{C}$ is $(11,51,107,95,30)$, and its boundary $\partial\mathbf{C}$ contains the complete 1-skeleton of $\mathbf{C}$. In particular, $\mathbf{C}$ is vertex minimal since the boundary $\mathbb{R}P^3$ requires already at least 11 vertices [138]. □

Remark 2.4.3. By starting with the product polytope of two $3k$-gons containing a $3k \times 3k$-grid torus, $k \geq 1$, one can similarly obtain a simplicial version of the Hopf map from the lens space $L(k,1)$ to S^2. Furthermore, the same procedure as above can be carried out for the corresponding mapping cylinder. We will come back to this observation in Section 2.6.

2.4.2 Simplicial blowups

Let us now construct a PL version of the Hopf σ-process from Section 2.3. We start by cutting out the star of one of the singular vertices, which is nothing but a cone over a triangulated $\mathbb{R}P^3$. This corresponds to the space A_i above. The boundary of the resulting space is a triangulated $\mathbb{R}P^3$ and is therefore PL homeomorphic to the boundary of the triangulated mapping cylinder $\mathbf{C}$ from Section 2.4.1 which corresponds to the space B_i. We then complete the blowup by gluing in B_i by an appropriate PL homeomorphism ϕ.

Definition 2.4.4. (Resolution of singularities in PL topology)
Let v be a singular vertex of a PL 4-pseudomanifold M with a compact neighborhood A of the type "cone over an $\mathbb{R}P^3$" and let $\phi : \partial A \to \partial\mathbf{C}$ be a PL homeomorphism. A *PL resolution of the singularity* v is given by the following construction:

$$M \mapsto \widetilde{M} := (M \setminus A^\circ) \cup_\phi \mathbf{C}. \tag{2.4.2}$$

We will refer to this operation as a *PL blowup* of v.

However, for a combinatorial version of a blowup with explicit triangulations we face the problem that in general it is not clear how to obtain the PL homeomorphism ϕ to glue the triangulations together. Hence, before cutting out and gluing in we have to modify the triangulations by bistellar moves until their boundaries are isomorphic. For details about how this is done see the proof of Theorem 2.4.6 or the source code of the GAP-package simpcomp [40, 41, 42]. Once the boundary complexes are combinatorially isomorphic, we can apply the following version of a blowup.

Definition 2.4.5. (Simplicial resolution of singularities)
Let v be a vertex of a combinatorial 4-pseudomanifold M the link of which is isomorphic to the particular 11-vertex triangulation of $\mathbb{R}P^3$ given by the boundary complex of $\mathbf{C}$ above. Let $\psi : \mathrm{lk}(v) \to \partial\mathbf{C}$ denote such an isomorphism. A *simplicial resolution of the singularity* v is given by the following construction:

$$M \mapsto \widetilde{M} := (M \setminus \mathrm{star}(v)^\circ) \cup_\psi \mathbf{C}. \tag{2.4.3}$$

We will refer to this operation as a *simplicial blowup* or just a *blowup* of v.

Since in either case both parts are glued together along their PL homeomorphic boundaries, the resulting complex is closed, and the construction of $\widetilde{M}$ is well defined. $\widetilde{M}$ is a closed pseudomanifold and the number of singular points in $\widetilde{M}$ is the number of singular points in M minus one. In particular, we can apply this to $M = (K^4)_{16}$ and then iterate the procedure on the resulting spaces until the last singularity disappears. We can then prove the following main result.

Theorem 2.4.6. *There is a 17-vertex triangulation of the $K3$ surface $(K3)_{17}$ with the standard PL structure. It can be constructed from $(K^4)_{16}$ by a sequence of bistellar moves and, in between, by 16 simplicial blowups.*

The proof of the theorem is constructive and will be given in the form of an algorithm. From the construction it is clear that the resulting PL manifold is PL homeomorphic to the classical $K3$ surface.

Let $\widetilde{K}_i$, $0 \leq i \leq 16$, be the 4-dimensional Kummer variety after the ith blowup. Since we have to modify its combinatorial type repeatedly, our notation will not distinguish here between two different complexes after bistellar moves. Furthermore, let $\mathbf{C}$ be the bounded complex from Section 2.4.1 and Q_i the intersection form of $\widetilde{K}_i$.

We start with a singular vertex $v \in \widetilde{K}_i$. In general, its link is not isomorphic to $\partial\mathbf{C}$. Thus, we have to modify $\widetilde{K}_i \backslash \mathrm{star}(v)$ in a suitable way to yield a complex which allows a simplicial blowup with the space $\mathbf{C}$. This is accomplished by modifying $\partial(\widetilde{K}_i \backslash \mathrm{star}(v)) = \mathrm{lk}(v)$ with respect to the combinatorial structure of the complex $\widetilde{K}_i$ using bistellar moves. We call this approach *simultaneous bistellar moves*. The source code of the algorithm is available within the GAP-package simpcomp [40, 41, 42]. Even though in general we cannot claim that this must be possible for arbitrary complexes, in this particular case we were able to find fairly short sequences of bistellar moves realizing this modification at any of the 16 steps.

Once $\partial(\widetilde{K}_i \backslash \mathrm{star}(v))$ is isomorphic to $\partial\mathbf{C}$, we can perform the simplicial blowup and obtain $\widetilde{K}_{i+1}$ as described above. Note that in each step we can perform the blowup in two significantly different ways corresponding to the choice of orientation of $\mathbf{C}$. For the verification of the choice of the right orientation we compute the intersection form Q_{i+1} of $\widetilde{K}_{i+1}$ and check that $|\mathrm{sign}(Q_{i+1})| = |\mathrm{sign}(Q_i) + 1|$ holds (note that $Q(K^4) = 0$ and $Q(K3) = \pm 16$).

Due to the various modifications above, the resulting complex $\widetilde{K}_{i+1}$ will be considerably bigger than $\widetilde{K}_i$. Thus, we use bistellar moves to reduce its size before repeating the same operation for the remaining singularities. In every step the signature of the intersection form and the second Betti number will increase by one and the number of singularities will decrease by one. Also, the torsion part of $H_2(K^4)$ will gradually decline. This, however, depends on the order of the blowing up process of the singularities. It follows that the resulting complex is a triangulation of the $K3$ surface with the expected intersection form and with the standard PL structure. As already mentioned, the algorithm is available within the GAP-package simpcomp (see Appendix A.4.4 or [40, 41, 42]). The correctness of the computations was additionally verified with the software `polymake` [50].

The smallest complex (with respect to the f-vector) we were able to obtain by bistellar moves is a 17-vertex version of the $K3$ surface which will be denoted by $(K3)_{17}$. Its facets as well as some basic properties are listed in Table 2.1.

Further data as well as all 16 steps of the dilatation process are listed in Appendix B. The source code itself is part of the GAP-package simpcomp [40, 41, 42].

So far, we were not able to prove PL equivalence to $(K3)_{16}$. However, since the given complex has only 17 vertices this is most likely to be true and further experiments will probably prove the following conjecture.

Conjecture 2.4.7. *$(K3)_{17}$ is PL homeomorphic to $(K3)_{16}$.*

Remark 2.4.8. If Conjecture 3.5.6 is false, this would imply that $(K3)_{16}$ is exotic. In this case $(K3)_{16}$ would to our knowledge be the first explicit triangulation with few vertices of a non-standard combinatorial 4-manifold.

2.5 Other simply connected combinatorial 4-manifolds

In each of the 16 steps of the dilatation process of the Kummer variety we have the choice between two orientations. Consequently, for the resulting non-singular manifold there are a number of different topological types which are possible at the end. One can describe these by the intersection form.

Proposition 2.5.1. *One can construct combinatorial 4-manifolds realizing any of the intersection forms of rank 22 and signature $2n$, $n \in \{0, \dots, 8\}$ from the triangulated 4-dimensional Kummer variety K^4 by 16 simplicial blowups, except for $19(\mathbb{C}P^2)\#3(-\mathbb{C}P^2)$ and, possibly, $11(S^2 \times S^2)$.*

$(K3)_{17} =$ ⟨⟨1 2 3 8 13⟩, ⟨1 2 3 8 14⟩, ⟨1 2 3 12 13⟩, ⟨1 2 3 12 15⟩, ⟨1 2 3 14 15⟩, ⟨1 2 4 7 13⟩,
⟨1 2 4 7 15⟩, ⟨1 2 4 13 15⟩, ⟨1 2 5 6 9⟩, ⟨1 2 5 6 14⟩, ⟨1 2 5 9 17⟩, ⟨1 2 5 14 15⟩,
⟨1 2 5 15 17⟩, ⟨1 2 6 8 14⟩, ⟨1 2 6 8 15⟩, ⟨1 2 6 9 16⟩, ⟨1 2 6 15 17⟩, ⟨1 2 6 16 17⟩,
⟨1 2 7 8 11⟩, ⟨1 2 7 8 15⟩, ⟨1 2 7 11 13⟩, ⟨1 2 8 10 11⟩, ⟨1 2 8 10 13⟩, ⟨1 2 9 16 17⟩,
⟨1 2 10 11 13⟩, ⟨1 2 12 13 15⟩, ⟨1 3 4 5 8⟩, ⟨1 3 4 5 17⟩, ⟨1 3 4 6 10⟩, ⟨1 3 4 6 12⟩,
⟨1 3 4 8 9⟩, ⟨1 3 4 9 10⟩, ⟨1 3 4 12 17⟩, ⟨1 3 5 7 11⟩, ⟨1 3 5 7 14⟩, ⟨1 3 5 8 16⟩,
⟨1 3 5 11 16⟩, ⟨1 3 5 14 15⟩, ⟨1 3 5 15 17⟩, ⟨1 3 6 10 12⟩, ⟨1 3 7 8 11⟩, ⟨1 3 7 8 14⟩,
⟨1 3 8 9 12⟩, ⟨1 3 8 11 16⟩, ⟨1 3 8 12 13⟩, ⟨1 3 9 10 12⟩, ⟨1 3 12 15 17⟩, ⟨1 4 5 8 16⟩,
⟨1 4 5 11 16⟩, ⟨1 4 5 11 17⟩, ⟨1 4 6 7 12⟩, ⟨1 4 6 7 15⟩, ⟨1 4 6 10 15⟩, ⟨1 4 7 12 13⟩,
⟨1 4 8 9 16⟩, ⟨1 4 9 10 14⟩, ⟨1 4 9 14 16⟩, ⟨1 4 10 14 16⟩, ⟨1 4 10 15 16⟩, ⟨1 4 11 16 17⟩,
⟨1 4 12 13 17⟩, ⟨1 4 13 15 16⟩, ⟨1 4 13 16 17⟩, ⟨1 5 6 9 13⟩, ⟨1 5 6 13 14⟩, ⟨1 5 7 10 12⟩,
⟨1 5 7 10 14⟩, ⟨1 5 7 11 12⟩, ⟨1 5 9 11 13⟩, ⟨1 5 9 11 17⟩, ⟨1 5 10 12 14⟩, ⟨1 5 11 12 13⟩,
⟨1 5 12 13 14⟩, ⟨1 6 7 8 14⟩, ⟨1 6 7 8 15⟩, ⟨1 6 7 10 12⟩, ⟨1 6 7 10 16⟩, ⟨1 6 7 14 16⟩,
⟨1 6 9 11 13⟩, ⟨1 6 9 11 14⟩, ⟨1 6 9 14 16⟩, ⟨1 6 10 15 16⟩, ⟨1 6 11 13 14⟩, ⟨1 6 15 16 17⟩,
⟨1 7 10 14 16⟩, ⟨1 7 11 12 13⟩, ⟨1 8 9 10 11⟩, ⟨1 8 9 10 12⟩, ⟨1 8 9 11 16⟩, ⟨1 8 10 12 13⟩,
⟨1 9 10 11 14⟩, ⟨1 9 11 16 17⟩, ⟨1 10 11 13 14⟩, ⟨1 10 12 13 14⟩, ⟨1 12 13 15 16⟩, ⟨1 12 13 16 17⟩,
⟨1 12 15 16 17⟩, ⟨2 3 4 6 12⟩, ⟨2 3 4 6 16⟩, ⟨2 3 4 7 14⟩, ⟨2 3 4 7 16⟩, ⟨2 3 4 12 14⟩,
⟨2 3 5 6 12⟩, ⟨2 3 5 6 16⟩, ⟨2 3 5 12 16⟩, ⟨2 3 7 14 16⟩, ⟨2 3 8 13 14⟩, ⟨2 3 9 10 13⟩,
⟨2 3 9 10 14⟩, ⟨2 3 9 11 14⟩, ⟨2 3 9 11 16⟩, ⟨2 3 9 12 13⟩, ⟨2 3 9 12 16⟩, ⟨2 3 10 13 14⟩,
⟨2 3 11 14 16⟩, ⟨2 3 12 14 15⟩, ⟨2 4 5 7 10⟩, ⟨2 4 5 7 11⟩, ⟨2 4 5 8 10⟩, ⟨2 4 5 8 12⟩,
⟨2 4 5 11 12⟩, ⟨2 4 6 11 12⟩, ⟨2 4 6 11 16⟩, ⟨2 4 7 9 14⟩, ⟨2 4 7 9 15⟩, ⟨2 4 7 10 13⟩,
⟨2 4 7 11 16⟩, ⟨2 4 8 10 12⟩, ⟨2 4 9 10 13⟩, ⟨2 4 9 10 14⟩, ⟨2 4 9 13 15⟩, ⟨2 4 10 12 14⟩,
⟨2 5 6 7 10⟩, ⟨2 5 6 7 12⟩, ⟨2 5 6 8 10⟩, ⟨2 5 6 8 14⟩, ⟨2 5 6 9 16⟩, ⟨2 5 7 11 12⟩,
⟨2 5 8 12 14⟩, ⟨2 5 9 15 16⟩, ⟨2 5 9 15 17⟩, ⟨2 5 12 14 15⟩, ⟨2 5 12 15 16⟩, ⟨2 6 7 10 13⟩,
⟨2 6 7 11 12⟩, ⟨2 6 7 11 13⟩, ⟨2 6 8 10 15⟩, ⟨2 6 10 11 13⟩, ⟨2 6 10 11 17⟩, ⟨2 6 10 15 17⟩,
⟨2 6 11 16 17⟩, ⟨2 7 8 11 15⟩, ⟨2 7 9 11 14⟩, ⟨2 7 9 11 15⟩, ⟨2 7 11 14 16⟩, ⟨2 8 10 11 15⟩,
⟨2 8 10 12 14⟩, ⟨2 8 10 13 14⟩, ⟨2 9 11 15 17⟩, ⟨2 9 11 16 17⟩, ⟨2 9 12 13 16⟩, ⟨2 9 13 15 16⟩,
⟨2 10 11 15 17⟩, ⟨2 12 13 15 16⟩, ⟨3 4 5 8 17⟩, ⟨3 4 6 8 9⟩, ⟨3 4 6 8 11⟩, ⟨3 4 6 9 15⟩,
⟨3 4 6 10 15⟩, ⟨3 4 6 11 13⟩, ⟨3 4 6 13 16⟩, ⟨3 4 7 12 14⟩, ⟨3 4 7 12 17⟩, ⟨3 4 7 13 16⟩,
⟨3 4 7 13 17⟩, ⟨3 4 8 11 17⟩, ⟨3 4 9 10 15⟩, ⟨3 4 11 13 17⟩, ⟨3 5 6 10 12⟩, ⟨3 5 6 10 15⟩,
⟨3 5 6 13 15⟩, ⟨3 5 6 13 16⟩, ⟨3 5 7 11 15⟩, ⟨3 5 7 14 15⟩, ⟨3 5 8 16 17⟩, ⟨3 5 9 10 15⟩,
⟨3 5 9 10 17⟩, ⟨3 5 9 15 17⟩, ⟨3 5 10 12 16⟩, ⟨3 5 10 16 17⟩, ⟨3 5 11 15 16⟩, ⟨3 5 13 15 16⟩,
⟨3 6 7 8 9⟩, ⟨3 6 7 8 15⟩, ⟨3 6 7 9 15⟩, ⟨3 6 8 11 15⟩, ⟨3 6 11 13 15⟩, ⟨3 7 8 9 13⟩,
⟨3 7 8 11 15⟩, ⟨3 7 8 13 14⟩, ⟨3 7 9 13 17⟩, ⟨3 7 9 15 17⟩, ⟨3 7 12 14 15⟩, ⟨3 7 12 15 17⟩,
⟨3 7 13 14 16⟩, ⟨3 8 9 12 13⟩, ⟨3 8 10 11 16⟩, ⟨3 8 10 11 17⟩, ⟨3 8 10 16 17⟩, ⟨3 9 10 11 14⟩,
⟨3 9 10 11 16⟩, ⟨3 9 10 12 16⟩, ⟨3 9 10 13 17⟩, ⟨3 10 11 13 14⟩, ⟨3 10 11 13 17⟩, ⟨3 11 13 14 16⟩,
⟨3 11 13 15 16⟩, ⟨4 5 7 8 13⟩, ⟨4 5 7 8 16⟩, ⟨4 5 7 10 13⟩, ⟨4 5 7 11 16⟩, ⟨4 5 8 10 15⟩,
⟨4 5 8 11 12⟩, ⟨4 5 8 11 17⟩, ⟨4 5 8 13 15⟩, ⟨4 5 9 10 13⟩, ⟨4 5 9 10 15⟩, ⟨4 5 9 13 15⟩,
⟨4 6 7 9 12⟩, ⟨4 6 7 9 15⟩, ⟨4 6 8 9 14⟩, ⟨4 6 8 11 14⟩, ⟨4 6 9 12 14⟩, ⟨4 6 11 12 14⟩,
⟨4 6 11 13 17⟩, ⟨4 6 11 16 17⟩, ⟨4 6 13 16 17⟩, ⟨4 7 8 13 16⟩, ⟨4 7 9 12 14⟩, ⟨4 7 12 13 17⟩,
⟨4 8 9 14 16⟩, ⟨4 8 10 11 12⟩, ⟨4 8 10 11 15⟩, ⟨4 8 11 14 15⟩, ⟨4 8 13 15 16⟩, ⟨4 8 14 15 16⟩,
⟨4 10 11 12 15⟩, ⟨4 10 12 14 16⟩, ⟨4 10 12 15 16⟩, ⟨4 11 12 14 15⟩, ⟨4 12 14 15 16⟩, ⟨5 6 7 10 12⟩,
⟨5 6 8 10 15⟩, ⟨5 6 8 13 14⟩, ⟨5 6 8 13 15⟩, ⟨5 6 9 13 16⟩, ⟨5 7 8 9 13⟩, ⟨5 7 8 9 17⟩,
⟨5 7 8 16 17⟩, ⟨5 7 9 10 13⟩, ⟨5 7 9 10 17⟩, ⟨5 7 10 14 16⟩, ⟨5 7 10 16 17⟩, ⟨5 7 11 15 16⟩,
⟨5 7 14 15 16⟩, ⟨5 8 9 12 13⟩, ⟨5 8 9 12 17⟩, ⟨5 8 11 12 17⟩, ⟨5 8 12 13 14⟩, ⟨5 9 11 12 13⟩,
⟨5 9 11 12 17⟩, ⟨5 9 13 15 16⟩, ⟨5 10 12 14 16⟩, ⟨5 12 14 15 16⟩, ⟨6 7 8 9 16⟩, ⟨6 7 8 14 16⟩,
⟨6 7 9 12 16⟩, ⟨6 7 10 13 17⟩, ⟨6 7 10 16 17⟩, ⟨6 7 11 12 13⟩, ⟨6 7 12 13 17⟩, ⟨6 7 12 16 17⟩,
⟨6 8 9 14 16⟩, ⟨6 8 11 13 14⟩, ⟨6 8 11 13 15⟩, ⟨6 9 11 12 13⟩, ⟨6 9 11 12 14⟩, ⟨6 9 12 13 16⟩,
⟨6 10 11 13 17⟩, ⟨6 10 15 16 17⟩, ⟨6 12 13 16 17⟩, ⟨7 8 9 12 16⟩, ⟨7 8 9 12 17⟩, ⟨7 8 12 16 17⟩,
⟨7 8 13 14 16⟩, ⟨7 9 10 13 17⟩, ⟨7 9 11 14 15⟩, ⟨7 9 12 14 15⟩, ⟨7 9 12 15 17⟩, ⟨7 11 14 15 16⟩,
⟨8 9 10 11 16⟩, ⟨8 9 10 12 16⟩, ⟨8 10 11 12 17⟩, ⟨8 10 12 13 14⟩, ⟨8 10 12 16 17⟩, ⟨8 11 13 14 15⟩,
⟨8 13 14 15 16⟩, ⟨9 11 12 14 15⟩, ⟨9 11 12 15 17⟩, ⟨10 11 12 15 17⟩, ⟨10 12 15 16 17⟩, ⟨11 13 14 15 16⟩⟩.

Table 2.1: 17-vertex triangulation of the $K3$ surface $(K3)_{17}$ with the standard PL structure. Its f-vector is $f(K3) = (17, 135, 610, 780, 312)$. Note that the complex is not 2-neighborly.

Proof. The case $n = 8$ was already considered in Section 2.4. In this case the orientation was uniquely determined in every step by the one in the first step. Therefore, the construction is essentially unique (up to the orientation in the first blowup) and leads to the $K3$ surface. Note that in particular, the manifold $19(\mathbb{C}P^2)\#3(-\mathbb{C}P^2)$ cannot be obtained in this way.

By Rohlin's Theorem (see [48]) the signature of an even intersection form of a simply connected PL 4-manifold is divisible by 16. It follows that for $n \in \{1,\dots,7\}$ we have an odd intersection form. In these cases the manifold is homeomorphic to

$$k(\mathbb{C}P^2)\#l(-\mathbb{C}P^2)$$

where $k - l = \pm 2n$, $n \in \{1,\dots,7\}$ and $k + l = 22$. In the case $n = 0$ the construction is not unique: The pattern of the orientations of all 16 blowups is not determined since there are 8 positive and 8 negative blowups distributed arbitrarily in K^4. One particular sequence of combinatorial blowups led to a manifold the intersection form of which was odd. This implies that the manifold is homeomorphic to $11(\mathbb{C}P^2)\#11(-\mathbb{C}P^2)$. □

The question whether or not the manifold $11(S^2 \times S^2)$ can also be obtained by this construction remains open at this point. It must also be left open whether or not any of the other manifolds with a 22-dimensional second homology admits a triangulation with only 16 vertices. Such a 16-vertex triangulation would have to be 3-neighborly by [77, Theorem 4.9] and would, by the Dehn-Sommerville equations, have the same f-vector as $(K3)_{16}$ and thus would give a solution to Problem 1 stated in the introduction. Further experiments in this direction could possibly produce such an example. This is work in progress.

In the case of 10 vertices and $\chi = 4$ the combinatorial data corresponds to three topological types of simply connected 4-manifolds, namely $S^2 \times S^2$, $\mathbb{C}P^2\#\mathbb{C}P^2$ and $\mathbb{C}P^2\#(-\mathbb{C}P^2)$. These would be candidates for a solution to Problem 2. However, it was shown in [81] that in fact for none of the topological manifolds above there exists a combinatorial triangulation with only 10 vertices.

The process in more detail is described in Appendix A. Moreover, the algorithm used to perform simplicial blowups is available within the GAP-package simpcomp (cf. Appendix A.4.4 or [40, 41, 42]).

2.6 Resolution of other types of singularities

In principle, the procedure of a simplicial blowup can be altered in order to resolve arbitrary isolated singularities v of combinatorial 4-pseudomanifolds M: If we cut out a neighborhood U_v of v, then $B := \partial(M \setminus U_v)$ is a combinatorial 3-manifold. Since for any 3-manifold B there is a bounded 4-manifold N such that $B = \partial N$, the singularity v can be resolved by gluing together M and N along their boundaries. Of course, the topology of the resulting complex depends on the choice of N. However, in some cases this choice is more or less canonical. For example

- for singularities of link type $S^2 \times S^1$ or $S^2 \underline{\times} S^1$, a 1-handle $D^3 \times S^1$ or $D^3 \underline{\times} S^1$ would be an obvious option for N. In both cases handlebodies exist with no more than the minimum number of vertices needed to triangulate their boundaries,
- for $\mathrm{lk}_M(v) \cong L(k,1)$, the mapping cylinders from Remark 2.4.3 can be used. Again, at least for $k \leq 4$, there exist very small triangulations of these mapping cylinders with no interior vertices.

Nevertheless, we cannot expect the algorithm to terminate in all of these cases as it heavily uses the concept of bistellar moves.

Example 2.6.1. The 17-vertex complex $C_{17}^{3^{11}}$ from the series of 2-transitive combinatorial 4-pseudomanifolds from Section 5.8 has 17 singularities of link type $S^2 \underline{\times} S^1$. A few of its basic properties are as follows:

- $f(C_{17}^{3^{11}}) = (17, 136, 544, 680, 272)$,
- $\pi_1(C_{17}^{3^{11}}) = 1$,
- $H_*(C_{17}^{3^{11}}) = (\mathbb{Z}, 0, \mathbb{Z}^{16} \times \mathbb{Z}_5, \mathbb{Z}_2, 0)$,
- $\chi(C_{17}^{3^{11}}) = 17$,
- $\mathrm{Aut}(C_{17}^{3^{11}}) = \mathrm{AGL}(1, 17)$.

A sequence of 17 resolutions K_i, $1 \leq i \leq 17$, of the singularities of $C_{17}^{3^{11}}$, starting with $K_0 := C_{17}^{3^{11}}$, gradually reduces the 2nd Betti number. The complex K_{15} (a combinatorial 4-pseudomanifold with 2 singularities) is still simply connected and its integral homology groups are $H_\star(K_{15}) = (\mathbb{Z}, 0, \mathbb{Z} \times \mathbb{Z}_5, \mathbb{Z}_2, 0)$. After 16 of the 17 resolutions, a non trivial fundamental group occurs ($\pi_1(K_{16}) = \mathbb{Z}_{17}$) and the homology changes to $H_\star(K_{16}) = (\mathbb{Z}, \mathbb{Z}_{17}, \mathbb{Z}_5, \mathbb{Z}_2, 0)$.

Finally, the complete resolution K_{17} is a combinatorial 4-manifold with homology groups $H_\star(K_{17}) = (\mathbb{Z}, \mathbb{Z} \times \mathbb{Z}_{17}, \mathbb{Z}_5, \mathbb{Z}_2, 0)$ and Euler characteristic $\chi(K_{17}) = 0$. The f-vector of the smallest complex obtained by bistellar moves is $f(K_{17}) = (44, 691, 2324, 2795, 1118)$. Unfortunately, this complex does not have any non-trivial automorphisms. Its fundamental group can be represented as follows:

$$\begin{aligned}\pi_1(K_{17}) &= \langle a, b, c \,|\, ac = ca, bc = c^{-1}bcb^{-1}cb, ab = bab^{-1}aba^{-1}, \\ &\quad bab^{-1} = (c^{-1}b)a^{-1}(b^{-1}c), ac^{-1}bca^3b^{-1}c^{-1} = 1, cbacba^{-1}b^{-1}c^2ab^{-1} = 1\rangle.\end{aligned}$$

As expected, $\pi_1(K_{17})$ factorized by the commutator subgroup reduces to

$$\begin{aligned}H_1(K_{17}) &= \langle a, b, c \,|\, ab = ba, ac = ca, bc = cb, a = c^{-4}, c = a^4\rangle \\ &= \langle a, b \,|\, ab = ba, a^{17} = 1\rangle \\ &\cong \mathbb{Z} \times \mathbb{Z}_{17}.\end{aligned}$$

Since $C_{17}^{3^{11}}$ with 17 singularities is a highly symmetric minimal triangulation, it is very likely that the element of order 17 of $H_1(K_{17})$ corresponds to some self homeomorphism of the underlying manifold. However, as of today no further investigations on the manifold K_{17} and its fundamental group were carried out. This is work in progress. A list of simplices of K_{17} can be found in Appendix C.

The `simpcomp` routine `SCBlowup` to perform blowups of ordinary double points can also be used to resolve singularities of link type $S^2 \times S^1$, $S^2 \underline{\times} S^1$, $L(3,1)$ and $L(4,1)$ (of course only within the scope of the possibilities of bistellar moves). In addition, if a suitably bounded 4-manifold N is provided, arbitrary isolated singularities can be processed as well. For more information see Appendix A or [40, 41, 42].

Chapter 3

Slicings and discrete normal surfaces

We investigate slicings of combinatorial manifolds as properly embedded co-dimension 1 submanifolds[1]. Focus is given to the case of dimension 3, where slicings are discrete normal surfaces. For the cases of 2-neighborly 3-manifolds as well as quadrangulated slicings, lower bounds on the number of quadrilaterals of slicings depending on its genus g are presented. These are shown to be sharp for infinitely many values of g. Furthermore, we classify slicings of combinatorial 3-manifolds which are weakly neighborly polyhedral maps. Finally, we study slicings of higher dimensions and the changes of their topological types between critical points of an induced rsl-function. In this way, we obtain slicings through the $K3$ surface of various topological types.

3.1 Introduction

In this chapter we develop a combinatorial theory of discrete normal surfaces in combinatorial 3-manifolds. The chapter is organized as follows:

In Section 3.2 we investigate upon the possible topological types of discrete normal surfaces a given combinatorial 3-manifold admits. In particular, we present a minimal combinatorial Heegaard splitting of the 3-torus and a slicing through a cylinder of type $S^2 \times [0, 1]$ with only 4 quadrilaterals.

In Section 3.3 we discuss the local combinatorial structure of slicings of combinatorial 3-manifolds and 3-pseudomanifolds by presenting a variety of observations on the different

[1] The results of this chapter are contained in [127].

roles of triangles and quadrilaterals.

In Section 3.4 we summarize how bistellar moves in a surrounding 3-manifold affect the combinatorial and topological structure of an embedded discrete normal surface.

In Section 3.5 we present a lower bound on the number of quadrilaterals of a slicing depending on its genus and assuming certain properties such as 2-neighborliness of the surrounding 3-manifold. Furthermore, we discuss some cases of equality.

In Section 3.6 we examine slicings which are weakly neighborly polyhedral maps. A condition for the weakly neighborliness is given, as well as a classification of all weakly neighborly slicings of combinatorial manifolds.

3.2 The genus of discrete normal surfaces

For embedded orientable surfaces $S \subset M$ which decompose a 3-manifold M into two pieces we have the following statement:

Proposition 3.2.1. *Let M be a connected compact orientable 3-manifold and $S \subset M$ a properly embedded connected surface decomposing M into two bounded 3-manifolds M^- and M^+ with common boundary S. Then*

$$\beta_1(M^-;\mathbb{Z}) - \beta_2(M^-;\mathbb{Z}) = g(S) = \beta_1(M^+;\mathbb{Z}) - \beta_2(M^+;\mathbb{Z}) \tag{3.2.1}$$

holds, where $g(S)$ denotes the genus of S and $\beta_i(M^\pm;\mathbb{Z})$ the ith integral Betti number of $M^\pm$.

Proof. Since M^+ (M^-) is a bounded manifold with orientable boundary S (the orientability of S follows from the orientability of M and the fact that S is separating M^+ from M^-), we can glue M^+ (M^-) with a copy of itself along its boundary S obtaining a closed 3-manifold M. Recall that $\chi(M) = 0$ since M is a closed 3-manifold. By the additivity of the Euler characteristic it now follows that

$$0 = \chi(M) = 2\chi(M^+) - \chi(S) = 2(1 - \beta_1(M^+) + \beta_2(M^+)) - (2 - \beta_1(S)), \tag{3.2.2}$$

which simplifies to

$$\beta_1(M^+) - \beta_2(M^+) = \frac{1}{2}\beta_1(S) = g(S). \tag{3.2.3}$$

The calculation for M^- is the same. □

Since the embedding of S is arbitrary, the genus $g(S)$ does not depend on any properties of M. In particular, any 3-manifold M admits an embedding of a connected orientable surface S of any genus $g(S)$.

But of course, if M is a combinatorial manifold and S is a discrete normal surface the situation is somewhat different. Due to the finite number of tetrahedra, the genus of an embedded discrete normal surface S is always bounded. In fact, the only topological type of discrete normal surfaces that occurs in any given combinatorial manifold M is the 2-sphere (for example as the vertex figure of an arbitrary vertex). In the following we will investigate the restrictions on the genus of S given by the combinatorial properties of M.

Proposition 3.2.2. *Let M be an orientable connected combinatorial 3-manifold and $V = V_1 \dot\cup V_2$ a partition of the set of vertices. Assume furthermore that $|V| \in \{2n, 2n+1\}$, $n \in \mathbb{N}$, such that $S_{(V_1,V_2)}$ is connected. Then*

$$g(S_{(V_1,V_2)}) \leq \binom{n-1}{2}. \tag{3.2.4}$$

Proof. Let $M = M^- \cup M^+$, $V_1 \subset M^+$, $V_2 \subset M^-$, be the decomposition of M with common boundary $S_{(V_1,V_2)}$. From Proposition 3.2.1 it follows that

$$\beta_1(M^+) - \beta_2(M^+) = g(S_{(V_1,V_2)}).$$

Now let e be the number of edges, t the number of triangles, Δ the number of tetrahedra of $\mathrm{span}(V_1)$ and w.l.o.g. $m := |V_1| \leq |V_2|$. Then

$$\begin{aligned} g(S_{(V_1,V_2)}) &= 1 - 1 + \beta_1(M^+) - \beta_2(M^+) \\ &= 1 - m + e - t + \Delta \end{aligned} \tag{3.2.5}$$

holds by the Euler-Poincaré formula. Since $t \geq 2\Delta$, $e \leq \binom{m}{2}$ and $m \leq n$, the right hand side of (3.2.5) is maximal for $t = \Delta = 0$, $e = \binom{m}{2}$ and $m = n$, and thus $g(S_{(V_1,V_2)}) \leq 1 - n + \binom{n}{2} = \binom{n-1}{2}$. □

See Figure 3.5.2 (right) for an example attaining equality in (3.2.4) in the case $n = 5$ with $g(S_{(V_1,V_2)}) = 6$.

In order to make a closer connection between the genus of S and the topology of M, we will restrict ourselves to special kinds of decompositions in the following: i) Heegaard splittings (cf. Section 1.4), ii) Heegaard splittings of minimal genus (cf. Section 1.4), iii) slicings $S_{(V_1,V_2)}$ induced by a vertex splitting $V = V_1 \dot\cup V_2$ such that the underlying set of $\mathrm{span}(V_i)$, $i \in \{1,2\}$, defines bounded 3-manifolds. These are called *separating surfaces.* Examples of such discrete normal surfaces do not put any restrictions on the topology of M in both M^+ and M^-. In particular, M can be extended to a closed combinatorial 3-manifold $\hat{M}$ in which $S_{(V_1,V_2)}$ is no longer separating $\hat{M}$ into two pieces $\hat{M}^\pm$. This gives rise to a family of examples of discrete normal surfaces that are no slicings (in $\hat{M}$).

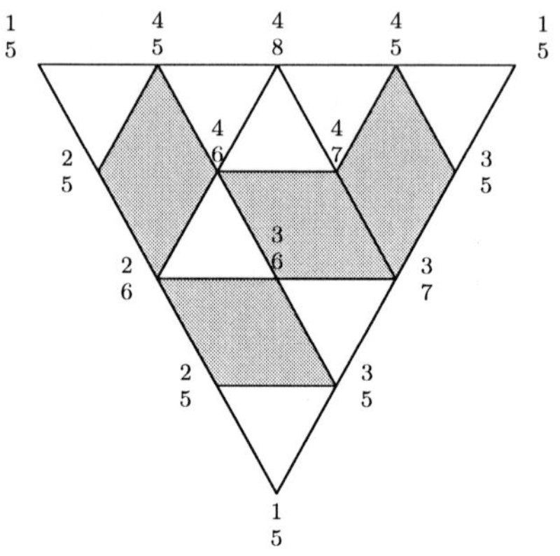

Figure 3.2.1: Slicing of C_1 in between $\{1,2,3,4\}$ and $\{5,6,7,8\}$ with the minimum number of facets.

Example 3.2.3 (Separating surfaces). Consider a combinatorial version of the cylinder $\mathrm{Cyl}_g := S_g \times [0,1]$ where S_g is a triangulated oriented surface of genus g. Cyl_g is a bounded 3-manifold which can be modified into a closed triangulated manifold by gluing suitably bounded triangulated manifolds to each boundary component. Thus, a slicing through Cyl_g which is disjoint to its boundary can be seen as a separating surface of genus g. In particular, the combinatorial properties of such a slicing do not depend on topological properties of the surrounding manifold, as handles may be attached to the surrounding manifold which do not alter the slicing.

In the case $g = 0$ we can consider the cylinder

$$C_1 = \langle\langle 1\,2\,3\,5\rangle,\langle 1\,2\,4\,5\rangle,\langle 1\,3\,4\,5\rangle,\langle 2\,3\,4\,6\rangle,\langle 2\,3\,5\,6\rangle,\langle 2\,4\,5\,6\rangle,$$
$$\langle 3\,4\,5\,7\rangle,\langle 3\,4\,6\,7\rangle,\langle 3\,5\,6\,7\rangle,\langle 4\,5\,6\,8\rangle,\langle 4\,5\,7\,8\rangle,\langle 4\,6\,7\,8\rangle\rangle$$

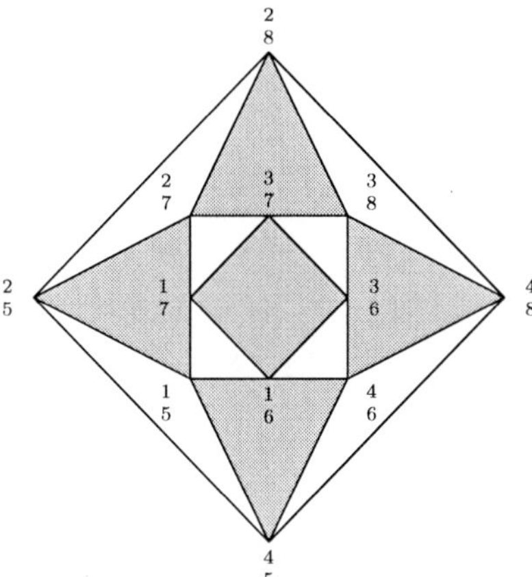

Figure 3.2.2: Schlegel diagram of the slicing of C_2 in between $\{1,2,3,4\}$ and $\{5,6,7,8\}$ (a cuboctahedron).

with boundary

$$\partial C_1 = \langle\langle 1\,2\,3\rangle,\langle 1\,2\,4\rangle,\langle 1\,3\,4\rangle,\langle 2\,3\,4\rangle,\\ \langle 5\,6\,7\rangle,\langle 5\,6\,8\rangle,\langle 5\,7\,8\rangle,\langle 6\,7\,8\rangle\rangle.$$

Since we need at least 8 vertices for a triangulation of the boundary of Cyl_0, we have $f_0(\mathrm{Cyl}_0) \geq 8$. Barnette's Lower Bound Theorem (see [9, 10] or Theorem 1.3.2 in the introduction) tells us that Cyl_0 needs at least $3 \cdot 8 - 10 - 2 = 12$ tetrahedra. Now let us consider a slicing S through Cyl_0 with $\partial(\mathrm{Cyl}_0) \cap S = \emptyset$. If there exist a set of tetrahedra $\Delta \subset \mathrm{Cyl}_0$ disjoint to S, we define $\tilde{\mathrm{Cyl}}_0 = \mathrm{Cyl}_0 \setminus \Delta$. By the arguments above, $\tilde{\mathrm{Cyl}}_0$ again has at least 12 tetrahedra and all of them intersect with S. Thus, S must have at least 12 facets, 8 of them being triangles. Figure 3.2.1 shows a slicing of C_1 with 8 triangles and a minimum number of 4 quadrilaterals. Note that the slicing is a subdivided tetrahedron.

A slightly different cylinder on 8 vertices

$$C_2 = \langle\langle 1\,2\,3\,7\rangle,\langle 1\,2\,4\,5\rangle,\langle 1\,2\,5\,7\rangle,\langle 1\,3\,4\,6\rangle,\langle 1\,3\,6\,7\rangle,\langle 1\,4\,5\,6\rangle,\langle 1\,5\,6\,7\rangle,\\ \langle 2\,3\,4\,8\rangle,\langle 2\,3\,7\,8\rangle,\langle 2\,4\,5\,8\rangle,\langle 2\,5\,7\,8\rangle,\langle 3\,4\,6\,8\rangle,\langle 3\,6\,7\,8\rangle,\langle 4\,5\,6\,8\rangle\rangle$$

and necessarily the same boundary as C_1 leads to a more symmetric slicing in the form of a cuboctahedron with 14 facets, 8 triangles and 6 quadrilaterals (see Figure 3.2.2).

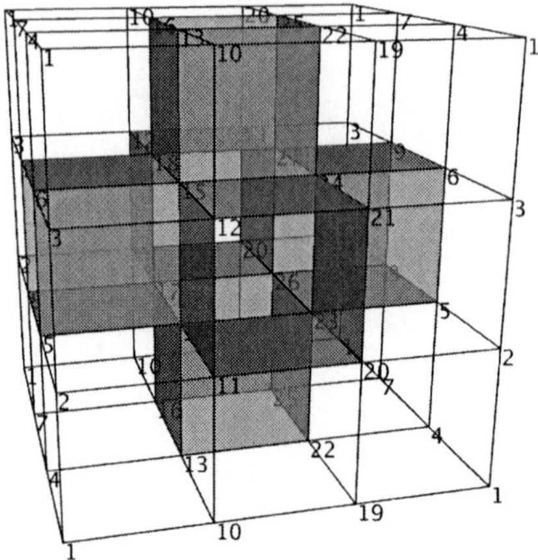

Figure 3.2.3: Combinatorial Heegaard splitting of the canonical $3 \times 3 \times 3$-cube subdivision of the 3-torus. Opposite faces are identified.

C_1 was obtained by canonically subdividing the prism complex $S_4^2 \times [0,1]$ into $3 \cdot 4 = 12$ tetrahedra. This procedure is available for any closed oriented surface S_g and gives rise to a slicing through $S_g \times [0,1]$ with $f_2(S_g)$ quadrilaterals and $2 \cdot f_2(S_g)$ triangles. Let S_g be a vertex minimal triangulation, g arbitrary. The question, whether or not this type of subdivision contains the minimum number of quadrilaterals needed for a slicing of $S_g \times [0,1]$ is interesting but does not seem to be answered yet. It is closely related to a generalization of Theorem 3.5.4. An equivalent reformulation would be a follows.

Question 3.2.4. Given a triangulation of Cyl_g such that no vertex lies in the interior of Cyl_g. What is the minimum number of tetrahedra that do not share a triangle with the boundary of Cyl_g?

Example 3.2.5 (Combinatorial Heegaard splittings). It is well known that the 3-torus $\mathbb{T}^3$ is a 3-manifold of Heegaard genus 3. By a *combinatorial Heegaard splitting* we mean the decomposition of a combinatorial 3-manifold into two polyhedral handlebodies such that the common boundary is a slicing. It seems likely that the minimal genus of a combinatorial Heegaard splitting equals the Heegaard genus of the underlying manifold, although this does not seem to be trivial. Hence, a visualization of a minimal combinatorial splitting of the 3-torus would be interesting.

The decomposition of $\mathbb{T}^3$ as a $3 \times 3 \times 3$ subdivided cube with a pairwise identification of opposite faces and $3^3 = 27$ vertices admits a splitting of (minimal) genus 3 as shown in Figure 3.2.3. We can easily transform the cube decomposition into a combinatorial manifold by subdividing each cube into six tetrahedra. The embedded surface of genus 3 is a sub-complex and thus not a slicing. However, by a slight perturbation such an example is obtained. Note that the perturbed slicing would be too complex to admit a useful visualization in this context.

3.3 The combinatorial structure of slicings

As we have already seen, it is difficult to link global properties of a slicing to its surrounding combinatorial manifold or pseudomanifold. However, we will observe in the following that its local combinatorial structure in fact depends on the manifold.

Remark 3.3.1. Slicings are very special polyhedral manifolds. In order to see this, let P be a facet of a slicing S of a combinatorial d-manifold or d-pseudomanifold M. Then

$$P \cong_{\text{comb.}} \Delta^{d-1-k} \times \Delta^k, \tag{3.3.1}$$

where $0 \leq k \leq \frac{d-1}{2}$. Hence, S contains at most $\lfloor \frac{d+1}{2} \rfloor$ different types of polytopes as facets (see Lemma 3.7.1 for a proof of Equation (3.3.1)). In particular, Equation (3.3.1) implies that a 2-dimensional slicing (or a discrete normal surface) only consists of triangles Δ^2 and quadrilaterals $\Delta^1 \times \Delta^1$. Hence, the search for relations between the number of triangles and quadrilaterals in the 2-dimensional case seems natural and has been investigated by Kalelkar in [65]. We will return to that question in Section 3.5.

If the slicing of a combinatorial 3-manifold or 3-pseudomanifold M is a vertex figure of M, it obviously contains no quadrilaterals. Hence, any triangulated sphere or surface can be seen as the vertex figure of a suitable combinatorial 3-manifold or 3-pseudomanifold. However, every slicing different from a disjoint union of vertex figures contains quadrilaterals as facets, since in this case both M^- and M^+ at least contain one edge. Thus, every connected slicing of a combinatorial 3-manifold different from the sphere has to contain quadrilaterals, whereas this is not true for combinatorial 3-pseudomanifolds.

However, not only the vertex figures of singular vertices of combinatorial pseudomanifolds are different from slicings of combinatorial manifolds: consider for example the slicing $S_{((1,4,5,6),(2,3,7,8))} \cong \mathbb{T}^2$ through the 8-vertex triangulation of a 3-dimensional (singular) Kummer variety K^3 (from [75]) shown in Figure 3.3.1. The 24 tetrahedra of the triangulation are completely determined by $S_{((1,4,5,6),(2,3,7,8))}$. Note that there exists a basis $\langle\alpha,\beta\rangle$ of $H_1(S_{((1,4,5,6),(2,3,7,8))})$ for which both cycles are quadrilaterals of the form $\langle\langle\binom{a}{c},\binom{a}{d}\rangle,\langle\binom{a}{d},\binom{b}{d}\rangle,\langle\binom{b}{d},\binom{b}{c}\rangle,\langle\binom{b}{c},\binom{a}{c}\rangle\rangle$. If we look at α and β in M^+, they both collapse to an edge and are thus contractible. The same holds for α and β in M^-. In fact, we have $\operatorname{span}(V_i) \cong S^2_4$, $i \in \{1,2\}$. This contradicts with Proposition 3.2.1. As a consequence, $S_{((1,4,5,6),(2,3,7,8))}$ cannot be a slicing of a combinatorial manifold.

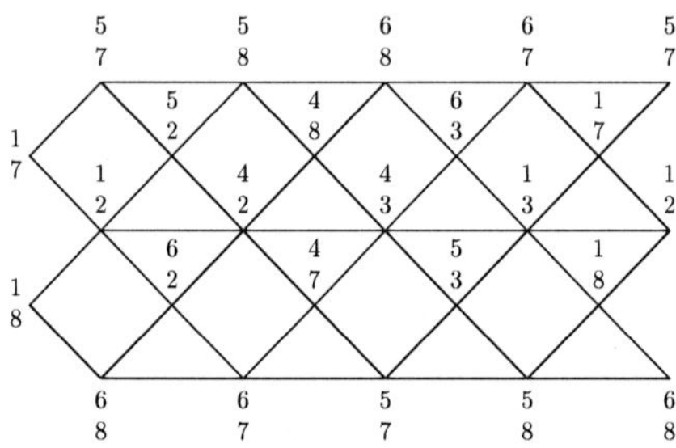

Figure 3.3.1: Highly symmetric centered slicing of genus 1 through an 8-vertex triangulation of a 3-dimensional Kummer variety. See Section 4.5.2 for a series of 3-manifolds obtained by a generalization of this slicing.

Despite the differences discussed above, the following definitions and lemmata hold for both combinatorial manifolds and pseudomanifolds. In the following, let $S_{(V_1,V_2)}$ be the slicing defined by the vertex splitting $V = V_1 \dot{\cup} V_2$ of a combinatorial 3-pseudomanifold.

Definition 3.3.2. Let $S_{(V_1,V_2)} \subset M$ be a slicing of a combinatorial 3-pseudomanifold M defined by $V = V_1 \dot{\cup} V_2$ and let $x \in V_1$ be a vertex. We define

$$C_2^x := \{\delta \in S_{(V_1,V_2)} \mid \delta = \langle\binom{x}{a},\binom{x}{b},\binom{x}{c}\rangle\,;\, a,b,c \in V_2\}, \tag{3.3.2}$$

$$C_1^x := \{\delta \in S_{(V_1,V_2)} \backslash C_2^x \mid \delta = \langle\binom{x}{a},\binom{x}{b}\rangle\,;\, a,b \in V_2\}, \tag{3.3.3}$$

$$C_0^x := \{\delta \in S_{(V_1,V_2)} \backslash (C_2^x \cup C_1^x) \mid \delta = \langle\binom{x}{a}\rangle\,;\, a \in V_2\} \tag{3.3.4}$$

and

$$C^x := \overline{C_2^x \cup C_1^x}.$$

We call C^x the *trace of* x *on* $S_{(V_1,V_2)}$. Analogously, we define C_y for any $y \in V_2$.

Note that $C^x \cup C_0^x = S_{(V_1,V_2)} \cap S_{(\{x\},V\setminus\{x\})}$ where $S_{(\{x\},V\setminus\{x\})}$ is the vertex figure of x which uniquely consists of triangles.

Lemma 3.3.3. *Let $S_{(V_1,V_2)} \subset M$ be a slicing of a combinatorial 3-pseudomanifold M, $x,y \in V_1$ and $a,b \in V_2$. If $\langle \binom{x}{a}, \binom{y}{b} \rangle$ is an edge of $S_{(V_1,V_2)}$, then either $a = b$ or $x = y$ holds.*

Proof. If $\langle \binom{x}{a}, \binom{y}{b} \rangle$ is an edge of a slicing $S_{(V_1,V_2)}$ in a 3-manifold M, it must be a subset of some triangle $\langle \alpha, \beta, \gamma \rangle \in M$. So $\binom{x}{a}$ and $\binom{y}{b}$ are both center points of edges in $\langle \alpha, \beta, \gamma \rangle$. It follows immediately that a, b, x and y cannot all be distinct. □

Corollary 3.3.4. *All quadrilaterals Q of a slicing $S_{(V_1,V_2)}$ of a combinatorial 3-pseudomanifold M are of the form*

$$Q = \left\langle \binom{x}{a}, \binom{x}{b}, \binom{y}{a}, \binom{y}{b} \right\rangle$$

and all triangles T are of the form

$$T = \left\langle \binom{x}{a}, \binom{x}{b}, \binom{x}{c} \right\rangle \quad \text{or} \quad T = \left\langle \binom{x}{a}, \binom{y}{a}, \binom{z}{a} \right\rangle,$$

with $x,y,z \in V_1$, $a,b,c \in V_2$.

Proof. Recall that all facets of $S_{(V_1,V_2)}$ originate from proper sections with a tetrahedron. □

Lemma 3.3.5. *Let $S_{(V_1,V_2)} \subset M$ be a slicing of a combinatorial 3-pseudomanifold M defined by $V = V_1 \dot\cup V_2$ and C^x (C_a) the trace of $x \in V_1$ ($a \in V_2$). Then the following implications hold:*

$$i) \quad x,y \in V_1, x \neq y \quad \Rightarrow \quad C^x \cap C^y = \varnothing \tag{3.3.5}$$

$$ii) \quad x \in V_1, a \in V_2 \quad \Rightarrow \quad C^x \cap C_a = \begin{cases} \left\langle \binom{x}{a} \right\rangle & \text{if } \langle x,a \rangle \in M \\ \varnothing & \text{otherwise} \end{cases} \tag{3.3.6}$$

Proof. This follows immediately from the definition of C^x:

Since $x,y \in V_1$, the upper entries of all vertices in C^x (C^y) are equal to x (y). But $x \neq y$ and the intersection of C^x and C^y must be empty. This proves i).

With the same argument we see that $C^x \cap C_a \subseteq \left\langle \binom{x}{a} \right\rangle$. However, the vertex $\binom{x}{a}$ is part of $S_{(V_1,V_2)}$ if and only if $\langle x,a \rangle$ is an edge of M. This shows ii). □

Lemma 3.3.6. *Let Q be a quadrilateral of a slicing $S_{(V_1,V_2)}$ of a combinatorial 3-pseudomanifold M, C^x the trace of a vertex $x \in V_1$ and C_a the trace of $a \in V_2$ in $S_{(V_1,V_2)}$. Then Q shares at most one edge with C^x and C_a, respectively.*

Proof. By Corollary 3.3.4, any quadrilateral Q of $S_{(V_1,V_2)}$ is of the form $Q = \langle \binom{x}{a}, \binom{x}{b}, \binom{y}{a}, \binom{y}{b} \rangle$. This implies that Q shares exactly one edge with C^x, C^y, C_a and C_b. □

The observations about the local combinatorial structures of slicings made above emphasize the fact that discrete normal surfaces without vertex linking components are completely determined by their quadrilaterals.

3.4 Bistellar moves in slicings

A given slicing can be modified by bistellar moves that are applied to the surrounding manifold. Here, the modification of the surface not only depends on the type of bistellar move performed, but also on the position of the surface in the altered domain. As a consequence, although we only have two distinct types of bistellar moves, we have nine different modifications of an embedded slicing in the 3-dimensional case. In particular, we have the following classification:

1. 0-moves:

A 0-move is the stellar subdivision of a tetrahedron, say $\langle 1,2,3,4 \rangle$. Up to relabeling, $\langle 1,2,3,4 \rangle$ can lie in a sliced 3-manifold $M_{(V_1,V_2)}$ in three essentially different ways:

i) $\{1,2,3,4\} \subset V_1$

ii) $\{1,2,3\} \subset V_1$ and $\{4\} \subset V_2$

iii) $\{1,2\} \subset V_1$ and $\{3,4\} \subset V_2$

For each of these cases, the center of the subdivision can be added either to V_1 or to V_2. See Figures 3.4.1 – 3.4.3 for all five cases.

In the first case (cf. Figure 3.4.1) the vertices of the tetrahedron are entirely contained in V_1. The transformation on the left hand side is the interesting one. The center vertex 5 is added to V_2. Thus, the complete vertex figure of vertex 5 is added to $S_{(V_1,V_2)}$. In particular, we have the topological modification $S_{(V_1,V_2)} \mapsto S_{(V_1,V_2)} \dot{\cup} S^2$. If vertex 5 is added to V_2, the whole setting is disjoint to $S_{(V_1,V_2)}$ and it remains unchanged.

In the second case (cf. Figure 3.4.2) we have the triangle $\langle \binom{1}{4}, \binom{2}{4}, \binom{3}{4} \rangle$ before applying the 0-move. If vertex 5 is added to V_1 (left side), the tetrahedra $\langle 1,2,4,5 \rangle$, $\langle 1,3,4,5 \rangle$ and

$\langle 2,3,4,5\rangle$ are cut in a triangle. $\langle 1,2,3,5\rangle$ is disjoint. If vertex 5 is added to V_2, the first three tetrahedra are sliced in a quadrilateral and the fourth in a triangle, resulting in the figure on the right side.

In the third case (cf. Figure 3.4.3) the tetrahedron $\langle 1,2,3,4\rangle$ is sliced symmetrically. As a consequence, it does not matter if vertex 5 is added to V_1 or to V_2. Two of the occurring tetrahedra are sliced symmetrically, the other two in a triangle.

2. 1-moves:

A 1-move transforms a double cone over a triangle into a bouquet of 3 tetrahedra joined along a common edge. 5 vertices $\{1,2,3,4,5\}$ are involved which split into two groups, the centered triangle $\langle 1,2,3\rangle$ and the missing diagonal $\{4,5\}$. This gives rise to six different cases:

i)	$\{3\}\subset V_1$	and	$\{1,2,4,5\}\subset V_2$
ii)	$\{4\}\subset V_1$	and	$\{1,2,3,5\}\subset V_2$
iii)	$\{2,3\}\subset V_1$	and	$\{1,4,5\}\subset V_2$
iv)	$\{1,2,3\}\subset V_1$	and	$\{4,5\}\subset V_2$
v)	$\{3,4\}\subset V_1$	and	$\{1,2,5\}\subset V_2$
vi)	$\{1,2,3,4,5\}\subset V_1$	or	$\{1,2,3,4,5\}\subset V_2$

The cases are described in detail in Figures 3.4.4 – 3.4.8. Since a 1-move adds no vertex, case vi) is disjoint to any slicing. Hence, this case is not shown.

In the first case (cf. Figure 3.4.4) one vertex of the triangle $\langle 1,2,3\rangle$ is isolated from the rest. Thus, $\langle 1,2,3,4\rangle$ and $\langle 1,2,3,5\rangle$ are cut in a triangle. After the 1-move we have three tetrahedra, namely $\langle 1,2,4,5\rangle$, $\langle 1,3,4,5\rangle$ and $\langle 2,3,4,5\rangle$. The first two are cut in a triangle, the third one is disjoint.

In the second case (cf. Figure 3.4.5) one vertex of the missing diagonal is isolated. Thus, an entire tetrahedron (here: $\langle 1,2,3,5\rangle$) is in V_2, and just one tetrahedron is sliced in a triangle. After the 1-move all tetrahedra are sliced in a triangle since all tetrahedra contain the vertex 4 (and 4 is the only vertex in V_1).

The third case (cf. Figure 3.4.6) shows two out of three vertices of the triangle $\langle 1,2,3\rangle$ isolated. As a consequence, we have 2 quadrilaterals before and one quadrilateral and two triangles after the 1-move. Note that on the left side two tetrahedra contain the triangle $\langle 1,4,5\rangle$ and that $\langle 2,3,4,5\rangle$ is sliced symmetrically.

In the fourth case (cf. Figure 3.4.7) the triangle and the missing diagonal are separated. Thus we have two disjoint triangles. Since on the left side all tetrahedra contain $\langle 4,5\rangle$, the

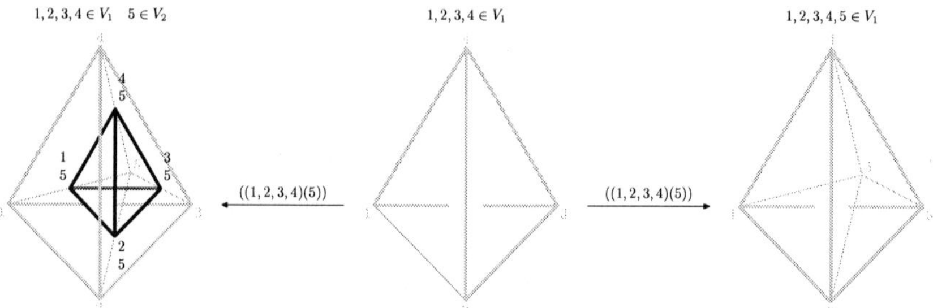

Figure 3.4.1: First type of modification by a 0-move: A connected component is added to the slicing – or not.

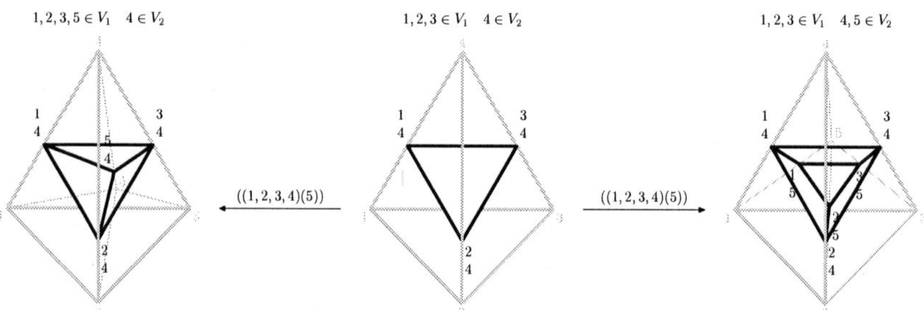

Figure 3.4.2: Second type of modification by a 0-move: The topological type of the slicing remains unchanged (see also Figure 3.4.5).

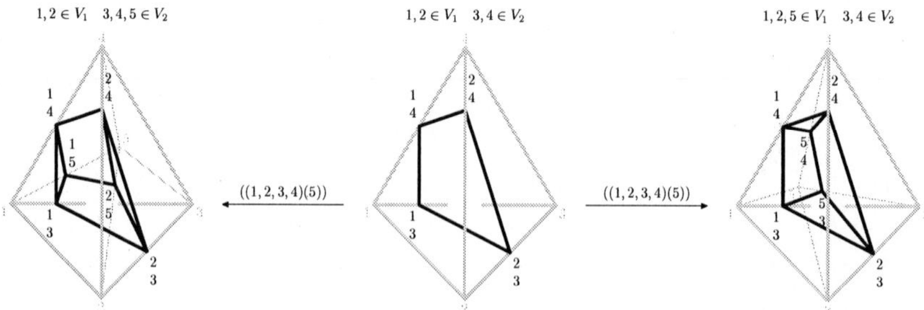

Figure 3.4.3: Third type of modification by a 0-move: The symmetric case.

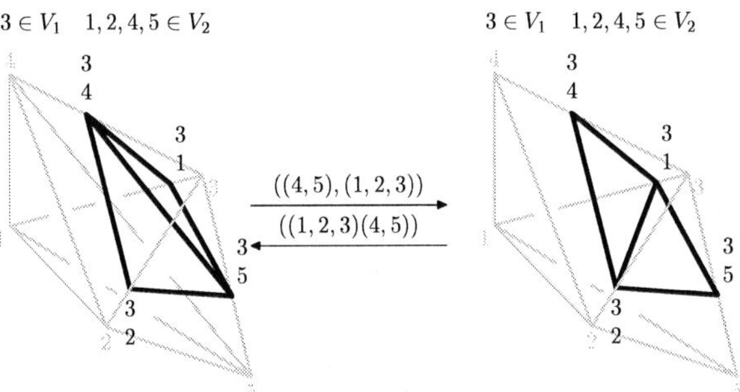

Figure 3.4.4: First type of modification by a 1-move.

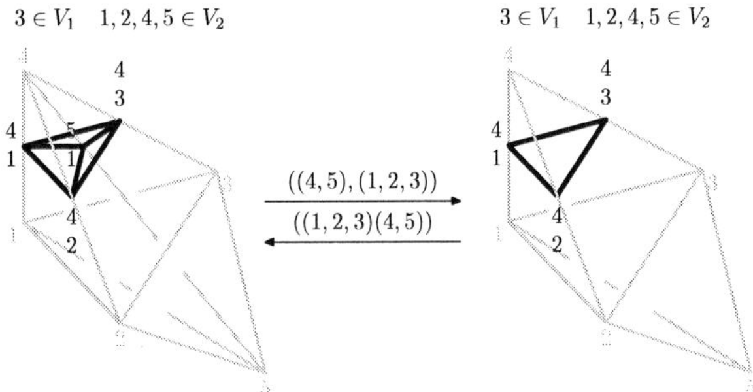

Figure 3.4.5: Second type of modification by a 1-move: A stellar subdivision of a triangle. This is the same modification as observed in the second type of a 0-move (see Figure 3.4.2)

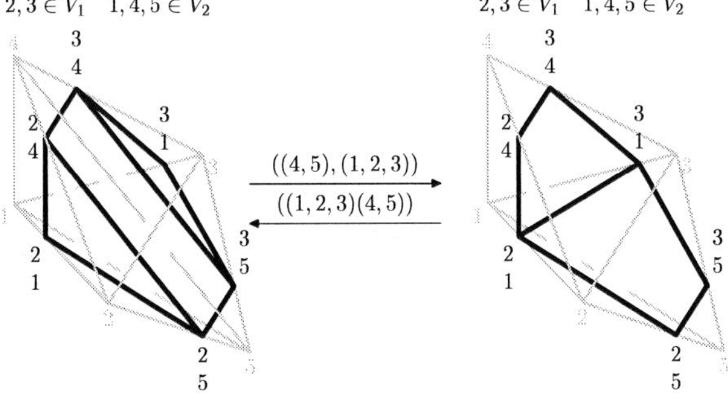

Figure 3.4.6: Third type of modification by a 1-move.

slicing locally transforms into a cylinder around the inserted edge, homotopy equivalent to its link. Topologically this operation is a handle addition that increases not only the genus of the slicing by one but also the number of quadrilaterals by three.

In the fifth case one vertex of $\langle 1,2,3\rangle$ and one of the missing edge are separated from the other vertices. One tetrahedron is cut in a triangle, one is cut in a quadrilateral. After the 1-move the two tetrahedra that contain the vertex of the triangle are sliced in the center, the third one is cut along a triangle.

Our main observation of the modifications of embedded slicings by bistellar moves is that Pachner's theorem does not hold for embedded slicings: The topological type of a slicing can be changed by the operations shown in Figure 3.4.2 and 3.4.7. In addition, it is an interesting fact that the handle addition shown in Figure 3.4.7 adds three quadrilaterals to the slicing. This is a first indication for the motivation of Conjecture 3.5.6.

3.5 Triangles vs. quadrilaterals

Discrete normal surfaces are polyhedral maps consisting of triangles and quadrilaterals. Although triangulated discrete normal surfaces exist, we already pointed out that any discrete normal surface of a combinatorial 3-manifold with non-trivial genus has to contain quadrilateral facets and thus is not simplicial. The relation between the genus of a normal surface and the number of quadrilaterals was already investigated by Kalelkar:

Theorem 3.5.1 (Kalelkar, [65]). *Let S be a closed, oriented, connected normal surface of a pseudotriangulation of a closed 3-manifold M, g its genus. Then we have for the number of quadrilaterals q of S*

$$g \le \frac{7}{2} q. \tag{3.5.1}$$

Although Theorem 3.5.1 holds for arbitrary values of g it merely assures the appearance of one quadrilateral for normal surfaces of genus less than or equal to 3 and two quadrilaterals in the case $3 < g \le 7$. However, a lot more quadrilaterals occur in practice. In particular, we have no non-trivial examples of discrete normal surfaces of combinatorial 3-manifolds satisfying equality in (3.5.1). In Chapter 3.3 we investigated some rules describing how strongly connected components of triangles determine areas of quadrilaterals in the case of slicings. Moreover, we know from Proposition 3.2.2 that an increasing genus of a slicing yields an increasing size of the manifold, thus an increasing minimum number of vertices, and in

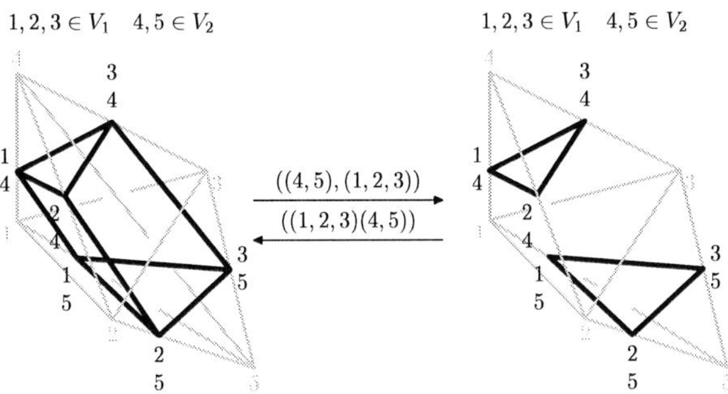

Figure 3.4.7: Fourth type of modification by a 1-move: A handle addition performed on the slicing.

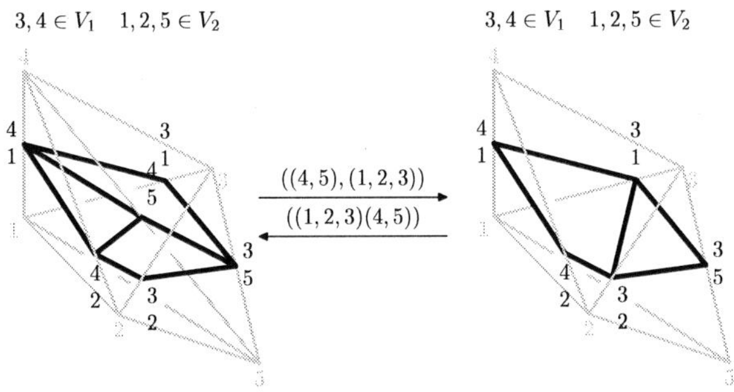

Figure 3.4.8: Fifth type of modification by a 1-move.

turn an increasing size of the slicing. In order to improve Inequality (3.5.1) in a combinatorial setting, we will start with some universal observations on combinatorial 3-manifolds before imposing some restrictions on the generic case.

Let M be an orientable, combinatorial 3-manifold with f-vector $f = (f_0, f_1, f_2, f_3)$ and V the set of vertices. From the Lower Bound Theorem (LBT, see Theorem 1.3.2 in the introduction) for combinatorial manifolds we get the following restrictions on the number of edges f_1:

$$4f_0 - 10 \leq f_1 \leq \binom{f_0}{2} \tag{3.5.2}$$

(here the rightmost inequality is just the trivial upper bound $f_1 \leq \binom{f_0}{2}$). In addition, the Dehn-Sommerville equations for 3-manifolds hold:

$$\begin{aligned} f_0 - f_1 + f_2 - f_3 &= 0 \\ 2f_2 - 4f_3 &= 0. \end{aligned}$$

Now, let $S_{(V_1,V_2)}$ be a slicing defined by the vertex partition $V = V_1 \dot\cup V_2$ and let n be the number of vertices of $S_{(V_1,V_2)}$. Then the obvious condition

$$n \leq |V_1||V_2| \tag{3.5.3}$$

holds, with equality whenever every vertex of V_1 is connected to every vertex of V_2 by an edge. Moreover, we have the following equalities which we call *Dehn-Sommerville equations for slicings*:

$$\begin{aligned} n - e + t + q &= 2 - 2g, \\ -2e + 3t + 4q &= 0, \end{aligned}$$

where e is the number of edges, t the number of triangles, q the number of quadrilaterals and g the genus of $S_{(V_1,V_2)}$. Note that in the following we will call the 4-tuple (v, e, t, q) the f-vector of a slicing despite the fact that the f-vector of $S_{(V_1,V_2)}$ seen as a polyhedral map would be the 3-tuple $(v, e, t + q)$.

Finally, we have another obvious relation between M and $S_{(V_1,V_2)}$: Since every tetrahedron of M contains at most one facet of $S_{(V_1,V_2)}$ one has

$$f_3 - t - q \geq 0. \tag{3.5.4}$$

More precisely, the number $f_3 - t - q \geq 0$ equals the number of tetrahedra in the span of V_1 and V_2. For a fixed value of n, the Dehn-Sommerville equations for 3-manifolds and slicings induce a linear system of equations of dimension 4×7 with rank 4:

f_1	f_2	f_3	v	e	t	q	
-1	1	-1					$-n$
	-2	4					0
			1	-1	1	1	$2-2g$
				-2	3	4	0

Solutions of these equations additionally have to fulfill the Inequalities (3.5.2), (3.5.3) and (3.5.4).

Note that we cannot expect useful information from the above linear system in the general case (a fact that is also implied by the separating surfaces from Section 3.2). However, if we restrict M to be 2-neighborly, the situation is different.

In the following, let $S_{(V_1,V_2)}$ be a slicing of a 2-neighborly combinatorial 3-manifold M such that $|V_1| + c = |V_2| - c$ for a $c = \frac{x}{2}$, $x \in \mathbb{N}$. In this case we get a 4×4 system of the form

f_3	e	t	q	
1				$\binom{n}{2} - n$
	-1	1	1	$2 - 2g - \frac{n^2}{4} + c^2$
	-2	3	4	0
1		-1	-1	$\geq 0.$

As the system is of rank 4, n and c completely determine the combinatorial properties of the discrete normal surface up to the relation $f_3 \geq t + q$.

From this we can deduce our main result.

Theorem 3.5.2. *Let $S_{(V_1,V_2)}$, $|V_1| + c = |V_2| - c$, be a slicing of a 2-neighborly combinatorial 3-manifold M of genus g, n the number of vertices of M and let q be the number of quadrilaterals of $S_{(V_1,V_2)}$. Then we have*

$$q \geq 4g + \frac{3n}{2} - (4 + 2c^2). \tag{3.5.5}$$

For $c = 0$, Inequality (3.5.5) is sharp for all values $g = \binom{\frac{n}{2}-1}{2}$.

Proof. The row echelon form of the linear system above is

f_3	e	t	q	
1				$\binom{n}{2} - n$
	-1	1	1	$2 - 2g - \frac{n^2}{4} + c^2$
		1	2	$4g - 4 + \frac{n^2}{2} - 2c^2$
			1	$\geq 4g + \frac{3n}{2} - (4 + 2c^2)$.

This proves Inequality (3.5.5).

It remains to show that Inequality (3.5.5) is sharp in the case $c = 0$ and $g = \binom{\frac{n}{2}-1}{2}$. The fact that $c = 0$ implies that $n = 2k$ for some $k \in \mathbb{N}$ and thus $g = \binom{k-1}{2}$ is well defined. By replacing the value of g with $\binom{k-1}{2}$ in Inequality (3.5.5) we get

$$q \geq 4\binom{k-1}{2} + 3k - 4 = \binom{2k}{2} - 2k \geq f_3$$

for the number of tetrahedra of a combinatorial 3-manifold with $2k$ vertices. The last step follows from the Dehn-Sommerville equations for combinatorial 3-manifolds and Inequality (3.5.4), which implies that $q = 4g + 3k - 4$ and $t = 0$. Hence, a slicing satisfying equality in Inequality (3.5.5) has to be embedded in a 3-manifold with the maximum number of facets and has to slice all of them symmetrically in a quadrilateral.

Now consider the boundary complex $\partial C_4(2k)$ of the cyclic 4-polytope $C_4(2k)$ (cf. [53, 83]). For any k, the boundary complex $\partial C_4(2k)$ can be constructed as follows: Given the dihedral group in the following permutation representation

$$D_{2k} = \langle (1, \ldots, 2k), (2k, 2)(2k-1, 3) \ldots (k+2, k) \rangle$$

with $4k$ elements, take the union of the following $k - 2$ orbits of length $2k$

$$(1, 2, 3, 4)_{2k}, (1, 2, 4, 5)_{2k}, \ldots, (1, 2, k, k+1)_{2k}$$

and the orbit $(1, 2, k+1, k+2)_k$ of length k. The resulting complex $\partial C_4(2k)$ is a 2-neighborly combinatorial 3-sphere with the maximum number of $2k(k-2) + k = \binom{2k}{2} - 2k$ facets.

As one easily deduces from the group action, all of the facets of $\partial C_4(2k)$ contain exactly two even and exactly two odd vertex labels. This property is also known as *Gale's evenness condition* of cyclic polytopes (see for example [53, p. 62]). If one defines $V_1 = \{1, 3, \ldots, 2k-1\}$ and $V_2 = \{2, 4, \ldots, 2k\}$, neither $\mathrm{span}(V_1)$ nor $\mathrm{span}(V_2)$ contains a triangle, and the induced slicing intersects with all tetrahedra of $\partial C_4(2k)$ in a quadrilateral and hence is an example of equality in Inequality (3.5.5). □

Example 3.5.3. The simplest case of equality in Inequality (3.5.5) is the equilibrium set in the boundary complex of the join of two triangles:

$$\begin{aligned}\partial C_4(6) = \langle \langle 1\,2\,3\,4\rangle, \quad &\langle 1\,2\,3\,6\rangle, \quad \langle 1\,2\,4\,5\rangle, \\ \langle 1\,2\,5\,6\rangle, \quad &\langle 1\,3\,4\,6\rangle, \quad \langle 1\,4\,5\,6\rangle, \\ \langle 2\,3\,4\,5\rangle, \quad &\langle 2\,3\,5\,6\rangle, \quad \langle 3\,4\,5\,6\rangle\rangle.\end{aligned}$$

By separating the odd from the even vertex labels as above, one obtains the slicing

$$\begin{aligned}S_{(\{1,3,5\},\{2,4,6\})} = \langle \langle \tbinom{1}{2}, \tbinom{1}{4}, \tbinom{3}{2}, \tbinom{3}{4}\rangle, \quad &\langle \tbinom{1}{2}, \tbinom{1}{6}, \tbinom{3}{2}, \tbinom{3}{6}\rangle, \quad \langle \tbinom{1}{2}, \tbinom{1}{4}, \tbinom{5}{2}, \tbinom{5}{4}\rangle, \\ \langle \tbinom{1}{2}, \tbinom{1}{6}, \tbinom{5}{2}, \tbinom{5}{6}\rangle, \quad &\langle \tbinom{1}{4}, \tbinom{1}{6}, \tbinom{3}{4}, \tbinom{3}{6}\rangle, \quad \langle \tbinom{1}{4}, \tbinom{1}{6}, \tbinom{5}{4}, \tbinom{5}{6}\rangle, \\ \langle \tbinom{3}{2}, \tbinom{3}{4}, \tbinom{5}{2}, \tbinom{5}{4}\rangle, \quad &\langle \tbinom{3}{2}, \tbinom{3}{6}, \tbinom{5}{2}, \tbinom{5}{6}\rangle, \quad \langle \tbinom{3}{4}, \tbinom{3}{6}, \tbinom{5}{4}, \tbinom{5}{6}\rangle\rangle,\end{aligned}$$

which by construction is the standard 3×3-grid torus; see Figure 3.5.1.

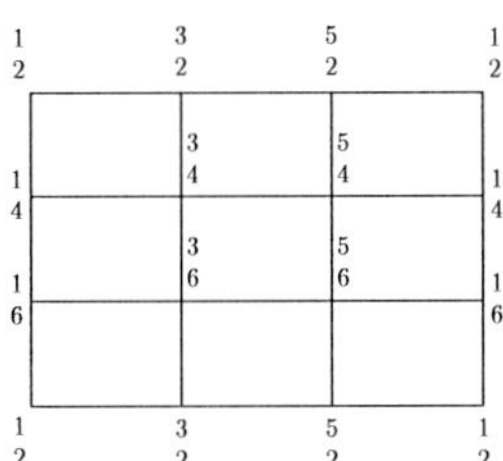

Figure 3.5.1: The standard 3×3-grid torus as a quadrangulated discrete normal surface of genus 1 of the combinatorial 3-sphere $\partial C_4(6)$.

Looking at slicings $S_{(V_1,V_2)}$ where either $\dim(\mathrm{span}(V_1)) = 1$ or $\dim(\mathrm{span}(V_2)) = 1$, we can prove the following theorem for arbitrary combinatorial 3-manifolds.

Theorem 3.5.4. *Let M be a combinatorial 3-manifold, $S_{(V_1,V_2)}$ a connected slicing of genus g of M, q its number of quadrilaterals, $\dim(\mathrm{span}(V_i)) = 1$ and $n := |V_i|$ for $i = 1$ or $i = 2$. Then*

$$q \geq 3(n + g - 1). \tag{3.5.6}$$

In particular, this applies to all quadrangulated slicings.

Proof. To obtain a connected slicing $S_{(V_1,V_2)}$ of genus g, both the span of V_1 and of V_2 must be connected and contain a graph of genus g. The Euler characteristic tells us that a graph of genus g with n vertices needs exactly $g+n-1$ edges ($n \geq k$ for $\binom{k-2}{2} < g \leq \binom{k-1}{2}$). In addition, every edge of the manifold is surrounded by at least three tetrahedra. Since $\dim(\mathrm{span}(V_i)) = 1$ for $i = 1$ or $i = 2$, each of the at least $3(g+n-1)$ tetrahedra surrounding the at least $n+g-1$ edges of the graph $\mathrm{span}(V_i)$ must be distinct from all the others. In particular, exactly one edge of each of these tetrahedra lies in $\mathrm{span}(V_i)$. Thus, the intersection of $S_{(V_1,V_2)}$ with each of the tetrahedra appears as a quadrilateral in the slicing.

If a slicing is quadrangulated, then $\dim(\mathrm{span}(V_1)) = \dim(\mathrm{span}(V_2)) = 1$ and the conditions of Theorem 3.5.4 are fulfilled. □

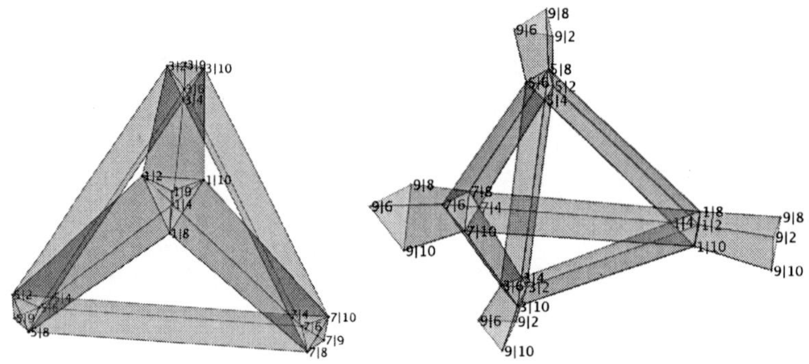

Figure 3.5.2: Slicings of the nearly neighborly centrally symmetric 10-vertex sphere $^{3}10_{1}^{22}$ (from [92], see also [53, Section 6.4]). On the left: Slicing behind the complete 1-skeleton of $\langle 1,3,5,7\rangle$ - a surface of genus 3. On the right: Slicing between the vertices $V_1 := \{1,3,5,7,9\}$ and $V_2 := \{2,4,6,8,10\}$. Both V_1 and V_2 span a complete graph $K_{5,5}$, generating a quadrangulated surface of genus 6.

For some examples of slicings that attain equality in Inequality (3.5.6), consider the nearly neighborly centrally symmetric 4-polytope P with 10 vertices, 40 edges and transitive automorphism group

$$\mathrm{Aut}(P) = \langle (1,4,7,6,9,2)(3,10)(5,8), (1,7,3,9)(2,8,4,6)\rangle \cong C_2 \times \sigma_5$$

due to Grünbaum (cf. [53, Section 6.4]). The boundary $\partial P = S$ is a combinatorial 3-sphere (see [92, ${}^{3}10^{22}_{1}$]) and equals the Aut(P)-orbit

$$S = (1,2,3,4)_{30}$$

of length 30. S admits an rsl-function f such that every slicing induced by f attains equality in Inequality (3.5.6):

slicing	g	n	# quadrilaterals
$\{1\},\{2,3,4,5,6,7,8,9,10\}$	0	1	0
$\{1,3\},\{2,4,5,6,7,8,9,10\}$	0	2	3
$\{1,3,5\},\{2,4,6,7,8,9,10\}$	1	3	9
$\{1,3,5,7\},\{2,4,6,8,9,10\}$	4	3	18
$\{1,3,5,7,9\},\{2,4,6,8,10\}$	5	6	30

For an illustration of some of the slicings see Figure 3.5.2.

In addition, there is a 15-vertex triangulation C of $S^2 \times S^1$ (see [92, ${}^{3}15^{1}_{17}$]) with a cyclic automorphism group

$$\mathrm{Aut}(C) = \langle (1,\dots,15) \rangle \cong \mathbb{Z}_{15},$$

generated by the $\mathbb{Z}_{15}$-orbits

$$C = (1,2,3,5)_{15} \cup (1,2,3,12)_{15} \cup (1,2,4,6)_{15} \cup (1,2,5,7)_{15} \cup (1,2,6,7)_{15},$$

which admits a number of examples of non-orientable surfaces that are obtained via slicings. All these fulfill equality in Inequality (3.5.6):

slicing	g	n	# quadrilaterals
$\{1,4,7\}$, $\{2,3,5,6,8,\dots,15\}$	1	3	9
$\{1,4,7,10\}$, $\{2,3,5,6,8,9,11,12,13,14,15\}$	3	4	18
$\{1,4,7,10,13\}$, $\{2,3,5,6,8,9,11,12,14,15\}$	6	5	30

The examples were found by an enumerative search using the support for slicings of the GAP-package simpcomp [40, 41, 42]. The complexes that were used to construct the slicings are included in the built-in library of simpcomp.

Remark 3.5.5. It seems natural to assume that Inequality (3.5.6) also holds in the generic case of arbitrary combinatorial 3-manifolds since the condition $\dim(\mathrm{span}(V_1)) = 1$ only holds for small examples. However, this is not true:

An analogue of Theorem 3.5.4 in the general case would state that at least 39 quadrilaterals are needed for a surface of genus 8. But there exist orientable and non-orientable slicings of genus 8 with only 35 quadrilaterals (slicing $(\{1,3,5,7,9,11,13\},\{2,4,6,8,10,12,14\})$ of $^{3}14^{1}_{3}$ and $(\{1,3,5,7,9,11,13\},\{2,4,6,8,10,12,14\})$ of $^{3}14^{1}_{32}$, notations from [92]). Moreover, there is an orientable slicing of genus 2 with only 14 quadrilaterals, whereas the lower bound in Theorem 3.5.4 would be 15 $((\{1,2,16,21\},\{3,4,5,6,7,8,9,10,11,12,13,14,15,17,18,19,20,22\})$ of a triangulation of $(S^2 \times S^1)^{\#14}$ from [91]).

Nevertheless, we assume that the following holds.

Conjecture 3.5.6. *Let M be a combinatorial 3-manifold and S a slicing of M with q quadrilaterals. Then*

$$q \geq 3 - \frac{3}{2}\chi(S).$$

Purely combinatorial methods do not seem to be suitable to prove Conjecture 3.5.6. However, at least for low genera, a geometric or algebraic approach could lead to new results.

Furthermore, it remains to investigate whether the stated theorems and conjectures can be extended to discrete normal surfaces. Although most proofs found upon the fact that a slicing splits the surrounding combinatorial manifold into two pieces, it is believed that Theorem 3.5.2 and Theorem 3.5.4 are also true for discrete normal surfaces that do not define such a splitting.

3.6 Weakly neighborly slicings

The famous Heawood inequality (cf. Inequality (1.3.1)) provides a lower bound on the number of vertices needed to triangulate an abstract surface with prescribed Euler characteristic. The inequality is sharp whenever the triangulation is 2-neighborly. Note that not all topological types of surfaces admit 2-neighborly triangulations, but whenever they do this triangulation is minimal with respect to the number of vertices (these are the so-called *regular cases* of the Heawood inequality). In contrast to the simplicial case, the notion of a neighborly complex has to be defined differently for a polyhedral map since a polyhedral map which is not simplicial can never be neighborly. This directly leads us to the following.

Definition 3.6.1 (Weakly neighborly polyhedral map, from [19]). We call a polyhedral map *weakly neighborly* if for any two vertices there is a face containing both of them.

Together with the above definition we can give a lower bound on the number of vertices for polyhedral m-gon maps (cf. Definition 1.4.2).

Proposition 3.6.2 (Lemma 3 in [19]). *Let S be a polyhedral m-gon map with Euler characteristic $\chi(S)$, then S needs at least*

$$n \geq \frac{1}{2}\left((2m+1) + \sqrt{(2m+1)^2 - 8m\chi(S)}\right) \tag{3.6.1}$$

vertices, with equality if and only if S is weakly neighborly.

Proof. Proposition 3.6.2 is a special case of [19, Lemma 3] with $n = V$ and $m = k_i$ for all $1 \leq i \leq p = F$, where $p = \frac{2}{m-2}(n - \chi(S))$ denotes the number of m-gons of S. □

In the case $m = 4$ we have

$$n \geq \frac{1}{2}\left(9 + \sqrt{81 - 32\chi(S)}\right). \tag{3.6.2}$$

Thus, the 9-vertex grid torus from Figure 3.5.1 is an example for a weakly neighborly quadrangulation of the torus.

Let us now consider the case of polyhedral maps consisting of triangles and quadrilaterals.

Lemma 3.6.3 (cf. Lemma 2 in [19]). *Let S be a polyhedral map with n vertices consisting of t triangles and q quadrilaterals, then the following statements are equivalent:*

a) S is weakly neighborly,

b) S has $e = \binom{n}{2} - 2q$ edges,

c) $n = \frac{1}{2}\left(7 + \sqrt{49 + 8q - 24\chi(S)}\right)$,

d) $q = \binom{n-3}{2} + 3\chi(S) - 6$.

Proof. *a)* $\Leftrightarrow$ *b)* Every quadrilateral in S decreases the maximal number of edges in S by 2. In particular, S has exactly $e = \binom{n}{2} - 2q$ edges. This can be seen as a special case of [19, Lemma 2 (5)] with $n = V$, t times $k_i = 3$ and q times $k_i = 4$.

b) $\Leftrightarrow$ *c)* By the weak pseudomanifold property we have $2e = 3t + 4q$ and with the Euler characteristic $\chi(S) = n - e + \frac{2e-4q}{3} + q$ and $e = \binom{n}{2} - 2q$ we get

$$\chi(S) = n - \frac{1}{3}\left(\binom{n}{2} - 3q\right),$$

which is equivalent to

$$0 = n^2 - 7n - 2q + 6\chi(S).$$

It follows that

$$n = \frac{1}{2}\left(7 \pm \sqrt{49 + 8q - 24\chi(S)}\right), \qquad (3.6.3)$$

where clearly only the greater one of the two solutions of the quadratic equation is valid (see also [19, Lemma 2 (6)] for a similar statement).

c) $\Leftrightarrow$ *d)* Given $\chi(S)$ and n we have at most one non-negative integer q solving Equation (3.6.3). Thus, we get

$$\begin{aligned} n &= \frac{1}{2}(2n) \\ &= \frac{1}{2}\left(7 + \sqrt{(2n-7)^2}\right) \\ &= \frac{1}{2}\left(7 + \sqrt{49 + 4(n^2 - 7n) + 24\chi(S) - 24\chi(S)}\right) \\ &= \frac{1}{2}\left(7 + \sqrt{49 + 8\left(\frac{n^2 - 7n}{2} + 6 - 6 + 3\chi(S)\right) - 24\chi(S)}\right) \\ &= \frac{1}{2}\left(7 + \sqrt{49 + 8\left(\binom{n-3}{2} - 6 + 3\chi(S)\right) - 24\chi(S)}\right). \end{aligned}$$

This directly leads to

$$q = \binom{n-3}{2} - 6 + 3\chi(S).$$

□

In a series of papers (see [18, 6, 19]), Altshuler and Brehm classified weakly neighborly maps on surfaces. It is obvious (for example by looking at the examples given in [18]) that not all weakly neighborly polyhedral maps can be realized as discrete normal surfaces or slicings of combinatorial pseudomanifolds. In fact, this case is a rare exception. In order to see that, just note that every edge of a discrete normal surface corresponds to a triangle of the surrounding 3-manifold and the number of triangles of a combinatorial 3-manifold is restricted by the Dehn-Sommerville equations (cf. Lemma 3.3.3). More precisely we have the following condition.

Lemma 3.6.4. *Let* $S_{(V_1,V_2)}$ *be a weakly neighborly slicing of a combinatorial 3-pseudomanifold* M *induced by the vertex partition* $V_1 \dot{\cup} V_2 = V$, n_i *the number of vertices that lie in the boundary of* $\operatorname{span}(V_i)$, $i \in \{1,2\}$, *then*

$$n_1 n_2 (15 - n_1 n_2 - n_1 - n_2) = 12\chi(S_{(V_1,V_2)}). \tag{3.6.4}$$

Proof. Let us first verify that any pair of vertices $a \in V_1$, $y \in V_2$ in the boundary of $\operatorname{span}(V_1)$, $\operatorname{span}(V_2)$ occurs as a vertex $\binom{a}{y}$ of $S_{(V_1,V_2)}$: Since a (y) lies in the boundary of $\operatorname{span}(V_1)$ ($\operatorname{span}(V_2)$), there exist two vertices $b \in V_1$ and $x \in V_2$ such that $\langle a,x\rangle$, $\langle b,y\rangle \in M$. If $a = b$ or $x = y$ then $\binom{a}{y}$ is a vertex of $S_{(V_1,V_2)}$. Now let $a \neq b$ and $x \neq y$, then $\binom{a}{x}$ and $\binom{b}{y}$ are both vertices of $S_{(V_1,V_2)}$. By Lemma 3.3.3 they cannot form an edge of $S_{(V_1,V_2)}$. However, since $S_{(V_1,V_2)}$ is weakly neighborly, $\binom{a}{x}$ and $\binom{b}{y}$ lie in one facet and thus $\langle\binom{a}{x},\binom{a}{y},\binom{b}{x},\binom{b}{y}\rangle$ must be one of its quadrilaterals, and $\binom{a}{y}$ is a vertex of $S_{(V_1,V_2)}$. As a consequence the slicing has $n_1 n_2$ vertices.

Now we show that the boundary of $\operatorname{span}(V_i)$, $i \in \{1,2\}$, is 2-neighborly. Let $a, b \in V_1$ be two vertices in the boundary of $\operatorname{span}(V_1)$, $a \neq b$. Then there exist two edges $\langle a,x\rangle$ and $\langle b,y\rangle$ with $x, y \in V_2$. If $x \neq y$, then $\langle\binom{a}{x},\binom{b}{y}\rangle$ cannot be an edge of $S_{(V_1,V_2)}$ by Lemma 3.3.3. Since $S_{(V_1,V_2)}$ is weakly neighborly $\langle\binom{a}{x},\binom{a}{y},\binom{b}{x},\binom{b}{y}\rangle$ must be a quadrilateral, $\langle a,b,x,y\rangle$ a tetrahedron of M and $\langle a,b\rangle$ an edge of $\operatorname{span}(V_1)$. If $x = y$, then $\langle\binom{a}{x},\binom{b}{x}\rangle$ must be an edge of $S_{(V_1,V_2)}$ since it is weakly neighborly. Hence, $\{a,b,x\}$ spans a triangle in M and $\langle a,b\rangle$ is an edge. Altogether, the boundary of $\operatorname{span}(V_i)$, $i \in \{1,2\}$, must be 2-neighborly. From this and the fact that $S_{(V_1,V_2)}$ has $n_1 n_2$ vertices we can deduce that $S_{(V_1,V_2)}$ must have exactly

$$n_1\binom{n_2}{2} + n_2\binom{n_1}{2} = \binom{n_1 n_2}{2} - 2\left(\binom{n_1 n_2 - 3}{2} + 3\chi(S) - 6\right) \tag{3.6.5}$$

edges, where the right hand side follows from Lemma 3.6.3 *d*). This yields Equation (3.6.4). □

If we furthermore restrict M to be a combinatorial manifold, we can classify all such weakly neighborly slicings.

Theorem 3.6.5. *The only weakly neighborly polyhedral maps that are slicings of a combinatorial 3-manifold M are the boundary of the 3-simplex $\partial\Delta^3$, the boundary of the triangular prism $\partial(\Delta^2 \times \Delta^1)$ and the 3×3-grid torus $\partial\Delta^2 \times \partial\Delta^2$ shown in Figure 3.5.1.*

Proof. Since M is a 3-manifold with $\chi(M) = 0$, $\chi(S)$ must be an even number.

In the case $\chi(S) = 2$, the only integer solutions of Equation (3.6.4) with $\chi(S) = 2$ are $n_1 = 1$, $n_2 = 4$ and $n_1 = 2$, $n_2 = 3$. In the first case we have a (triangulated) vertex figure which has to be 2-neighborly. But the only triangulated 2-neighborly 2-sphere is the boundary of the 3-simplex. The second case must occur as a slicing of a closed combinatorial manifold with $2 + 3 = 5$ vertices (since the span of three vertices cannot have interior faces in dimension 3). The only combinatorial 3-manifold with five vertices is the boundary of the 4-simplex. The slicing between an edge and a triangle yields a weakly neighborly 2-sphere consisting of three quadrilaterals and two triangles: this has to be the boundary of a 3-dimensional prism $\Delta^1 \times \Delta^2$.

In the case $\chi(S) = 0$, since M is a combinatorial 3-manifold, it follows from Proposition 3.2.2 that $n_1, n_2 \geq 3$. Then the only solution for Equation (3.6.4) is $n_1 = n_2 = 3$. Since the span of three vertices cannot have interior faces in dimension 3, the surrounding 3-manifold M must have six vertices (which necessarily forms a triangulated 3-sphere by virtue of Theorem 1.3.7 from Section 1.3) and at least two disjoint empty triangles. This determines the combinatorial type of M to be the boundary of the cyclic 4-polytope with 6 vertices $\partial C_4(6)$ where only one slicing is of genus 1: the 3×3-grid torus shown in Figure 3.5.1. In particular, the Klein bottle cannot appear as a weakly neighborly slicing (by the orientability of S^3).

In the case $\chi(S) \leq -2$ the lower bounds on n_1 and n_2 given in Proposition 3.2.2 and the asymptotic behavior of the left hand side of Equation (3.6.4) do not admit further solutions. □

3.7 Slicings and critical point theory in higher dimensions

In contrast to the theory of discrete normal surfaces we can apply polyhedral Morse theory and the concept of slicings to combinatorial manifolds in arbitrary dimensions. Following Remark 3.3.1, we have the following lemma regarding the polytopal facets of slicings in arbitrary dimensions.

Lemma 3.7.1. *Let M be a combinatorial d-manifold, S a slicing of M, Δ^k the k-simplex and P a facet of S. We then have*

$$P \cong_{\text{comb.}} \Delta^{d-1-k} \times \Delta^k. \tag{3.7.1}$$

In particular, S contains at most $\lfloor \frac{d}{2} \rfloor$ different types of polytopes as facets.

Proof. Let $\Delta^d = \langle 0, \dots, d \rangle$ be the d-simplex. Δ^d can be described by barycentric coordinates as follows:

$$\Delta^d = \left\{ (x_1, \dots, x_{d+1}) \in [0,1]^{d+1} \,\middle|\, \sum_{i=1}^{d+1} x_i = 1 \right\}.$$

A facet P of a slicing S is given by a partition of $V = \{0, \dots, d\} = V_1 \dot{\cup} V_2$. The vertices of P are all center points c_j, $1 \le j \le |V_1||V_2|$, of edges of Δ^d with one vertex in V_1 and the other in V_2 and we have $P = \text{conv}(\{c_1, \dots c_{|V_1||V_2|}\})$. Thus, after resorting the barycentric coordinates, P is given by

$$\begin{aligned} P &= \left\{ (x_1, \dots, x_{d+1}) \in [0,1]^{d+1} \,\middle|\, \sum_{i=1}^{|V_1|} x_i = \frac{1}{2}; \sum_{i=|V_1|+1}^{d+1} x_i = \frac{1}{2} \right\} \\ &= \left\{ (x_1, \dots, x_{|V_1|}) \in [0,1]^{|V_1|} \,\middle|\, \sum_{i=1}^{|V_1|} x_i = \frac{1}{2} \right\} \times \left\{ (x_{|V_1|+1}, \dots, x_{d+1}) \in [0,1]^{|V_2|} \,\middle|\, \sum_{i=|V_1|+1}^{d+1} x_i = \frac{1}{2} \right\} \\ &= \Delta^{|V_1|-1} \times \Delta^{|V_2|-1}. \end{aligned}$$

□

In particular, 2-dimensional slicings consist only of triangles Δ^2 and quadrilaterals $\Delta^1 \times \Delta^1$, 3-dimensional slicings consist of tetrahedra Δ^3 and prisms $\Delta^2 \times \Delta^1$. In the following we will take a closer look at the 16-vertex triangulation of the $K3$ surface $(K3)_{16}$ as well as the 4-dimensional Kummer variety with 16 ordinary double points $(K^4)_{16}$ from Chapter 2 using Morse theory and 3-dimensional slicings.

Combinatorial Morse analysis on $(K3)_{16}$

Let us first recall that the 16-vertex triangulation of the $K3$ surface is a very special object in combinatorial topology:

1. It satisfies equality in the generalized Heawood inequality (cf. Inequality (1.3.8)) for the number n of vertices of a 4-manifold M with Euler characteristic $\chi(M)$.
2. It is 3-neighborly (sometimes called super-neighborly, cf. [77]) and thus tight. See [133, Theorem 5.8] for a characterization of all possible g-vectors of a triangulated $K3$ surface, starting with the minimum $(g_0, g_1, g_2) = (1, 10, 55)$.
3. It is the only known triangulation of a 4-manifold admitting an automorphism group acting 2-*transitively* on the set of vertices (besides the trivial case of the 6-vertex 4-sphere, cf. Section 5.6).

From the view-point of Morse theory, this has the following consequence.

Proposition 3.7.2. *Any rsl-function f defined on $(K3)_{16}$ is a perfect function in the sense that the total number of critical points is* 24. *More precisely, we have $\mu_0(f,\mathbb{F}) = \mu_4(f,\mathbb{F}) = 1$, $\mu_1(f,\mathbb{F}) = \mu_3(f,\mathbb{F}) = 0$ and $\mu_2(f,\mathbb{F}) = 22$. This holds for any choice of a field $\mathbb{F}$.*

Proof. This follows from the fact that the 16-vertex triangulation of the $K3$ surface is a tight triangulation in the sense of [77, 86]. The reason is that the triangulation is 3-neighborly which implies that there are no critical points of index 1: Any subset of vertices spans a connected and simply connected subset. The 2-neighborliness implies that every rsl-function has exactly one critical point of index 0. The rest follows from the duality $\mu_i(f,\mathbb{F}) = \mu_{4-i}(f,\mathbb{F})$ and the Poincaré relation $\mu_0 - \mu_1 + \mu_2 - \mu_3 + \mu_4 = \chi(K3) = 24$. □

Corollary 3.7.3. *Any rsl-function f defined on $(K3)_{16}$ has a critical point of index* 2 *with a multiplicity higher than* 2. *More precisely,* 10 *possible critical vertices have to build up the second Betti number* 22. *This holds for any choice of a field F. Moreover, any slicing of an rsl-function on $(K3)_{16}$ is a connected 3-manifold.*

Proof. The first part is obvious from the Morse inequality $\mu_2(f) \geq b_2(M) = 22$ and the fact that by the 3-neighborliness only the middle vertices (i. e. all but the three on top and the three on bottom) can be critical of index 2. For examples of multiplicity vectors see Table 3.1 below. The second part follows from the fact that there is no critical point of index 1. If there were a disconnected level it would have to be modified into a connected level later, and this procedure requires a critical point of index 1 in between. □

It may be interesting to see how the levels of such a function change when passing through a critical level. It does not seem to be known from differential topology what the possible levels can be for smooth perfect functions on the $K3$ surface. The standard embedding $(z_0, z_1, z_2, z_3) \mapsto (z_i \bar{z}_j)_{ij}$ of a quartic surface in projective 3-space $K3 \to \mathbb{C}P^3 \to S^{14} \to \mathbb{R}^{15}$ induces smooth Morse functions by linear projections from 15-space to $\mathbb{R}$. However, in general these won't be perfect. It is well known that there is no tight smooth embedding or immersion of the $K3$ surface into any Euclidean space [135]. Not too much seems to be known about possible slicings of perfect smooth Morse functions defined on the $K3$ surface. In the PL case we have the following special situation:

An rsl-function on $(K3)_{16}$

$$f_\Omega : (K3)_{16} \to [0,1]$$

is essentially determined by a fixed ordering on the set of vertices $\Omega := \{v_1, \dots, v_{16}\}$ determining the function f_Ω by the condition $0 = f_\Omega(v_1) < \dots < f_\Omega(v_{16}) = 1$. Any slicing $f_\Omega^{-1}(\alpha)$ of a 4-manifold only consists of tetrahedra and 3-dimensional prisms of type $\Delta^2 \times \Delta^1$, induced by proper sections with the 4-simplices of $(K3)_{16}$. In many cases the topological type of $f_\Omega^{-1}(\alpha)$ can be identified using standard techniques. Some of the topological types of the slicings can be predicted:

- **The 3-torus:** Obviously there is a 3-torus as a slicing of the 4-torus. It can be arranged that such a slicing avoids all the 16 fixed points of the involution. Hence, we have the same slicing in the Kummer variety and, by the purely local resolution procedure, also in the $K3$ surface.

- **The real projective 3-space:** The link of any singular point in K^4 is a real projective 3-space. By resolving the singularities we only change a neighborhood of these points. Thus, there are slicings in $(K3)_{16}$ separating such a neighborhood. These are homeomorphic with $\mathbb{R}P^3$.

- **The Poincaré homology sphere Σ^3:** There is a surgery description of the $K3$ surface showing the Poincaré homology sphere as a possible slicing (see [120]). Even though this does not tell about the number of vertices which will be needed it turned out that a certain slicing of the 16-vertex triangulation is this manifold Σ^3, see below.

Proposition 3.7.4. *As slicings of $(K3)_{16}$ we obtain at least the manifolds*

$$S^3, \mathbb{R}P^3, L(3,1), L(4,1), L(5,1), \Sigma^3$$

and a number of other space forms: The 3-torus, the cube space, the octahedron space, the truncated cube space and the prism space $P(3)$. Here Σ^3 denotes the Poincaré homology sphere with a fundamental group of order 120.

Proof. Let $\{1,\dots,16\}$ be the set of vertices of $(K3)_{16}$. The permutation

$$(1,16)(2,15)(3,14)(4,13)(5,12)(6,11)(7,10)(8,9)$$

on the 16 vertices is an automorphism of $(K3)_{16}$, and we have $f^{-1}_{\{1,\dots,16\}}(\frac{1}{2}) \cong \mathbb{T}^3$. We use this slicing as a starting point and analyze all possible slicings of $(K3)_{16}$ around this 3-torus in the middle.

Since $(K3)_{16}$ is 3-neighborly, all slicings with three or less vertices on one side are trivial (i. e. the slicing is a 3-sphere). With four vertices on one side we have two possible situations. Either the tetrahedron formed by the four vertices is contained in the complex (in this case the slicing is clearly trivial), or it is not. In the latter case we have a slicing behind an empty tetrahedron. This type of slicing is a real projective 3-space. Therefore, a simplicial decomposition of the set $|f^{-1}_{\Omega}([0,\alpha])|$, $\frac{3}{15} < \alpha < \frac{4}{15}$ is PL homeomorphic to the mapping cylinder $\mathbf{C}$ of the Hopf map $\tilde{h}$ in Section 2.4. Hence, we can find topological copies of $\mathbf{C}$ in $(K3)_{16}$ (which is not surprising).

Neither the span of $\{1,\dots,8\}$ nor the span of $\{9,\dots,16\}$ contains a 4-simplex of $(K3)_{16}$. Thus, five vertices on one side cannot induce a trivial slicing but such PL homeomorphic to a lens space of type L(4,1), L(3,1) or L(2,1) = $\mathbb{R}P^3$. In the case of six or seven vertices we have the cube space, the octahedron space or the Poincaré homology sphere. As already mentioned, eight vertices on each side result in the 3-torus, the only non-spherical 3-manifold in this series. For a complete list of the topological types of these slicings see Table 3.2.

	$f_{\{1,\ldots,5,7,6,8,9,11,10,12,\ldots,16\}}$		$f_{\{2,\ldots,7,1,8,9,16,10,\ldots,15\}}$		$f_{\{1,\ldots,5,7,6,8,9,11,10,12,\ldots,16\}}$	
$f_i(v)$	v	$\mathbf{m}(v,\mathbb{F}_2)$	v	$\mathbf{m}(v,\mathbb{F}_2)$	v	$\mathbf{m}(v,\mathbb{F}_2)$
0	1	$(1,0,0,0,0)$	2	$(1,0,0,0,0)$	1	$(1,0,0,0,0)$
$\frac{1}{15}$	2	$(0,0,0,0,0)$	3	$(0,0,0,0,0)$	2	$(0,0,0,0,0)$
$\frac{2}{15}$	3	$(0,0,0,0,0)$	4	$(0,0,0,0,0)$	3	$(0,0,0,0,0)$
$\frac{3}{15}$	4	$(0,0,1,0,0)$	5	$(0,0,0,0,0)$	4	$(0,0,1,0,0)$
$\frac{4}{15}$	5	$(0,0,2,0,0)$	6	$(0,0,1,0,0)$	5	$(0,0,1,0,0)$
$\frac{5}{15}$	7	$(0,0,3,0,0)$	7	$(0,0,3,0,0)$	7	$(0,0,2,0,0)$
$\frac{6}{15}$	6	$(0,0,2,0,0)$	1	$(0,0,4,0,0)$	6	$(0,0,4,0,0)$
$\frac{7}{15}$	8	$(0,0,3,0,0)$	8	$(0,0,3,0,0)$	8	$(0,0,3,0,0)$
$\frac{8}{15}$	9	$(0,0,3,0,0)$	9	$(0,0,3,0,0)$	9	$(0,0,3,0,0)$
$\frac{9}{15}$	11	$(0,0,2,0,0)$	16	$(0,0,4,0,0)$	11	$(0,0,4,0,0)$
$\frac{10}{15}$	10	$(0,0,3,0,0)$	10	$(0,0,3,0,0)$	10	$(0,0,2,0,0)$
$\frac{11}{15}$	12	$(0,0,2,0,0)$	11	$(0,0,1,0,0)$	12	$(0,0,1,0,0)$
$\frac{12}{15}$	13	$(0,0,1,0,0)$	12	$(0,0,0,0,0)$	13	$(0,0,1,0,0)$
$\frac{13}{15}$	14	$(0,0,0,0,0)$	13	$(0,0,0,0,0)$	14	$(0,0,0,0,0)$
$\frac{14}{15}$	15	$(0,0,0,0,0)$	14	$(0,0,0,0,0)$	15	$(0,0,0,0,0)$
1	16	$(0,0,0,0,1)$	15	$(0,0,0,0,1)$	16	$(0,0,0,0,1)$

Table 3.1: Multiplicity vectors of the critical points of $f_{\{1,\ldots,5,7,6,8,9,11,10,12,\ldots,16\}}$, $f_{\{2,\ldots,7,1,8,9,16,10,\ldots,15\}}$, $f_{\{1,\ldots,5,7,6,8,9,11,10,12,\ldots,16\}} : (K3)_{16} \to [0,1]$.

Besides the symmetric slicings of Table 3.2 we found a number of other 3-dimensional spherical space forms like the truncated cube space, the prism space $P(3)$ or the lens space $L(5,1)$ as well as some orientable flat manifolds. Triangulations of such spaces were found in [93]. These can be used for comparison by bistellar moves. □

Combinatorial Morse analysis on $(K^4)_{16}$

In this section we will use the field $F := \mathbb{F}_2$ because the Kummer variety has 2-torsion in the homology, see Equation (2.2.1). Since $(K^4)_{16}$ is not a combinatorial manifold, we cannot apply critical point theory as easily as for the $K3$ surface. The reason is that now all vertex links are distinct from combinatorial 3-spheres. This implies that duality no longer holds. Moreover, it has the following consequence in Morse theory which is somehow against our intuition: Slicings below a non-critical vertex do not necessarily have to be homeomorphic to

α	$f_\Omega^{-1}(\alpha)$	slicing in between
$\frac{1}{30}$	S^3	$\{1\}$ and $\{2,\dots,16\}$
$\frac{1}{10}$	S^3	$\{1,2\}$ and $\{3,\dots,16\}$
$\frac{1}{6}$	S^3	$\{1,2,3\}$ and $\{4,\dots,16\}$
$\frac{7}{30}$	S^3	$\{2,3,4,5\}$ and $\{1,6,\dots,16\}$
	$\mathbb{R}P^3$	$\{1,2,3,4\}$ and $\{5,\dots,16\}$
$\frac{3}{10}$	$\mathrm{L}(4,1)$	$\{1,\dots,5\}$ and $\{6,\dots,16\}$
	$\mathrm{L}(3,1)$	$\{2,\dots,6\}$ and $\{1,7,\dots,16\}$
	$\mathbb{R}P^3$	$\{1,2,3,5,6\}$ and $\{4,7,\dots,16\}$
$\frac{11}{30}$	$\mathbf{C}^3$	$\{2,\dots,7\}$ and $\{1,8,\dots,16\}$
	$\mathbf{O}^3$	$\{1,\dots,5,7\}$ and $\{6,8,\dots,16\}$
$\frac{13}{30}$	Σ^3	$\{1,\dots,7\}$ and $\{8,\dots,16\}$
$\frac{1}{2}$	$\mathbb{T}^3$	$\{1,\dots,8\}$ and $\{9,\dots,16\}$

Table 3.2: Topological types of slicings of $(K3)_{16}$. Here Σ^3 denotes the Poincaré homology sphere, $\mathbf{C}^3$ the cube space and $\mathbf{O}^3$ the octahedron space.

the ones above the same non-critical vertex. Moreover, $(K^4)_{16}$ is not a tight triangulation. A tight triangulation of a simply connected space (manifold or not) must be 3-neighborly, but this does not hold for $(K^4)_{16}$ because of $f_2 = 400 < \binom{16}{3}$. In particular, not all rsl-functions on $(K^4)_{16}$ are perfect functions.

Looking at the $\mathbb{F}_2$-Betti numbers $b_0 = 1$, $b_1 = 0$, $b_2 = 11$, $b_3 = 5$, $b_4 = 1$ of the Kummer variety, we expect that any rsl-functions has 18 or more critical points, counted with multiplicity. The question is whether there is a perfect rsl-function on this triangulation which in addition respects the symmetry of the complex. This would be an excellent candidate for visualizing the space $(K^4)_{16}$ by various 3-dimensional slicings.

Proposition 3.7.5. *As slicings associated with perfect functions on $(K^4)_{16}$ we obtain at least the manifolds*

$$\mathbb{R}P^3, \mathbb{R}P^3\#\mathbb{R}P^3, \mathbb{R}P^3\#\mathbb{R}P^3\#\mathbb{R}P^3, S^2\times S^1\#\mathbb{R}P^3\#\mathbb{R}P^3$$

and the 3-*torus.*

Proof. There is the following perfect rsl-function $f_{\{1,\ldots,16\}}$ given by

$$f_{\{1,\ldots,16\}} : (K^4)_{16} \to [0,1]; \quad i \mapsto \tfrac{i-1}{15}.$$

As we already know, the first and the last slicing represent the link of the vertex 1 (or 16, resp.) and are, therefore, combinatorial real projective 3-spaces. Furthermore, the middle slicing $f^{-1}_{\{1,\ldots,16\}}(\frac{1}{2})$ is homeomorphic to the 3-torus which is more or less immediate from the construction of $(K^4)_{16}$ as the 4-torus modulo central involution. The other slicings are connected sums of $\mathbb{R}P^3$ and $S^2 \times S^1$. They are listed in Table 3.3, the multiplicity vectors are shown in Table 3.4.

level of $f_{\{1,\ldots,16\}}$	type	slicing in between
$\frac{1}{30}$	$\mathbb{R}P^3$	$\{1\}$ and $\{2,\ldots,16\}$
$\frac{1}{10}$	$\mathbb{R}P^3\#\mathbb{R}P^3$	$\{1,2\}$ and $\{3,\ldots,16\}$
$\frac{1}{6}$	$\mathbb{R}P^3\#\mathbb{R}P^3\#\mathbb{R}P^3$	$\{1,\ldots,3\}$ and $\{4,\ldots,16\}$
$\frac{7}{30}$	$(S^2 \times S^1)\#2(\mathbb{R}P^3)$	$\{1,\ldots,4\}$ and $\{5,\ldots,16\}$
$\frac{1}{2}$	$\mathbb{T}^3$	$\{1,\ldots,8\}$ and $\{9,\ldots,16\}$

Table 3.3: Slicings of $(K^4)_{16}$ by the perfect and symmetric rsl-function $f_{\{1,\ldots,16\}}$.

□

An example of an rsl-function which is not a perfect function is the function

$$f_{\{1,4,6,2,3,5,7,\ldots,16\}} : (K^4)_{16} \to [0,1].$$

This admits an empty triangle on one side leading to a critical point of index 1. In fact $f_{\{1,4,6,2,3,5,7,\ldots,16\}}$ has precisely 20 critical points, counted with multiplicity, see Table 3.4.

	$f_{\{1,\ldots,16\}}$		$f_{\{1,4,6,2,3,5,7,\ldots,16\}}$	
level	v	$\mathbf{m}(v,\mathbb{F}_2)$	v	$\mathbf{m}(v,\mathbb{F}_2)$
0	1	$(1,0,0,0,0)$	1	$(1,0,0,0,0)$
$\frac{1}{15}$	2	$(0,0,0,0,0)$	4	$(0,0,0,0,0)$
$\frac{2}{15}$	3	$(0,0,0,0,0)$	6	$(0,1,0,0,0)$
$\frac{3}{15}$	4	$(0,0,1,0,0)$	2	$(0,0,1,0,0)$
$\frac{4}{15}$	5	$(0,0,0,0,0)$	3	$(0,0,1,0,0)$
$\frac{5}{15}$	6	$(0,0,1,0,0)$	5	$(0,0,1,0,0)$
$\frac{6}{15}$	7	$(0,0,1,0,0)$	7	$(0,0,1,0,0)$
$\frac{7}{15}$	8	$(0,0,1,1,0)$	8	$(0,0,1,1,0)$
$\frac{8}{15}$	9	$(0,0,0,0,0)$	9	$(0,0,0,0,0)$
$\frac{9}{15}$	10	$(0,0,1,0,0)$	10	$(0,0,1,0,0)$
$\frac{10}{15}$	11	$(0,0,1,0,0)$	11	$(0,0,1,0,0)$
$\frac{11}{15}$	12	$(0,0,1,1,0)$	12	$(0,0,1,1,0)$
$\frac{12}{15}$	13	$(0,0,1,0,0)$	13	$(0,0,1,0,0)$
$\frac{13}{15}$	14	$(0,0,1,1,0)$	14	$(0,0,1,1,0)$
$\frac{14}{15}$	15	$(0,0,1,1,0)$	15	$(0,0,1,1,0)$
1	16	$(0,0,1,1,1)$	16	$(0,0,1,1,1)$

Table 3.4: Multiplicity vectors of two rsl-functions $f_{\{1,\ldots,16\}}$ and $f_{\{1,4,6,2,3,5,7,\ldots,16\}}$ on $(K^4)_{16}$.

CHAPTER 4

Simply transitive combinatorial 2- and 3-manifolds

This chapter focuses on combinatorial 2- and 3-manifolds with simply transitive automorphism group[1]. In particular, we present a decomposition of the 2-skeleton of the k-dimensional cross polytope β^k into centrally symmetric transitive surfaces of genus $g \leq 1$ together with an explicit description of all of these surfaces in terms of a 3-parameter family of transitive triangulations. In addition, a partition of the 2-skeleton of the $(k-1)$-simplex in a union of transitive tori and Möbius strips for each $k \equiv 1, 5(6)$ is described. Furthermore, we use the theory of discrete normal surfaces to construct new infinite series of transitive 3-dimensional sphere bundles over the circle.

4.1 Introduction

In geometric topology the concept of symmetries is discussed in terms of self homeomorphisms of a topological space X, where isotopic homeomorphisms are considered equivalent. The set of such isotopy classes of homeomorphisms forms the so-called mapping class group $\mathrm{MCG}(X)$ of X (cf. Section 1.4) which, in general, is not easy to understand. A lot of attention has been payed to the case where X is an oriented surface. The results that were obtained, allowed a better understanding of 3-manifolds and, in particular, the nature of Heegaard splittings (cf. Section 1.4). However, even the mapping class groups of orientable surfaces are not completely understood as of today.

[1]The content of the Sections 4.2, 4.3 and 4.4 is contained in [125].

In combinatorial topology on the other hand symmetries of a simplicial complex C seem to be easier to handle: A symmetry or automorphism of C is a permutation on the set of vertices which does not change the complex as a whole (cf. 1.1). Thus, any automorphism group of a complex on n vertices is a permutation group isomorphic to a subgroup of the symmetric group σ_n. Thus, it is a rather easy task (though it needs a lot of computational effort) to compute the automorphism group of a given complex which makes the discussion about symmetries of simplicial complexes seem to be trivial. However, the geometric interpretation of a given permutation is not obvious in general. Moreover, isotopic but combinatorially different automorphisms (such as rotations of a torus) are not considered equivalent and, most importantly, the automorphism group does not contain any information about the symmetries of the underlying topological object but just about one particular combinatorial structure.

Nonetheless, the study of highly symmetric simplicial complexes, in particular combinatorial manifolds, has been very important to the field of combinatorial topology. In particular, transitive complexes follow certain rules which makes them easy to find (cf. Appendix A.4.2). The regularity involved admits a more efficient description in terms of orbit representatives and allows less experimental but more intuitive proofs.

First sporadic examples of highly symmetric (regular) combinatorial 2-spheres have been known for thousands of years: The tetrahedron, the octahedron and the icosahedron. The first well known infinite family of symmetric combinatorial $(d-1)$-spheres is given by the series of cyclic d-polytopes which has been described for the first time by Carathéodory at the beginning of the 20th century (see [27]). A more recent development is the study of highly symmetric combinatorial manifolds different from the boundary complex of a convex polytope. Here again, some examples (like the 7-vertex Möbius torus) have been known for a long time. In 1971, Altshuler discovered transitive tori in the boundary of the cyclic 4-polytopes (see [3]). About 15 years later, Kühnel discovered a number of transitive total spaces of bundles over the circle in increasing dimension with dihedral symmetry (see [74]). At the same time Kühnel and Lassmann started a first systematical (computer aided) approach to classify 3-dimensional complexes with dihedral symmetry up to 19 vertices. In addition, they constructed infinite series of highly symmetric 3-dimensional Klein bottles (see [83]). In 2003, a classification of all transitive combinatorial manifolds with up to 15 vertices was given by Lutz in [92]. See Appendix A.4.2 for further details about the computer aided classification of transitive complexes. For further articles on highly symmetric combinatorial manifolds see Brehm and Kühnel [22], Casella and Kühnel [28], Kühnel [75, 78], Kühnel and Lassmann [84, 85], Lassmann and Sparla [88] and Lutz [96].

In addition, a lot of research was done in the more general field of design theory regarding for example triple or quadruple systems. Although the focus given in this context is neither geometric nor topological, there are a number of one to one correspondences between designs and combinatorial pseudomanifolds as pointed out by Kühnel in [79]. Thus, some of the work done in this field can be translated into the language of combinatorial topology (see for example the book by Huber [62]).

Example 4.1.1 (A geometric interpretation of the automorphisms of chiral and regular maps). A geometric interpretation of the action of a given automorphism on a simplicial complex (other than the explanation of the signum and the order of its underlying permutation of the vertex labels) is not obvious in general. For orientable surfaces S_g of genus g, one step towards a solution of this problem could be to look at the action of an automorphism $h \in \mathrm{Aut}(S_g)$ on a basis of $H_1(S_g,\mathbb{Z})$ and thus a representation

$$\phi : \mathrm{Aut}(S_g) \to \mathrm{GL}(2g,\mathbb{Z}).$$

Here, case $g = 0$ always results in the trivial representation. For $g = 1$ it is interesting to see that for every regular (chiral) triangulation $\{3,6\}_{(p,q)}$ with automorphism group $\mathrm{Aut}(\{3,6\}_{(p,q)}) = \langle \tau,\rho,\sigma\rangle$ ($\mathrm{Aut}(\{3,6\}_{(p,q)}) = \langle \tau,\rho\rangle$), where τ is any translation, ρ the rotation by $\frac{\pi}{3}$ around the origin and σ the reflection at the diagonal of a fundamental domain, we always have

$$\phi : \{3,6\}_{(p,q)} \to \mathrm{GL}(2g,\mathbb{Z}); \quad \tau \mapsto \begin{pmatrix} 1 & 0 \\ 0 & 1 \end{pmatrix}, \quad \rho \mapsto \begin{pmatrix} 0 & -1 \\ 1 & 1 \end{pmatrix}, \quad \sigma \mapsto \begin{pmatrix} 0 & -1 \\ -1 & 0 \end{pmatrix}. \tag{4.1.1}$$

with $\phi(\mathrm{Aut}(\{3,6\}_{(p,q)})) \cong D_6$ ($\phi(\mathrm{Aut}(\{3,6\}_{(p,q)})) \cong C_6$) of order 12 (6). See Figure 4.1.1 for a visualization of this representation in the case $q = p = 3$. For a more general discussion of regular, chiral and equivelar tori see [22].

For arbitrary values of g it follows by the Lefschetz fixed point formula (see [89]) that whenever $\phi(h) = E_{2g}$ for an orientation preserving automorphism $\mathbf{Id} \neq h \in \mathrm{Aut}(S_g)$ (which has a finite number of fixed points $F(\phi(h))$ in this case, see [98]) we have

$$\mathrm{Tr}(\phi(h)) = 2g = 2 - F(\phi(h)). \tag{4.1.2}$$

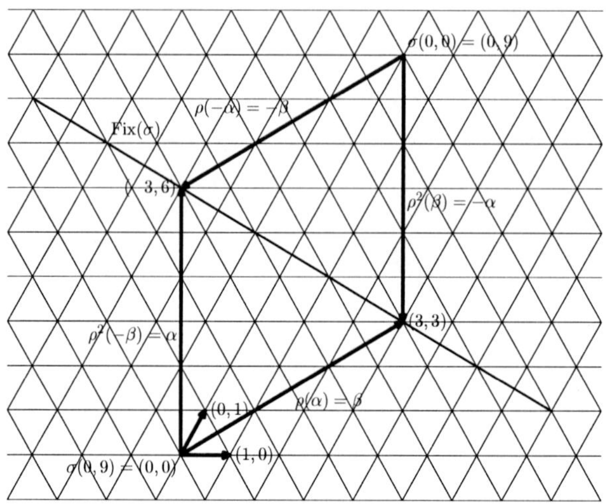

Figure 4.1.1: Fundamental domain of the regular torus $\{3,6\}_{(3,3)}$ with homology basis $H_1(\{3,6\}_{(3,3)}) = \langle\alpha,\beta\rangle$ and axis of reflexion $\mathrm{Fix}(\sigma)$.

Thus, we have the following result.

Corollary 4.1.2. *Let G be the group of all orientation preserving automorphisms of a triangulated orientable surface S_g of genus $g > 1$ and*

$$\phi : G \to \mathrm{GL}(2g,\mathbb{Z})$$

the representation of G induced by the action of G on the first homology of S_g. Then ϕ is faithful.

Example 4.1.3 (3.1 in [45]). As an example let us take a closer look at Dyck's regular map ($\{3,8\}_6$ in Coxeter notation). $\{3,8\}_6$ is a triangulated orientable surface of genus 3 with 12 vertices and regular automorphism group. It was first described in 1880 by Dyck in [35], see Figure 4.1.2 for an illustration. Polyhedral realizations were found by Schulte and Wills [121], Bokowski [16] and Brehm [17] more than 100 years later. The first homology group $H_1(\{3,8\}_6,\mathbb{Z})$ is generated by the eight simple closed curves a_i, $1 \le i \le 8$, shown in Figure 4.1.2. Note that $a_7 = -(a_1 + a_2 + a_3)$ and $a_8 = -(a_4 + a_5 + a_6)$. Hence, the cycles a_i, $1 \le i \le 6$,

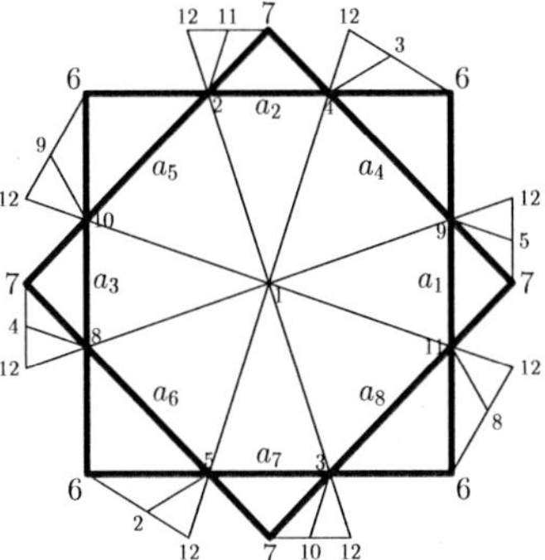

Figure 4.1.2: Fundamental domain of Dyck's map $\{3,8\}_6$ together with generators $\{a_i \mid 1 \leq i \leq 8\}$ of its first homology group.

form a base of $H_1(\{3,8\}_6, \mathbb{Z}) \cong \mathbb{Z}^6$.

The automorphism group $\text{Aut}(\{3,8\}_6) = \langle \sigma, \tau, \rho \rangle$ is generated by $\sigma = (6,7)(4,2,10,8,5,3,11,9)$ – a rotation around vertex 1, $\tau = (1,2)(3,12)(6,8)(7,9)$ – a rotation around edge $\langle 1,2 \rangle$ and $\rho = (1,4,2)(3,12,5)(6,11,8)(7,10,9)$ – a rotation about triangle $\langle 1,2,4 \rangle$. This results in a faithful representation

$$\phi : \text{Aut}(\{3,8\}_6) \to \text{GL}(6, \mathbb{Z})$$

given by

$$\phi(\sigma) = \begin{pmatrix} 0 & 0 & 0 & 0 & 0 & -1 \\ 0 & 0 & 0 & 1 & 0 & -1 \\ 0 & 0 & 0 & 0 & 1 & -1 \\ 1 & 0 & 0 & 0 & 0 & 0 \\ 0 & 1 & 0 & 0 & 0 & 0 \\ 0 & 0 & 1 & 0 & 0 & 0 \end{pmatrix}, \quad \phi(\tau) = \begin{pmatrix} 1 & 0 & 0 & 0 & 0 & 0 \\ 1 & 0 & 0 & 0 & 1 & -1 \\ 1 & 0 & -1 & 0 & 0 & -1 \\ -1 & 0 & 0 & -1 & 0 & 1 \\ -1 & 1 & 0 & 0 & 0 & 1 \\ 0 & 0 & 0 & 0 & 0 & 1 \end{pmatrix}$$

and

$$\phi(\rho) = \begin{pmatrix} -1 & 0 & 1 & 0 & 0 & 1 \\ -1 & 0 & 0 & 0 & -1 & 1 \\ -1 & 0 & 0 & 0 & 0 & 0 \\ 1 & -1 & 0 & 0 & 0 & 0 \\ 1 & 0 & 0 & 1 & 0 & 0 \\ 0 & 0 & 0 & 0 & 0 & 1 \end{pmatrix}.$$

For more information about representations of the automorphism group of regular maps as well as a more detailed description of the representation ϕ see [45].

Before looking at the partition of the 2-skeleton of the k-dimensional cross polytope β^k into symmetric closed surfaces, let us first summarize what is known about surfaces as 2-dimensional subcomplexes of polytopes in general.

4.2 Surfaces as subcomplexes of regular polytopes

A priori we can state that every triangulated surface with k vertices is a sub-complex of the $(k-1)$-simplex Δ^{k-1} and that every centrally symmetric surface with $2k$ vertices lies in the 2-skeleton of the k-dimensional cross polytope β^k, i. e. the convex hull of $2k$ points $x_i^{\pm} = (0,\dots,0,\pm 1,0,\dots,0) \in \mathbb{R}^k$, $1 \le i \le k$. Betke, Schulz and Wills were able to prove that every orientable 2-manifold is contained in the 2-skeleton of infinitely many 4-polytopes (see [12]). Hence, if we talk about surfaces as subcomplexes of convex polytopes it seems useful to impose some further restrictions on the embedded surface in order to get interesting examples. One possibility is to only consider surfaces which in addition contain the full edge graph of an ambient polytope P. These surfaces are referred to as 1-*Hamiltonian in* P (see for example [4] for triangulated 1-Hamiltonian surfaces in the 2-skeleton of stacked polytopes).

One possibility to approach the search for 1-Hamiltonian surfaces in convex polytopes is to start with the edge graph of a certain polytope and embed it into triangulated surfaces (so-called *triangular embeddings*). This was done extensively in the course of the proof of Heawood's Map Color Theorem (see [115] or [116, Chapter 4]). If a graph G is embeddable into a surface of genus g but not into a surface of smaller genus, then g is called the *genus of* G. Recall that we are interested in graph embeddings that cover the full edge graph of the ambient triangulated surface. For the complete graph or, equivalently, the n-simplex these are exactly the 2-neighborly surfaces. For the complete n-partite graph with two vertices in each partition corresponding to the k-dimensional cross polytope, Jungerman and Ringel were able to show the following theorem.

Theorem 4.2.1 (Regular cases in [63]). *For any orientable surface M and any $k \not\equiv 2\,(3)$ satisfying the equality*

$$2(k-1)(k-3) = 3(2-\chi(M)),$$

there exists a triangulation of M that contains the 1-skeleton of β^k.

There is a series of centrally symmetric 1-Hamiltonian surfaces $S(n)$, $n \geq 0$, with $12n+8$ vertices in β^{6n+4} of genus $g(S(n)) = 12n^2+8n+1$ (see [63] and [78, Example 3.6] for an explicit list of triangles). In particular, $S(0)$ is the 8-vertex Altshuler torus in the decomposition of β^4 described below.

A closely related question is whether or not the i-skeleton $\operatorname{skel}_i(P)$, $1 \leq i \leq (d-2)$, of a d-polytope P is *decomposable*, i. e. if there exist two (possibly bounded) PL i-manifolds M_1 and M_2 with $M_1 \cup M_2 = \operatorname{skel}_i(P)$ such that $M_1 \cap M_2 \subset \operatorname{skel}_{i-1}(P)$.

Grünbaum and Malkevitch [54] as well as Martin [100] investigated the case $i = 1$. The case $i = 2$ was settled by Betke, Schulz and Wills for simplicial polytopes.

Theorem 4.2.2 (Betke, Schulz, Wills, [12]). *There are exactly 5 simplicial polytopes with decomposable 2-skeletons:*

1. *The 2-skeleton of the 4-simplex Δ^4 is decomposable into 2 Möbius strips with cyclic symmetry, each with the minimum number of 5 vertices, 10 edges and 5 triangles.*

2. *The 18 triangles of the cyclic 4-polytope $C_4(6)$ with 6 vertices form the union of two Möbius strips where the triangulations equal the 6-vertex real projective plane with one triangle removed.*

3. *The triangles of the double pyramid over the 3-simplex can be partitioned into two Möbius strips on 6 vertices and 8 triangles each.*

4. *The 2-skeleton of the 4-dimensional cross polytope (i. e. the double pyramid over the octahedron) equals the union of two 8-vertex Altshuler tori.*

5. *The 20 triangles of the 5-simplex Δ^5 decompose into two copies of the 6-vertex real projective plane.*

Note that 1., 4. and 5. are highly symmetric. 1. and 4. occur as a part of the two series of decompositions presented below.

The proof of Theorem 4.2.2 relies on the fact that a decomposition of the 2-skeleton of a d-polytope P into two surfaces is only possible if each edge of P is contained in at most 4 triangles. Thus, we have $4 \leq d \leq 5$ and the number of vertices has to be bounded.

The idea of the proof of Theorem 4.2.2 shows that in general decompositions of a polytope P with more than 2 components are not as restrictive towards the local combinatorial structure of P. Here, we will focus on highly symmetric decompositions of the 2-skeleton of β^k and Δ^{k-1} with arbitrary many components.

Definition 4.2.3 (Difference cycle)**.** Let $a_i \in \mathbb{N}$, $0 \leq i \leq d$, $n := \sum_{i=0}^{d} a_i$ and $\mathbb{Z}_n = \langle(0,1,\ldots,n-1)\rangle$. The simplicial complex

$$(a_0 : \ldots : a_d) := \mathbb{Z}_n \cdot \langle 0, a_0, \ldots, \Sigma_{i=0}^{d-1} a_i \rangle$$

is called *difference cycle of dimension d on n vertices.* The number of its elements is referred to as the *length* of the difference cycle. If a complex C is a union of difference cycles of dimension d on n vertices and λ is a unit of $\mathbb{Z}_n$ such that the complex λC (obtained by multiplying all vertex labels modulo n by λ) equals C, then λ is called a *multiplier* of C.

Note that for any unit $\lambda \in \mathbb{Z}_n^{\times}$, the complex λC is combinatorially isomorphic to C. In particular, all $\lambda \in \mathbb{Z}_n^{\times}$ are multipliers of the complex $\bigcup_{\lambda \in \mathbb{Z}_n^{\times}} \lambda C$ by construction.

The definition of a difference cycle above is similar to the one of Kühnel and Lassmann given in [85]. For a more thorough introduction into the field of the more general difference sets and their multipliers see Chapter VI and VII in [11].

Throughout this work we will look at difference cycles as (1-)transitive abstract simplicial complexes.

Remark 4.2.4. It follows from Definition 4.2.3 that the set of difference cycles of dimension d on k vertices defines a partition of the d-skeleton of the $(k-1)$-simplex. Two $(d+1)$-tuples $(a_0, \ldots, a_d)$ and $(b_0, \ldots, b_d)$ with $\Sigma_{i=0}^{d} a_i = \Sigma_{i=0}^{d} b_i = k$ define the same difference cycle if and only if for a fixed $j \in \mathbb{Z}$ we have $a_{(i+j) \bmod (d+1)} = b_i$ for all $0 \leq i \leq d$.

Proposition 4.2.5. *Let $(a_0 : \ldots : a_d)$ be a difference cycle of dimension d on n vertices and $1 \leq k \leq d+1$ the smallest integer such that $k \mid (d+1)$ and $a_i = a_{i+k}$, $0 \leq i \leq d-k$. Then $(a_0 : \ldots : a_d)$ is of length $\sum_{i=0}^{k-1} a_i = \frac{nk}{d+1}$.*

Proof. We set $m := \frac{nk}{d+1}$ and compute

$$\begin{aligned}\langle 0+m, a_0+m, \ldots, (\Sigma_{i=0}^{d-1} a_i)+m\rangle &= \langle \Sigma_{i=0}^{k-1} a_i, \Sigma_{i=0}^{k} a_i, \ldots, \Sigma_{i=0}^{d-1} a_i, 0, a_1, \ldots, \Sigma_{i=0}^{k-2} a_i\rangle \\ &= \langle 0, a_0, \ldots, \Sigma_{i=0}^{d-1} a_i\rangle\end{aligned}$$

(all entries are computed modulo n). Hence, for the length l of $(a_0 : \ldots : a_d)$ we have $l \leq \frac{nk}{d+1}$ and since k is minimal with $k \mid (d+1)$ and $a_i = a_{i+k}$, the upper bound is attained. □

4.3 Centrally symmetric transitive surfaces and the decomposition of the 2-skeleton of β^k

In this section we will present a partition of the 2-skeleton $\mathrm{skel}_2(\beta^k)$ of the k-dimensional cross polytope into *centrally symmetric transitive surfaces* or *cst-surfaces* for short. Moreover, we present two infinite series of explicit triangulations of such surfaces (cf. Lemma 4.3.4 and Lemma 4.3.8).

In the sequel, we will look at the boundary of the k-dimensional cross polytope in terms of the abstract simplicial complex

$$\partial\beta^k = \{\langle a_1, \ldots, a_k\rangle \mid a_i \in \{0, \ldots, 2k-1\}, \{i, k+i\} \notin \{a_1, \ldots, a_k\}, \forall\, 0 \leq i \leq k-1\}. \quad (4.3.1)$$

In particular, the diagonals of β^k are precisely the edges $\langle i, k+i\rangle$, $1 \leq i \leq k$, and thus coincide with the difference cycle $(k{:}k)$. We can now state our main result.

Theorem 4.3.1. *The 2-skeleton of the k-dimensional cross polytope β^k can be decomposed into triangulated vertex transitive closed surfaces.*

More precisely, if $k \not\equiv 0\,(3)$, $\mathrm{skel}_2(\beta^k)$ decomposes into $\frac{(k-1)(k-2)}{3}$ triangulated vertex transitive closed surfaces of Euler characteristic 0 on $2k$ vertices and, if $k \equiv 0\,(3)$, into $\frac{k}{3}$ disjoint copies of $\partial\beta^3$ (on 6 vertices each) and $\frac{k(k-3)}{3}$ triangulated vertex transitive closed surfaces of Euler characteristic 0 on $2k$ vertices.

In the following, we will explicitly construct the transitive surfaces and determine their topological types for any given integer $k \geq 3$. The proof will consist of a number of consecutive lemmata.

Lemma 4.3.2. *The 2-skeleton of β^k can be written as the following set of difference cycles:*

$$(l : j : 2k - l - j), (l : 2k - l - j : j)$$

for $0 < l < j < 2k - l - j$, $k \notin \{l, j, l + j\}$, and

$$(j : j : 2(k - j))$$

for $0 < j < k$ with $2j \neq k$. If $k \not\equiv 0\,(3)$ all of them are of length $2k$, if $k \equiv 0\,(3)$ the difference cycle $(\frac{2k}{3} : \frac{2k}{3} : \frac{2k}{3})$ has length $\frac{2k}{3}$.

Proof. Let β^k be the k-dimensional cross polytope with vertices $\{0, \ldots, 2k-1\}$ and diagonals $\{j, k+j\}$, $0 \leq j \leq k-1$. It follows from the recursive construction of β^k as the double pyramid over β^{k-1} that it contains all 3-tuples of vertices as triangles except the ones including a diagonal. Thus, a difference cycle of the form $(a : b : c)$ lies in $\mathrm{skel}_2(\beta^k)$ if and only if $k \notin \{a, b, a + b\}$. In particular, $\mathrm{skel}_2(\beta^k)$ is a union of difference cycles.

Note that each ordered 3-tuple $0 < l < j < 2k - l - j$ defines exactly two distinct difference cycles on the set of $2k$ vertices, namely

$$(l : j : 2k - l - j) \text{ and } (l : 2k - l - j : j)$$

and it follows immediately that there is no other difference cycle $(a : b : c)$, $k \notin \{a, b, a + b\}$, on $2k$ vertices with a, b, c pairwise distinct.

For any positive integer $0 < j < k$ with $2j \neq k$ there is exactly one difference cycle

$$(j : j : 2(k - j)),$$

and since j must fulfill $0 < 2j < 2k$, there are no further difference cycles without diagonals with at most two different entries.

The length of the difference cycles follows directly from Proposition 4.2.5 with $d = 2$ and $n = 2k$. □

Lemma 4.3.3. *A closed 2-dimensional pseudomanifold S defined by m difference cycles of full length on the set of n vertices has Euler characteristic $\chi(S) = (1 - \frac{m}{2})n$.*

Proof. Since all difference cycles are of full length, S consists of n vertices and $m \cdot n$ triangles.

Additionally, the pseudomanifold property asserts that S has $\frac{3}{2}m \cdot n$ edges and thus

$$\chi(S) = n - \frac{3}{2}m \cdot n + m \cdot n = n(1 - \frac{m}{2}).$$

□

Lemma 4.3.4. *Let* $0 < l < j < 2k - l - j$, $k \notin \{l, j, l + j\}$ *and* $m := \gcd(l, j, 2k)$. *Then*

$$S_{l,j,2k} := \{(l : j : 2k - l - j), (l : 2k - l - j : j)\} \cong \{1, \ldots, m\} \times \mathbb{T}^2,$$

where all connected components of $S_{l,j,2k}$ *are combinatorially isomorphic to each other.*

Proof. The link of vertex 0 in $S_{l,j,2k}$ is equal to the cycle

$$\mathrm{lk}_{S_{l,j,2k}}(0) =$$

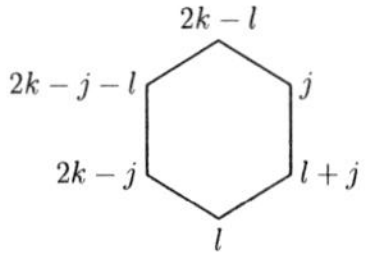

Since $0 < l < j < 2k - l - j$ and $k \notin \{l, j, l + j\}$, all vertices are distinct and $\mathrm{lk}_{S_{l,j,2k}}(0)$ is the boundary of a hexagon. By the vertex transitivity all other links are also hexagons and $S_{l,j,2k}$ is a surface.

Since l, j and $2k - l - j$ are pairwise distinct, both $(l : j : 2k - l - j)$ and $(l : 2k - l - j : j)$ have full length and by Lemma 4.3.3 the surface has Euler characteristic 0.

In order to see that $S_{l,j,2k}$ is oriented, we look at the (oriented) boundary of the triangles in $S_{l,j,2k}$ in terms of 1-dimensional difference cycles:

$$\begin{aligned}
\partial(l : j : 2k - l - j) &= (j : 2k - j) - (l + j : 2k - l - j) + (l : 2k - l) \\
\partial(l : 2k - l - j : j) &= (2k - l - j : l + j) - (2k - j : j) + (l : 2k - l) \\
&= (j : 2k - j) - (l + j : 2k - l - j) + (l : 2k - l)
\end{aligned}$$

and thus $\partial(l : j : 2k - l - j) - \partial(l : 2k - l - j : j) = 0$ and $S_{l,j,2k}$ is oriented.

Now consider

$$(l : j : 2k - l - j) = \mathbb{Z}_{2k} \cdot \langle 0, l, l + j \rangle$$

Clearly, the triangles $\langle (0 + i) \bmod 2k, (l + i) \bmod 2k, (l + j + i) \bmod 2k \rangle$, $0 \le i < 2k$, share at least one vertex if $i \in \{0, l, 2k - l, j, 2k - j, 2k - l - j, l + j\}$. For any other value of $i < 2k$, the intersection of the triangles is empty. By iteration it follows that $(l : j : 2k - l - j)$ has exactly $\gcd(0, l, 2k-l, j, 2k-j, 2k-l-j, l+j) = \gcd(l, j, 2k) = m$ connected components. The same holds for $(l : 2k - l - j : j)$. The complex $(0, \ldots, (2k-1))^i \cdot \langle 0, l, l+j \rangle$ is disjoint to $\langle 0, l, 2k-j \rangle$ for $i \notin \{0, l, 2k-l, j, 2k-j, 2k-l-j, l+j\}$. Together with the fact that $\mathrm{star}_{S_{l,j,2k}}(0)$ consists of triangles of both $(l : j : 2k-l-j)$ and $(l : 2k-l-j : j)$, it follows that $S_{l,j,2k}$ has m connected components and by a shift of the indices one can see that all of them must be combinatorially isomorphic. Altogether it follows that $S_{l,j,2k} \cong \{1, \ldots, m\} \times \mathbb{T}^2$. □

Remark 4.3.5. Some of the connected components of the surfaces presented above are combinatorially isomorphic to the so-called Altshuler tori

$$\{(1 : n-3 : 2), (1 : 2 : n-3)\}$$

with $n = \frac{2k}{m} \geq 7$ vertices (cf. [4, Theorem 4]). However, other triangulations of transitive tori are part of the decomposition as well: in the case $k = 6$, there are four different combinatorial types of tori. This is in fact the total number of combinatorial types of transitive tori on 12 vertices (cf. Table 4.1). The number of distinct combinatorial types of centrally symmetric transitive tori for $k \leq 30$ is listed below:

k	# comb. types	k	# comb. types	k	# comb. types	k	# comb. types
3	0	4	1	5	1	6	4
7	2	8	3	9	4	10	6
11	4	12	9	13	5	14	8
15	11	16	7	17	7	18	12
19	8	20	13	21	15	22	12
23	10	24	17	25	13	26	14
27	16	28	17	29	13	30	26

Remark 4.3.6. All centrally symmetric transitive surfaces $S_{l,j,2k}$ from Lemma 4.3.4 can be constructed using the function `SCSeriesCSTSurface(l,j,2k)` from the `GAP`-package simpcomp [40, 41, 42]. If the second parameter is not provided (`SCSeriesCSTSurface(l,2k)`), the surface $S_{l,2k}$ from Lemma 4.3.8 is generated.

Lemma 4.3.7. *Let*

$$\begin{aligned} M &:= \left\{(j:j:2(k-j)) \mid 0<j<k; 2j \neq k\right\}, \\ M_1 &:= \left\{(l:l:2(k-l)) \mid 1 \leq l \leq \left\lfloor \frac{k-1}{2} \right\rfloor\right\} \quad \textit{and} \\ M_2 &:= \left\{(k-l:k-l:2l)) \mid 1 \leq l \leq \left\lfloor \frac{k-1}{2} \right\rfloor\right\}. \end{aligned}$$

For all $k \geq 3$, the triple (M, M_1, M_2) defines a partition

$$M = M_1 \dot\cup M_2$$

into two sets of equal size. In particular, we have $\mid M \mid \equiv 0\,(2)$.

Proof. From $1 \leq l \leq \lfloor \frac{k-1}{2} \rfloor$ it follows that $k-l > l$ and $2l < k < 2(k-l)$. Thus, $M_1 \cap M_2 = \emptyset$ and $M_1 \cup M_2 \subseteq M$.

On the other hand let $\lfloor \frac{k-1}{2} \rfloor < j < k - \lfloor \frac{k-1}{2} \rfloor$. If k is odd, then $\frac{k-1}{2} < j < \frac{k+1}{2}$ which is impossible for $j \in \mathbb{N}$. If k is even, then $\frac{k}{2} - 1 < j < \frac{k}{2} + 1$, hence it follows that $j = k$ which is excluded in the definition of M. Altogether $M_1 \cup M_2 = M$ holds and

$$|M| = 2\left\lfloor \frac{k-1}{2} \right\rfloor = \begin{cases} k-1 & \text{if } k \text{ is odd} \\ k-2 & \text{else.} \end{cases}$$

□

Lemma 4.3.8. *The complex*

$$S_{l,2k} := \{(l:l:2(k-l)), (k-l:k-l:2l)\},$$

$1 \leq l \leq \lfloor \frac{k-1}{2} \rfloor$, is a disjoint union of $\frac{k}{3}$ copies of $\partial\beta^3$ if $k \equiv 0\,(3)$ and $l = \frac{k}{3}$ and a surface of Euler characteristic 0 otherwise.

Proof. We prove that $S_{l,2k}$ is a surface by looking at the link of vertex 0:

$\mathrm{lk}_{S_{j,2k}}(0) =$

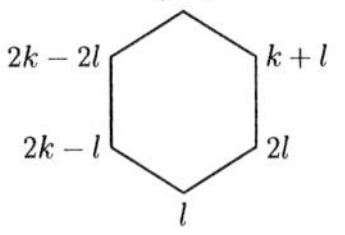

where $2l = k - l$ and $2k - 2l = k + l$ if and only if $l = \frac{k}{3}$. Thus, $\mathrm{lk}_{S_{l,j,2k}}(0)$ is either the boundary of a hexagon or, in the case $l = \frac{k}{3}$, the boundary of a quadrilateral and $S_{l,2k}$ is a surface.

Furthermore, if $l \neq \frac{k}{3}$ the surface $S_{l,2k}$ is a union of two difference cycles of full length and by Lemma 4.3.3 we have $\chi(S_{l,2k}) = 0$. If $l = \frac{k}{3}$, $(\frac{2k}{3} : \frac{2k}{3} : \frac{2k}{3})$ is of length $\frac{2k}{3}$ and it follows that

$$\chi(S_{\frac{k}{3},2k}) = 2k - \frac{8}{2}k + \frac{8}{3}k = \frac{2}{3}k.$$

By a calculation analogue to the one in the proof of Lemma 4.3.4, one obtains that $S_{\frac{k}{3},2k}$ consists of $\gcd(l,2k) = \frac{k}{3}$ isomorphic connected components of type $\{3,4\}$. Hence, $S_{\frac{k}{3},2k}$ is a disjoint union of $\frac{k}{3}$ copies of $\partial\beta^3$. □

Subsection 3.1 of [96] by Lutz contains a series of transitive tori and Klein bottles with $n = 8 + 2m$ vertices, $m \geq 0$. The series is given by

$$A_6(n) := \{(1:1:(n-2)), (2:(\frac{n}{2}-1):(\frac{n}{2}-1)\}.$$

$A_6(n)$ is a torus for m even and a Klein bottle for m odd.

Lemma 4.3.9. *Let $k \geq 3$, $1 \leq l \leq \lfloor\frac{k-1}{2}\rfloor$, $l \neq \frac{k}{3}$ and $n := \gcd(l,k)$. Then $S_{l,2k}$ is isomorphic to n copies of $A_6(\frac{2k}{n})$.*

Proof. Since $n = \gcd(l,k) = \gcd(l,k-l)$, we have $n = \min\{\gcd(l,2(k-l)), \gcd(2l,k-l)\}$ and either $\frac{l}{n} \in \mathbb{Z}^{\times}_{\frac{2k}{n}}$ or $\frac{k-l}{n} \in \mathbb{Z}^{\times}_{\frac{2k}{n}}$ holds. It follows by mulitplying with l or $k-l$ that $A_6(\frac{2k}{n})$ is isomorphic to $S_{\frac{l}{n},\frac{2k}{n}} = S_{\frac{k-l}{n},\frac{2k}{n}}$. The monomorphism

$$\mathbb{Z}_{\frac{2k}{n}} \to \mathbb{Z}_{2k} \quad j \mapsto (lj \mod 2k)$$

represents a relabeling of $S_{\frac{l}{n},\frac{2k}{n}}$ and a small computation shows that the relabeled complex is equal to the connected component of $S_{l,2k}$ containing 0. By a shift of the vertex labels we see that all other connected components of $S_{l,2k}$ are isomorphic to the one containing 0 what states the result. □

Let us now come to the proof of Theorem 4.3.1.

Proof. Lemma 4.3.2 and Lemma 4.3.7 describe $\mathrm{skel}_2(\beta^k)$ in terms of 2 series of pairs of difference cycles

$$\{(l : j : 2k - l - j), (l : 2k - l - j : j)\} \text{ and } \{(l : l : 2(k - l)), (k - l : k - l : 2l)\}$$

for certain parameters j and l. Lemma 4.3.4 determines the topological type of the first and the series $A_6(n)$ together with Lemma 4.3.8 and 4.3.9 determines the type of the second series.

Since $|\mathrm{skel}_2(\beta^k)| = \binom{2k}{3} - k(2k-2)$ and for $k \not\equiv 0\,(3)$ all surfaces have exactly $4k$ triangles, we get an overall number of $\frac{(k-1)(k-2)}{3}$ surfaces. If $k \equiv 0\,(3)$, all surfaces but one have $4k$ triangles, the last one has $\frac{8k}{3}$ triangles. Altogether this implies that there are $\frac{k(k-3)}{3}$ surfaces of Euler characteristic 0 and $\frac{k}{3}$ copies of $\partial\beta^3$. □

Table 4.1 shows the decomposition of $\mathrm{skel}_2(\beta^k)$ for $3 \leq k \leq 10$. The table was computed with the help of the GAP-package simpcomp [40, 41, 42]. For a complete list of the decomposition for $k \leq 100$ see [126].

4.4 The decomposition of the 2-skeleton of Δ^{k-1}

First, note that $\mathrm{skel}_2(\Delta^{k-1})$, $k \geq 3$, equals the set of all triangles on k vertices. By looking at its vertex links we can see that in the case that k is an even number the complex $\{(l : \frac{k}{2}-l : \frac{k}{2})\}$ cannot be part of a triangulated surface for any $0 < l < \frac{k}{2}$. Thus, the decomposition of $\mathrm{skel}_2(\beta^k)$ cannot be extended to a decomposition of $\mathrm{skel}_2(\Delta^{2k-1})$ in an obvious manner. However, for other numbers of vertices the situation is different.

Theorem 4.4.1. *Let $k > 1$, $k \equiv 1, 5(6)$. Then the 2-skeleton of Δ^{k-1} decomposes into $\frac{k-1}{2}$ collections of Möbius strips*

$$M_{l,k} := \{(l : l : k - 2l)\},$$

$1 \leq l \leq \frac{k-1}{2}$ each with $n := \gcd(l,k)$ isomorphic connected components on $\frac{k}{n}$ vertices and $\frac{k^2-6k+5}{12}$ collections of tori

$$S_{l,j,k} := \{(l : j : k - l - j), (l : k - l - j : j)\},$$

$1 \leq l < j < k - l - j$, with $m := \gcd(l,j,k)$ connected components on $\frac{k}{m}$ vertices each.

We first prove the following lemma.

Lemma 4.4.2. *The complex $M_{l,k}$ with $k \geq 5$, $k \not\equiv 0\,(3)$ and $k \not\equiv 0\,(4)$ is a triangulation of $n := \gcd(l,k)$ cylinders $[0,1] \times \partial\Delta^2$ if $\frac{k}{n}$ is even and of n Möbius strips if $\frac{k}{n}$ is odd.*

Proof. We first look at

$$M_{1,k} = \{\langle 0,1,2\rangle, \langle 1,2,3\rangle, \ldots, \langle k-2,k-1,0\rangle, \langle k-1,0,1\rangle\}$$

for $k \geq 5$ (see Figure 4.4.1). Every triangle has exactly two neighbors. Thus, the alternating sum

$$+\langle 0,1,2\rangle - \langle 1,2,3\rangle + \ldots - +(-1)^{k-1}\langle k-1,0,1\rangle$$

induces an orientation if and only if k is even and for any $l \in \mathbb{Z}_k^\times$ the complex $M_{l,k}$ is a cylinder if k is even and a Möbius strip if k is odd. Now suppose that $n = \gcd(l,k) > 1$. Since $k \not\equiv 0\,(3)$ and $k \not\equiv 0\,(4)$ we have $\frac{k}{n} \geq 5$ and by a relabeling we see that the connected components of $M_{l,k}$ are combinatorially isomorphic to $M_{\frac{l}{n},\frac{k}{n}} \cong M_{1,\frac{k}{n}}$. □

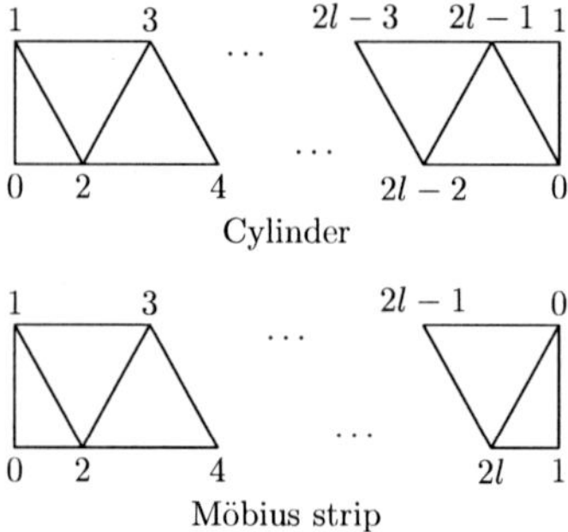

Figure 4.4.1: The cylinder $(1:1:2l-2)$ and the Möbius strip $(1:1:2l-3)$. The vertical boundary components $(\langle 0,1\rangle)$ are identified.

Remark 4.4.3. If $k \equiv 0\,(4)$, the connected components of $M_{\frac{k}{4},k} = \{(\frac{k}{4} : \frac{k}{4} : \frac{k}{2})\}$ are equal to $\{(1:1:2)\}$, the boundary of Δ^3. If $k \equiv 0\,(3)$, then $M_{\frac{k}{3},k}$ is a collection of disjoint triangles.

Lemma 4.4.4. *The complex $S_{l,j,k}$, $0 < l < j < k$, $k \not\equiv 0\,(2)$, is a triangulation of $m := \gcd(l,j,k)$ connected components of isomorphic tori on $\frac{k}{m}$ vertices.*

Proof. The link of vertex 0 equals

$$\mathrm{lk}_{S_{l,j,k}}(0) =$$

$k-l$
$k-j-l$ j
$k-j$ $l+j$
l

(cf. proof of Lemma 4.3.4). Since $0 < l < j < k-l-j$ and $k \not\equiv 0\,(2)$ the link is the boundary of a hexagon, $\frac{k}{m} \geq 7$ and $S_{l,j,k}$ is a surface. By Lemma 4.3.3, the complex $S_{l,j,k}$ is of Euler characteristic 0. The proof of the orientability and the number of connected components is analogue to the one given in the proof of Lemma 4.3.4. It follows that

$$S_{l,j,k} \cong \{1,\ldots,m\} \times \mathbb{T}^2.$$

□

Together with Lemma 4.4.2 and Lemma 4.4.4 in order to proof Theorem 4.4.1, it suffices to show that the two series presented above contain all triangles of Δ^{k-1}.

Proof. Let $\langle a,b,c\rangle \in \mathrm{skel}_2(\Delta^{k-1})$, $a<b<c$. Then $\langle a,b,c\rangle \in (b-a : c-b : k-(c-a))$. Now if $b-a$, $c-b$ and $k-(c-a)$ are pairwise distinct, we have

- $\langle a,b,c\rangle \in S_{b-a,c-b,k} = S_{b-a,k-(c-a),k}$ if $b-a < c-b, k-(c-a)$,
- $\langle a,b,c\rangle \in S_{c-b,b-a,k} = S_{c-b,k-(c-a),k}$ if $c-b < b-a, k-(c-a)$ or
- $\langle a,b,c\rangle \in S_{k-(c-a),b-a,k} = S_{k-(c-a),c-b,k}$ if $k-(c-a) < c-b, b-a$.

If on the other hand at least two of the entries are equal, then $(b-a : c-b : k-(c-a)) = (l : l : k-2l)$ for $1 \leq l \leq \frac{k-1}{2}$. Thus, the union of all Möbius strips $M_{l,k}$ and collections of tori $S_{l,j,k}$ equals the full 2-skeleton of the k-simplex $\mathrm{skel}_2(\Delta^{k-1})$.

Table 4.2 shows the decomposition of $\mathrm{skel}_2(\Delta^{k-1})$ for $k \in \{5,7,11,13,35\}$. For a complete list of the decomposition of $\mathrm{skel}_2(\beta^k)$ and $\mathrm{skel}_2(\Delta^{k-1})$ for $k \leq 100$ see [126]. □

Table 4.1: The decomposition of the 2-skeleton of β^k, $3 \leq k \leq 10$, into transitive surfaces.

k	$f_2(\beta^k)$	**topological type**	**difference cycles**
3	8	S^2	$\{(1{:}1{:}4),(2{:}2{:}2)\}$
4	32	$\mathbb{T}^2$	$\{(1{:}2{:}5),(1{:}5{:}2)\}$, $\{(1{:}1{:}6),(3{:}3{:}2)\}$

continued on next page –

Table 4.1 – continued from previous page

k	$f_2(\beta^k)$	**topological type**	**difference cycles**
5	80	$\mathbb{T}^2$	$\{(1{:}2{:}7),(1{:}7{:}2)\}$, $\{(1{:}3{:}6),(1{:}6{:}3)\}$
		$\mathbb{K}^2$	$\{(1{:}1{:}8),(4{:}4{:}2)\}$, $\{(2{:}2{:}6),(3{:}3{:}4)\}$
6	160	$\{1,2\}\times S^2$	$\{(2{:}2{:}8),(4{:}4{:}4)\}$
		$\mathbb{T}^2$	$\{(1{:}2{:}9),(1{:}9{:}2)\}$, $\{(1{:}3{:}8),(1{:}8{:}3)\}$, $\{(1{:}4{:}7),(1{:}7{:}4)\}$, $\{(2{:}3{:}7),(2{:}7{:}3)\}$, $\{(3{:}4{:}5),(3{:}5{:}4)\}$, $\{(1{:}1{:}10),(5{:}5{:}2)\}$
7	280	$\mathbb{T}^2$	$\{(1{:}2{:}11),(1{:}11{:}2)\}$,$\{(1{:}3{:}10),(1{:}10{:}3)\}$,$\{(1{:}4{:}9),(1{:}9{:}4)\}$, $\{(1{:}5{:}8),(1{:}8{:}5)\}$, $\{(2{:}3{:}9),(2{:}9{:}3)\}$, $\{(3{:}5{:}6),(3{:}6{:}5)\}$
		$\{1,2\}\times\mathbb{T}^2$	$\{(2{:}4{:}8),(2{:}8{:}4)\}$
		$\mathbb{K}^2$	$\{(1{:}1{:}12),(6{:}6{:}2)\}$, $\{(2{:}2{:}10),(5{:}5{:}4)\}$, $\{(3{:}3{:}8),(4{:}4{:}6)\}$
8	448	$\mathbb{T}^2$	$\{(1{:}2{:}13),(1{:}13{:}2)\}$,$\{(1{:}3{:}12),(1{:}12{:}3)\}$,$\{(1{:}4{:}11),(1{:}11{:}4)\}$, $\{(1{:}5{:}10),(1{:}10{:}5)\}$,$\{(1{:}6{:}9),(1{:}9{:}6)\}$, $\{(2{:}3{:}11),(2{:}11{:}3)\}$, $\{(2{:}5{:}9),(2{:}9{:}5)\}$, $\{(3{:}4{:}9),(3{:}9{:}4)\}$, $\{(3{:}6{:}7),(3{:}7{:}6)\}$, $\{(4{:}5{:}7),(4{:}7{:}5)\}$, $\{(1{:}1{:}14),(7{:}7{:}2)\}$, $\{(3{:}3{:}10),(5{:}5{:}6)\}$
		$\{1,2\}\times\mathbb{T}^2$	$\{(2{:}4{:}10),(2{:}10{:}4)\}$,$\{(2{:}2{:}12),(6{:}6{:}4)\}$
9	672	$\{1,2,3\}\times S^2$	$\{(3{:}3{:}12),(6{:}6{:}6)\}$
		$\mathbb{T}^2$	$\{(1{:}2{:}15),(1{:}15{:}2)\}$,$\{(1{:}3{:}14),(1{:}14{:}3)\}$,$\{(1{:}4{:}13),(1{:}13{:}4)\}$, $\{(1{:}5{:}12),(1{:}12{:}5)\}$,$\{(1{:}6{:}11),(1{:}11{:}6)\}$,$\{(1{:}7{:}10),(1{:}10{:}7)\}$, $\{(2{:}3{:}13),(2{:}13{:}3)\}$,$\{(2{:}5{:}11),(2{:}11{:}5)\}$,$\{(3{:}4{:}11),(3{:}11{:}4)\}$, $\{(3{:}5{:}10),(3{:}10{:}5)\}$,$\{(3{:}7{:}8),(3{:}8{:}7)\}$, $\{(5{:}6{:}7),(5{:}7{:}6)\}$
		$\{1,2\}\times\mathbb{T}^2$	$\{(2{:}4{:}12),(2{:}12{:}4)\}$,$\{(2{:}6{:}10),(2{:}10{:}6)\}$,$\{(4{:}6{:}8),(4{:}8{:}6)\}$
		$\mathbb{K}^2$	$\{(1{:}1{:}16),(8{:}8{:}2)\}$, $\{(2{:}2{:}14),(7{:}7{:}4)\}$, $\{(4{:}4{:}10),(5{:}5{:}8)\}$
10	960	$\mathbb{T}^2$	$\{(1{:}2{:}17),(1{:}17{:}2)\}$,$\{(1{:}3{:}16),(1{:}16{:}3)\}$,$\{(1{:}4{:}15),(1{:}15{:}4)\}$, $\{(1{:}5{:}14),(1{:}14{:}5)\}$,$\{(1{:}6{:}13),(1{:}13{:}6)\}$,$\{(1{:}7{:}12),(1{:}12{:}7)\}$, $\{(1{:}8{:}11),(1{:}11{:}8)\}$,$\{(2{:}3{:}15),(2{:}15{:}3)\}$,$\{(2{:}5{:}13),(2{:}13{:}5)\}$, $\{(2{:}7{:}11),(2{:}11{:}7)\}$,$\{(3{:}4{:}13),(3{:}13{:}4)\}$,$\{(3{:}5{:}12),(3{:}12{:}5)\}$, $\{(3{:}6{:}11),(3{:}11{:}6)\}$,$\{(3{:}8{:}9),(3{:}9{:}8)\}$, $\{(4{:}5{:}11),(4{:}11{:}5)\}$, $\{(4{:}7{:}9),(4{:}9{:}7)\}$, $\{(5{:}6{:}9),(5{:}9{:}6)\}$, $\{(5{:}7{:}8),(5{:}8{:}7)\}$, $\{(1{:}1{:}18),(9{:}9{:}2)\}$, $\{(3{:}3{:}14),(7{:}7{:}6)\}$
		$\{1,2\}\times\mathbb{T}^2$	$\{(2{:}4{:}14),(2{:}14{:}4)\}$,$\{(2{:}6{:}12),(2{:}12{:}6)\}$
		$\{1,2\}\times\mathbb{K}^2$	$\{(2{:}2{:}16),(8{:}8{:}4)\}$, $\{(4{:}4{:}12),(6{:}6{:}8)\}$

Table 4.2: The decomposition of the 2-skeleton of Δ^{k-1}, $k \in \{5, 7, 11, 13, 35\}$, by topological types.

k	**topological type**	**difference cycles**		
5	$\mathbb{M}^2$	{(1:1:3)},	{(2:2:1)}	
7	$\mathbb{M}^2$	{(1:1:5)},	{(2:2:3)},	{(3:3:1)}
	$\mathbb{T}^2$	{(1:2:4), (1:4:2)}		
11	$\mathbb{M}^2$	{(1:1:9)},	{(2:2:7)},	{(3:3:5)},
		{(4:4:3)},	{(5:5:1)}	
	$\mathbb{T}^2$	{(1:2:8), (1:8:2)},	{(1:3:7), (1:7:3)},	{(1:4:6), (1:6:4)},
		{(2:3:6), (2:6:3)},	{(2:4:5), (2:5:4)}	
13	$\mathbb{M}^2$	{(1:1:11)},	{(2:2:9)},	{(3:3:7)},
		{(4:4:5)},	{(5:5:3)},	{(6:6:1)}
	$\mathbb{T}^2$	{(1:2:10), (1:10:2)},	{(1:3:9), (1:9:3)},	{(1:4:8), (1:8:4)},
		{(1:5:7), (1:7:5)},	{(2:3:8), (2:8:3)},	{(2:4:7), (2:7:4)},
		{(2:5:6), (2:6:5)},	{(3:4:6), (3:6:4)}	
35	$\mathbb{M}^2$	{(1:1:33)},	{(2:2:31)},	{(3:3:29)},
		{(4:4:27)},	{(6:6:23)},	{(8:8:19)},
		{(9:9:17)},	{(11:11:13)},	{(12:12:11)},
		{(13:13:9)},	{(16:16:3)},	{(17:17:1)}
	$\{1,\dots,5\} \times \mathbb{M}^2$	{(5:5:25)},	{(10:10:15)},	{(15:15:5)}
	$\{1,\dots,7\} \times \mathbb{M}^2$	{(7:7:21)},	{(14:14:7)}	
	$\mathbb{T}^2$	{(1:2:32), (1:32:2)},	{(1:3:31), (1:31:3)},	{(1:4:30), (1:30:4)},
		{(1:5:29), (1:29:5)},	{(1:6:28), (1:28:6)},	{(1:7:27), (1:27:7)},
		{(1:8:26), (1:26:8)},	{(1:9:25), (1:25:9)},	{(1:10:24), (1:24:10)},
		{(1:11:23), (1:23:11)},	{(1:12:22), (1:22:12)},	{(1:13:21), (1:21:13)},
		{(1:14:20), (1:20:14)},	{(1:15:19), (1:19:15)},	{(1:16:18), (1:18:16)},
		{(2:3:30), (2:30:3)},	{(2:4:29), (2:29:4)},	{(2:5:28), (2:28:5)},
		{(2:6:27), (2:27:6)},	{(2:7:26), (2:26:7)},	{(2:8:25), (2:25:8)},
		{(2:9:24), (2:24:9)},	{(2:10:23), (2:23:10)},	{(2:11:22), (2:22:11)},
		{(2:12:21), (2:21:12)},	{(2:13:20), (2:20:13)},	{(2:14:19), (2:19:14)},
		{(2:15:18), (2:18:15)},	{(2:16:17), (2:17:16)},	{(3:4:28), (3:28:4)},
		{(3:5:27), (3:27:5)},	{(3:6:26), (3:26:6)},	{(3:7:25), (3:25:7)},
		{(3:8:24), (3:24:8)},	{(3:9:23), (3:23:9)},	{(3:10:22), (3:22:10)},
		{(3:11:21), (3:21:11)},	{(3:12:20), (3:20:12)},	{(3:13:19), (3:19:13)},
		{(3:14:18), (3:18:14)},	{(3:15:17), (3:17:15)},	{(4:5:26), (4:26:5)},
		{(4:6:25), (4:25:6)},	{(4:7:24), (4:24:7)},	{(4:8:23), (4:23:8)},
		{(4:9:22), (4:22:9)},	{(4:10:21), (4:21:10)},	{(4:11:20), (4:20:11)},

continued on next page –

Table 4.2 – continued from previous page

k	**topological type**	**difference cycles**
		$\{(4{:}12{:}19),(4{:}19{:}12)\}$, $\{(4{:}13{:}18),(4{:}18{:}13)\}$, $\{(4{:}14{:}17),(4{:}17{:}14)\}$, $\{(4{:}15{:}16),(4{:}16{:}15)\}$, $\{(5{:}6{:}24),(5{:}24{:}6)\}$, $\{(5{:}7{:}23),(5{:}23{:}7)\}$, $\{(5{:}8{:}22),(5{:}22{:}8)\}$, $\{(5{:}9{:}21),(5{:}21{:}9)\}$, $\{(5{:}11{:}19),(5{:}19{:}11)\}$, $\{(5{:}12{:}18),(5{:}18{:}12)\}$, $\{(5{:}13{:}17),(5{:}17{:}13)\}$, $\{(5{:}14{:}16),(5{:}16{:}14)\}$, $\{(6{:}7{:}22),(6{:}22{:}7)\}$, $\{(6{:}8{:}21),(6{:}21{:}8)\}$, $\{(6{:}9{:}20),(6{:}20{:}9)\}$, $\{(6{:}10{:}19),(6{:}19{:}10)\}$, $\{(6{:}11{:}18),(6{:}18{:}11)\}$, $\{(6{:}12{:}17),(6{:}17{:}12)\}$, $\{(6{:}13{:}16),(6{:}16{:}13)\}$, $\{(6{:}14{:}15),(6{:}15{:}14)\}$, $\{(7{:}8{:}20),(7{:}20{:}8)\}$, $\{(7{:}9{:}19),(7{:}19{:}9)\}$, $\{(7{:}10{:}18),(7{:}18{:}10)\}$, $\{(7{:}11{:}17),(7{:}17{:}11)\}$, $\{(7{:}12{:}16),(7{:}16{:}12)\}$, $\{(7{:}13{:}15),(7{:}15{:}13)\}$, $\{(8{:}9{:}18),(8{:}18{:}9)\}$, $\{(8{:}10{:}17),(8{:}17{:}10)\}$, $\{(8{:}11{:}16),(8{:}16{:}11)\}$, $\{(8{:}12{:}15),(8{:}15{:}12)\}$, $\{(8{:}13{:}14),(8{:}14{:}13)\}$, $\{(9{:}10{:}16),(9{:}16{:}10)\}$, $\{(9{:}11{:}15),(9{:}15{:}11)\}$, $\{(9{:}12{:}14),(9{:}14{:}12)\}$, $\{(10{:}11{:}14),(10{:}14{:}11)\}$, $\{(10{:}12{:}13),(10{:}13{:}12)\}$
	$\{1,\ldots,5\}\times\mathbb{T}^2$	$\{(5{:}10{:}20),(5{:}20{:}10)\}$

4.5 Transitive 3-dimensional combinatorial manifolds

In this section we will present a number of infinite series of cyclic combinatorial 3-manifolds homeomorphic to the total space of a bundle over the circle. By construction, these types of manifolds ($S^2 \times S^1$, $S^2 \underline{\times} S^1$, $S^1 \times S^1 \times S^1$, etc.) admit a cyclic structure induced by the S^1 component. Hence, it seems natural to conjecture the existence of such series. Indeed, there have been various publications concerning infinite families of combinatorial manifolds with increasing numbers of vertices and consequently a number of infinite series of combinatorial manifolds homeomorphic to bundles over the circle as well as some sporadic triangulations are already known (see [77, Theorem 5.5] or [84] for bundles over the circle in arbitrary dimensions and [83, 85, 92] for examples in dimension 3).

Throughout this section, we will call such a series *of order* l, if it contains triangulations with $n = m + l \cdot k$ vertices for positive integers $l, m \in \mathbb{N}$ and for all $k \geq 0$. If $l = 1$, the series is called *dense*. For arbitrary cyclic triangulations of d-manifolds the following proposition holds.

Proposition 4.5.1. *Let M be a combinatorial d-manifold with n vertices, given by the difference cycles*

$$M := \{d_1, \ldots, d_m\}$$

with $d_i := (a_{i,0} : \cdots : a_{i,d})$, such that $a_{i,d} > a_{i,j}$, $a_{i,d} > \frac{n}{2}$ for all $1 \leq i \leq m$, $0 \leq j \leq d-1$. Then

$$M_k := \{d_1^k, \ldots, d_m^k\} \tag{4.5.1}$$

with $d_i^k := (a_{i,0} : \cdots : a_{i,d} + k)$, $1 \leq i \leq m$, is a combinatorial d-manifold with $n + k$ vertices, $k \geq 0$. In other words, $(M_k)_{k \geq 0}$ is a dense series of combinatorial d-manifolds.

Proof. We choose $V_k := \{1, \ldots, n + k\}$ as vertex labeling for all complexes M_k, $k \geq 0$. Due to the cyclic symmetry, all vertex links of M_k, k fixed, must be combinatorially isomorphic. Hence, it suffices to look at the link of vertex 1 in M_k for any $k \geq 0$. The link $\mathrm{lk}_{M_0}(1)$ is a $(d-1)$-sphere since $M_0 = M$ is a combinatorial d-manifold. Now observe that the link $\mathrm{lk}_{M_k}(1)$ is equal to $\mathrm{lk}_{M_{k+1}}(1)$ up to the relabeling $v \mapsto v + 1$ of all vertices $v > \frac{n}{2}$. Hence, all vertex links are combinatorially isomorphic to $\mathrm{lk}_{M_0}(1)$ and M_k is a combinatorial manifold. □

Example 4.5.2 (cf. Theorem 5.5 in [77]). The complex $\mathbb{H}_k^d := \{(\overbrace{1 : \cdots : 1}^{d \times} : k)\}$ defines a

d-dimensional 1-handle for all $d \geq 2$, $k \geq d+3$. Thus, its boundary $\partial \mathbb{H}_k^d$ defines a $(d-1)$-dimensional sphere bundle over the circle with $d+k$ vertices.

In this section, we want to construct new series of such sphere bundles with the help of the theory of slicings developed in Chapter 3: The main idea is to look at slicings $S_{(V_1,V_2)}$ of existing sporadic examples of triangulated sphere bundles M with cyclic automorphism group satisfying the following properties:

1. $S_{(V_1,V_2)}$ defines a minimal Heegaard splitting of M. In the case that M is a sphere bundle over the circle of type $S^1 \times S^2$ or $S^1 \underline{\times} S^2$ this means that $S_{(V_1,V_2)}$ is a torus.

2. $S_{(V_1,V_2)}$ inherits the cyclic symmetry,

3. $|V_1| = |V_2|$ (this already follows from 2.),

4. the subset $K \subset M$ formed by all tetrahedra which are cut by $S_{(V_1,V_2)}$ (i. e. the subset of M which is given by $S_{(V_1,V_2)}$) can be extended to M in a canonical way.

Extending $S_{(V_1,V_2)}$ along its cyclic symmetry yields a series of slicings, which in turn defines a series of combinatorial 3-manifolds by the properties we imposed on $S_{(V_1,V_2)}$. In most cases it is easy to prove that the series in fact only consists of combinatorial manifolds. However, it is a little bit more work to determine the exact topological type (which does not always have to be the same for each member of the series). Some arguments can be deduced from the theory of slicings. A fact which emphasizes the convenience of the construction principle explained above. In some cases, collapsing arguments and discrete (polyhedral) Morse theory can be helpful. Additional information can be obtained by looking at cycles in different parts of the manifolds, together with arguments of 3-manifold theory (Dehn surgery and the decomposition theorem of 3-manifolds). For a brief overview of the research that has already been done in the field of transitive combinatorial manifolds see Section 4.1 above.

In the following, we will give a summary of all series that have been found so far using this method. Note that the topological types of some members of the series are not verified in a general setting. In these cases the topological type of the first members was checked by a computer. All checked complexes were homeomorphic to $S^1 \times S^2$, $S^1 \underline{\times} S^2$ or $\mathbb{T}^3$. Nonetheless, we believe that other types of manifolds occur in form of such series, too. Further conjectures based on the results of the computer experiments are listed in more detail in Table 4.3, Table 4.4 and Table 4.5. More precisely, we have the following results:

- The series Le_n of order 2 generalizes the complex of Figure 4.5.1 and will be discussed in Section 4.5.1.
- The series Ku_{4n} of order 4 generalizing Figure 3.3.1 with a non-cyclic transitive automorphism group is described in Section 4.5.2.
- The series C_{2n} and D_{2n}, both of order 2, will reappear in Section 5.8. Both series will thus be discussed in detail in Section 4.5.3.
- A short overview of a small family of further dense series (K_k^i, $1 \le i \le 14$) and series of order 2 (L_k^i, $1 \le i \le 18$) will be given in Section 4.5.4.
- Three series of neighborly combinatorial 3-manifolds $\mathrm{NSB_i}(k)$, $1 \le i \le 3$, where the conjectured topological type is a **N**eighborly **S**phere **B**undle (thus the name), will be presented in Section 4.5.5.

The series K_k^i and L_k^i complete the list of all series of order less or equal two having at least one member with 15 or less vertices. Note that this is not a complete classification of such series as series of higher order could exist. Also, there are series of 3-manifolds with an increasing number of difference cycles (for example series of 2-neighborly combinatorial 3-manifolds like the series $\mathrm{NSB_i}(k)$, $1 \le i \le 3$ from Section 4.5.5) which are not included in this classification.

Using the above construction principle, more series can easily be constructed in the same or a similar manner. As already mentioned, the search for infinite series of 3-manifolds distinct from bundles over the circle did not lead to further results as of today. However, this is work in progress.

Remark 4.5.3 (Technical note). Note that in order to ease the computation in the `GAP` environment, a vertex labeling $V = \{1, 2, \dots, n\}$ is used throughout this section. Moreover, the complexes are expressed in difference cycles (cf. Definition 4.2.3) whenever this is possible. In addition, all series are available from within the `GAP`-Package `simpcomp` [40, 41, 42], all with the function prefix `SCSeries` (see Table 4.6 for a list of all functions).

4.5.1 The series Le_n

Figure 4.5.1 shows the separating torus of a transitive 2-neighborly 14-vertex triangulation of the lens space $\mathrm{L}(3,1)$ (see [83, 3_{14}] or [92, ${}^{3}14_{6}^{1}$]).

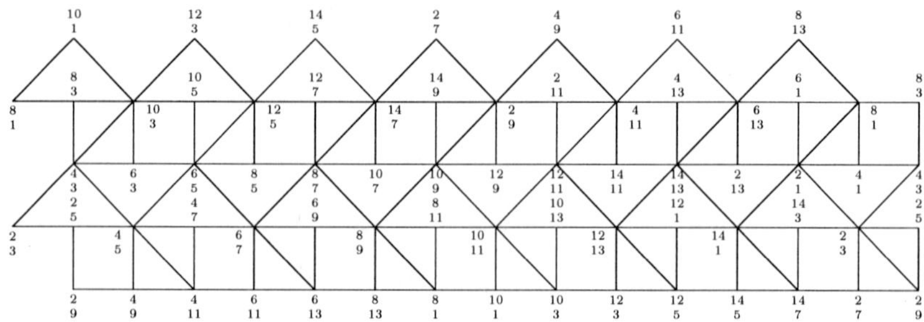

Figure 4.5.1: Slicing of genus 1 with 35 quadrilaterals, 28 triangles, $7 \cdot 7 = 49$ vertices and cyclic $\mathbb{Z}_{14}$-symmetry. It is obtained from a transitive 2-neighborly 14-vertex version of the lens space $\mathrm{L}(3,1)$.

As a matter of fact, this slicing coincides with slicings of two other transitive 2-neighborly 14-vertex triangulations. Namely with the 3-sphere [83, 1_{14}], [92, ${}^{3}14^{1}_{4}$] and the orientable sphere bundle over the circle [83, 2_{14}], [92, ${}^{3}14^{1}_{5}$].

The 35 quadrilaterals and the 28 triangles from Figure 4.5.1 induce a 3-dimensional simplicial complex K with $35 + 28 = 63$ tetrahedra which must be a subcomplex of ${}^{3}14^{1}_{4}$, ${}^{3}14^{1}_{5}$ and ${}^{3}14^{1}_{6}$. It is homeomorphic to the Cartesian product of the 2-dimensional torus with the unit interval and its boundary is isomorphic to two copies of the 7-vertex Möbius torus. The three different combinatorial 3-manifolds correspond to the three different fillings of the chiral 7-vertex Möbius torus (see [80]) and the transformation from one complex to another can be seen as a type of combinatorial Dehn surgery. In terms of difference cycles we have the complex:

$$K := \{(1{:}1{:}1{:}11),(1{:}2{:}4{:}7),(1{:}4{:}2{:}7),(1{:}4{:}7{:}2),(2{:}5{:}2{:}5)\} \qquad (4.5.2)$$

duch that

$$\begin{aligned} {}^{3}14^{1}_{4} &= K \cup \{(2{:}2{:}2{:}8)\} \\ {}^{3}14^{1}_{5} &= K \cup \{(2{:}4{:}2{:}6)\} \\ {}^{3}14^{1}_{6} &= K \cup \{(2{:}4{:}4{:}4)\}. \end{aligned}$$

In addition, the sphere bundle extends to a series of simplicial complexes Le_{2k} with $2k$

vertices, $k \geq 7$, given by the following difference cycles.

$$\begin{aligned} \mathrm{Le}_{2k} = \{ & (1{:}1{:}1{:}2k-3),(1{:}2{:}k-3{:}k),(1{:}k-3{:}2{:}k), \\ & (1{:}k-3{:}k{:}2),(2{:}k-2{:}2{:}k-2),(2{:}k-3{:}2{:}k-1) \}. \end{aligned} \qquad (4.5.3)$$

Figure 4.5.2 shows the vertex link of Le_{2k}. For $k \geq 7$ it is a combinatorial 2-sphere. Hence, Le_{2k} is a combinatorial 3-manifold for all $k \geq 7$. The topology is believed to be $S^2 \times S^1$ for all members of the series. For odd values of k this follows from the construction. For even numbers of k the situation is similar but not as clear. However, standard techniques should suffice to determine the topological type in this case as well.

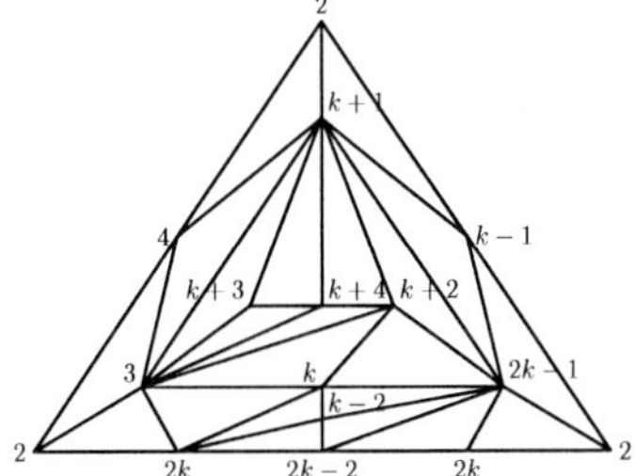

Figure 4.5.2: Link of vertex 1 in Le_{2k}, $k \geq 7$ – a combinatorial 2-sphere.

4.5.2 The series Ku_{4n}

Figure 3.3.1 from Section 3.3 can be extended to a series of slicings T_n as shown in Figure 4.5.3. The underlying 3-dimensional simplicial complexes Ku_{4n} can be closed in a canonical way yielding the following result.

Theorem 4.5.4. *There is an infinite series of combinatorial triangulations of* $\mathrm{Ku}_{4n} \cong S^2 \times S^1$, $n \geq 3$, *with* $4n$ *vertices given by the automorphism group*

$$\begin{aligned} G := & \langle (1,2n)(2,2n-1)\dots(n,n+1)(2n+1,4n)(2n+2,4n-1)\dots(3n,3n+1) \\ & (1,n+1,2,n+2,\dots,n,2n)(2n+1,3n+1,2n+2,3n+2,\dots 3n,4n) \\ & (1,4n)(2,4n-1)\dots(2n,2n+1) \rangle, \end{aligned} \qquad (4.5.4)$$

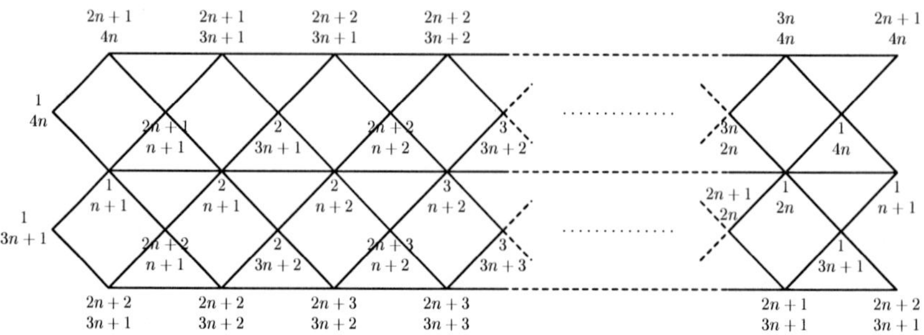

Figure 4.5.3: Slicing T_n with $4n$ quadrilaterals and $8n$ triangles: A polyhedral torus.

and the G-orbits

$$(1,2,n+1,2n+1)_{8n},(1,n+1,2n+1,4n)_{4n},(1,2,2n+1,2n+2)_{2n}.$$

In particular, Ku_{4n} *does not have a cyclic symmetry.*

Proof. T_n equals the slicing $S_{(V_1,V_2)}$ of a closed simplicial complex Ku_{4n} with vertex set $V := \{1,\dots,4n\}$ where $V_1 := \{1,\dots,n,2n+1,\dots,3n\}$ and $V_2 := \{n+1,\dots,2n,3n+1,\dots,4n\}$. As can be deduced from Figure 4.5.3, Ku_{4n} has the following symmetries generating its automorphism group

$$\begin{aligned} G &:= \langle (1,2n)(2,2n-1)\dots(n,n+1)(2n+1,4n)(2n+2,4n-1)\dots(3n,3n+1) \\ &\quad (1,n+1,2,n+2,\dots,n,2n)(2n+1,3n+1,2n+2,3n+2,\dots 3n,4n) \\ &\quad (1,4n)(2,4n-1)\dots(2n,2n+1)\rangle \\ &\cong \mathbb{Z}_2 \times D_{2n}, \end{aligned}$$

which acts transitively on the set of vertices. It follows that Ku_{4n} is generated by the G-orbits

$$(1,2,n+1,2n+1)_{8n},(1,n+1,2n+1,4n)_{4n},(1,2,2n+1,2n+2)_{2n}.$$

The f-vector is $f = (4n,18n,28n,14n)$, the vertex link of vertex 1 in Ku_{4n} is the 9-vertex 2-sphere shown in Figure 4.5.4. Thus, Ku_{4n} is a combinatorial 3-manifold for $n \geq 3$.

As every element of the automorphism group either preserves V_i or sends V_i to V_j, $i,j \in$

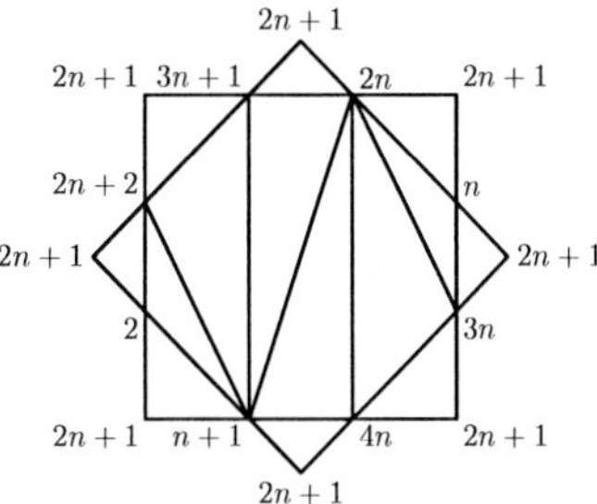

Figure 4.5.4: Vertex link of vertex 1 in Ku_{4n} with C_2-symmetry. A 2-sphere for $n \geq 3$, a 7-vertex real projective plane for $n = 2$ (cf. Figure 3.3.1 from Section 3.3). The non-trivial automorphism is given by the permutation $(2,n)(n+1,2n)(2n+1)(2n+2,3n)(3n+1,4n)$.

$\{1,2\}$, $i \neq j$, the span of V_1 and V_2 coincides with the G-orbit $(1,2,2n+1,2n+2)_{2n}$ of length $2n$. Hence, we have

$$\mathrm{span}_{V_1}(\mathrm{Ku}_{4n}) = \{\langle 1,2,2n+1,2n+2\rangle, \ldots, \langle n-1,n,3n-1,3n\rangle, \langle n,1,3n,2n+1\rangle\}$$

and

$$\mathrm{span}_{V_2}(\mathrm{Ku}_{4n}) = \{\langle n+1,n+2,3n+1,3n+2\rangle, \ldots, \langle 2n-1,2n,4n-1,4n\rangle, \langle 2n,n+1,4n,3n+1\rangle\},$$

which obviously both collapse to circles c_1 and c_2. Thus, the complex T_n defines a handlebody decomposition of genus 1 of Ku_{4n}. Furthermore, it follows directly from Figure 4.5.3 that $[c_1] = [c_2]$ in $H_1(T_n,\mathbb{Z})$ and thus $\mathrm{Ku}_{4n} \cong S^2 \times S^1$. □

4.5.3 The series C_{2n} and D_{2n}

The series C_{2n} and D_{2n} most likely coincide with the connected components of the vertex links of the series of 2-transitive combinatorial 4-pseudomanifolds from Section 5.8. Hence, it seems rewarding to pay a little more attention to these complexes.

Theorem 4.5.5. *Let $n \geq 8$. The simplicial complex given by the difference cycles*

$$C_{2n} = \{(1:1:n-5:n+3),(1:1:n+3:n-5),(1:n-5:1:n+3),(2:n-5:2:n+1),(2:n-3:2:n-1)\}$$

is homeomorphic to $S^2 \times S^1$ if n is odd and homeomorphic to $S^1 \times S^2$ if n is even. The

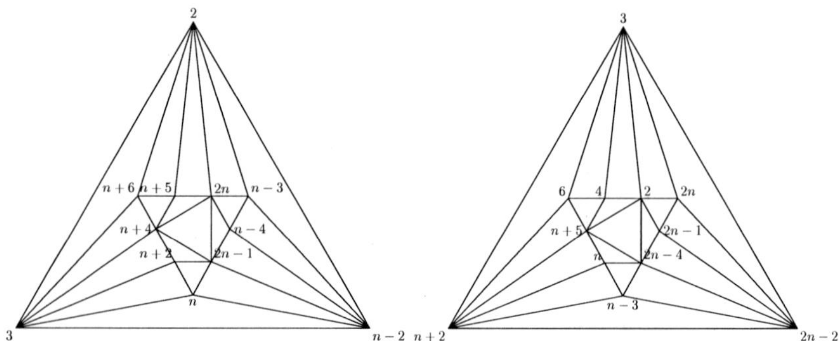

Figure 4.5.5: Links of vertex 1 in C_{2n} on the left and in D_{2n} on the right side. Note that the links are combinatorially isomorphic to the edge link of $C_p^{x^l}$.

complex

$$D_{2n} = \{(1{:}1{:}1{:}2n-3),(1{:}2{:}2n-5{:}2),(3{:}n-4{:}5{:}n-4),(2{:}3{:}n-4{:}n-1),(2{:}n-1{:}n-4{:}3)\}$$

is homeomorphic to $S^1 \times S^2$ *for all* $n \geq 8$, $n \neq 9$. *For even numbers* n *the complexes* C_{2n} *and* D_{2n} *are combinatorially isomorphic.*

Both C_{2n} *and* D_{2n} *are invariant under multiplication by* -1.

Proof. First, let us verify that C_{2n}, $n \geq 8$, and D_{2n}, $n \geq 8$, $n \neq 9$, are infinite series of combinatorial 3-manifolds: Since both series of simplicial complexes have transitive symmetry by construction, it suffices to look at the link of vertex 1 in each complex. The vertex links of vertex 1 in C_{2n} and D_{2n} are shown in Figure 4.5.5. For $n \geq 8$, and in the case of D_{2n} additionally $n \neq 9$, all vertices of the link are distinct and all links are 2-spheres what states the assumption.

In order to show that C_{2n} and D_{2n} are combinatorially isomorphic whenever n is an even number, consider the following relabeling of the vertices of C_{2n}:

$$\phi : \{1, \ldots, 2n\} \to \{1, \ldots, 2n\}; \quad i \mapsto \begin{cases} (n+i \mod 2n) + 1 & i \equiv 1\,(2) \\ (i \mod 2n) + 1 & i \equiv 0\,(2) \end{cases}$$

The mapping ϕ is an isomorphism if and only if n is even. Applying ϕ to the orbit representatives of C_{2n} yields a system of representatives of the orbits of D_{2n}.

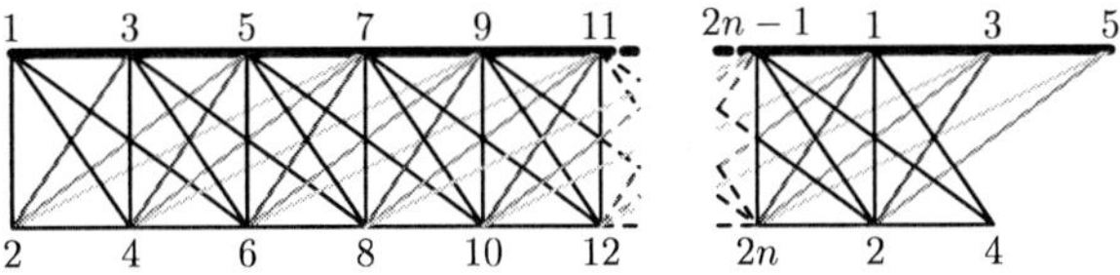

Figure 4.5.6: Handlebody $\tilde{D}_-$. The thick line $\langle 1,3,5,\ldots,2n-1,1\rangle$ represents a generator of the first homology group $H_1(\tilde{D}_-,\mathbb{Z})$.

In the following, we will determine the topological type of D_{2n}. In order to do this, consider the partition of D_{2n} into the following bounded manifolds

$$D_+ := \{(3{:}n-4{:}5{:}n-4),(2{:}3{:}n-4{:}n-1),(2{:}n-1{:}n-4{:}3)\}$$

and

$$D_- := \{(1{:}1{:}1{:}2n-3),(2{:}1{:}2{:}2n-5)\}$$

with common boundary

$$T := \partial D_+ = \partial D_- = \{(2{:}3{:}2n-5),(2{:}2n-5{:}3)\},$$

which is a triangulated torus by Lemma 4.3.4 ($l = 2$, $j = 3$) for $n \geq 6$. D_- is shown in Figure 4.5.6. As we can deduce from the drawing, D_- collapses onto the cycle $\langle 1,3,5,\ldots,2n-1,1\rangle$ and thus D_- is a handlebody of genus 1.

A set of orbit representatives of D_+ equals $\{\langle 1,4,6,n+5\rangle,\langle 1,4,n,n+5\rangle,\langle 1,4,n,2n-1\rangle\}$. It collapses onto the two triangles $\{\langle 1,n,n+5\rangle,\langle 4,n,n+5\rangle\}$ which form a system of orbit representatives of the torus $S_{5,n-4,2n}$ for $n \geq 8$, $n \neq 9$, from Lemma 4.3.4. For $n = 9$, the complex does not generate a manifold. See Figure 4.5.7 for a visualization of the collapsing process.

The projection of the boundary torus T onto $S_{5,n-4,2n}$ is a twofold covering. Thus, D_+ is homeomorphic to the total space of a non-trivial line bundle and hence its topological type is a product of a Möbius strip with a circle. The rest of the proof for D_{2n} consists of an exact copy of the proof of Theorem 2 in [83, pages 154, 155]. It follows that D_{2n} is a 3-dimensional Klein bottle for all $n \geq 8$, $n \neq 9$.

Now let us have a closer look at C_{2n}. Again, we start by decomposing the complex into

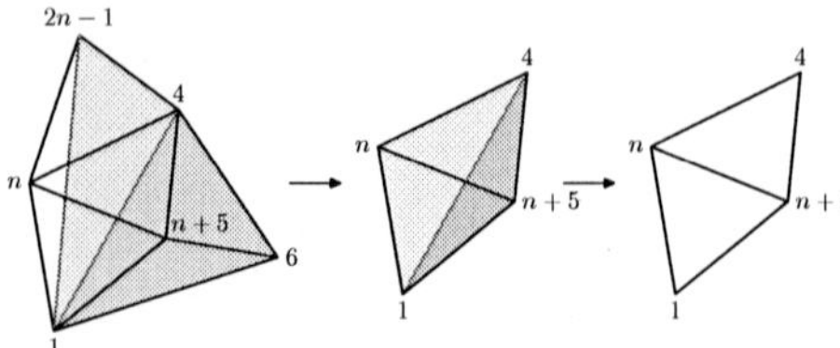

Figure 4.5.7: Collapsing procedure from a generating subset of D_+ onto a generating set of the transitive torus $S_{5,n-4,2n}$ in the case $n \geq 8$, $n \neq 9$.

two bounded manifolds:

$$C_+ := \{(1:1:n-5:n+3),(1:1:n+3:n-5),(1:n-5:1:n+3)\}$$

and

$$C_- := \{(2:n-5:2:n+1),(2:n-3:2:n-1)\}.$$

Both parts are bounded by

$$S := \partial C_+ = \partial C_- = \{(2:n-5:n+3),(2:n+3:n-5)\}.$$

By virtue of Lemma 4.3.4 ($l = 2$, $j = n-5$), the complex S is a torus if n is even and the disjoint union of two tori if n is odd.

Since C_{2n} and D_{2n} are combinatorially isomorphic whenever n is even, it suffices to look at C_{2n} in the case that $n := 2k+1$ is odd. Hence, we have

$$C_- = \{(2:2k-4:2:2k+2),(2:2k-2:2:2k)\},$$

and the complex C_- has two connected components. One is spanned by all even labeled vertices, the other by all odd labeled vertices. Both are combinatorially isomorphic to

$$\tilde{C}_- = \{(1:k-2:1:k+1),(1:k-1:1:k)\}$$

with boundary

$$\partial\tilde{C}_- = \{(1:k-2:k+2),(1:k+2:k-2)\}$$

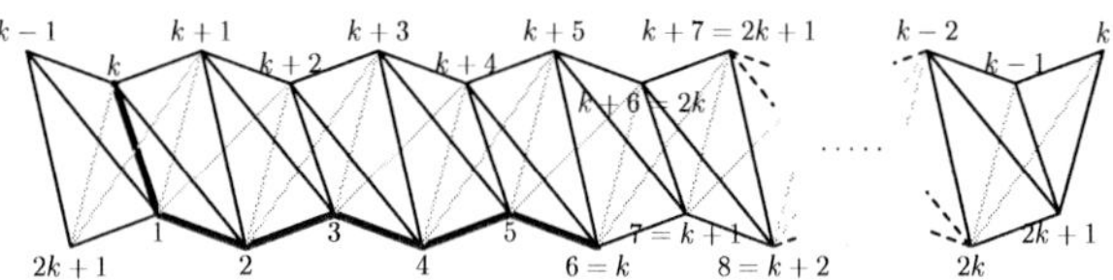

Figure 4.5.8: Handlebody $\tilde{C}_-$. The thick line $\langle 1,2,\ldots,k,1\rangle$ represents a generator of the first homology group $H_1(\tilde{C}_-,\mathbb{Z})$. In this picture, we have $k=6$ to illustrate a particular basis element.

which is a torus by Theorem 4.4.1 ($l=1$, $j=k-2,k\mapsto 2k+1$). From Figure 4.5.8 we can see that $\tilde{C}_-$ collapses onto the cycle $\langle 1,2,\ldots,k,1\rangle$ and, hence, $\tilde{C}_-$ is a handlebody $\mathbb{B}^2\times S^1$.

Since $\partial C_+ = \partial C_-$, the complex

$$C_+ = \{(1{:}1{:}2k-4{:}2k+4),(1{:}1{:}2k+4{:}2k-4),(1{:}2k-4{:}1{:}2k+4)\}$$

has two toroidal boundary components, one spanned by all even labeled vertices, the other spanned by all odd labeled vertices. In particular, no vertices lie in the interior of C_+. Thus, a simplexwise linear Morse function $f:|C_+|\to[0,1]$ where the images of all of the odd labeled vertices $v_1\in V_1$ are greater than $\frac{2}{3}$ and the images of all of the vertices with an even labeling $v_2\in V_2$ are smaller than $\frac{1}{3}$ cannot have any critical points in $[\frac{1}{3},\frac{2}{3}]$. In addition, any pre-image $f^{-1}(\alpha)$, $\alpha\in[\frac{1}{3},\frac{2}{3}]$, is connected, orientable and homeomorphic to the slicing $S_{(V_1,V_2)}$ of C_+ with $8k+4$ vertices, $20k+10$ edges, $8k+4$ triangles and $4k+2$ quadrilaterals. Hence, $f^{-1}(\alpha)$ is homeomorphic to $\mathbb{T}^2$ as well. Altogether, C_+ is a bundle of tori over the (contractible) unit interval $[0,1]$ and thus $C_+\cong\mathbb{T}^2\times[0,1]$.

To see that C_{4k+2} is actually homeomorphic to $S^2\times S^1$, it suffices to check that the cores of both connected components of C_- are homologous in C_{4k+2}: As we can see in Figure 4.5.8, the cores of the connected components of C_- are the two cycles $b_1:=\langle 1,3,\ldots,2k-1,1\rangle$ and $b_2:=\langle 2,4,\ldots,2k,2\rangle$. By looking at the orbit $(1:1:2k-4:2k+4)=\{\langle 1,2,3,2k-1\rangle,\langle 2,3,4,2k\rangle,\langle 3,4,5,2k+1\rangle,\ldots\}\subset C_+$ we see that b_1 as well as b_2 are contained in C_+ and can be transformed into each other by a simple homotopy, what proves the result.

Multiplication of C_{2n} by -1 interchanges the two difference cycles $(1{:}1{:}n-5{:}n+3)$ and $(1{:}1{:}n+3{:}n-5)$ and leaves the three difference cycles $(1{:}n-5{:}1{:}n+3)$, $(2{:}n-5{:}2{:}n+1)$ and $(2{:}n-3{:}2{:}n-1)$ invariant.

Multiplication of D_{2n} by -1 leaves the three difference cycles $(1{:}1{:}1{:}2n-3)$, $(1{:}2{:}2n-5{:}2)$

and $(3{:}n-4{:}5{:}n-4)$ invariant and interchanges $(2{:}3{:}n-4{:}n-1)$ and $(2{:}n-1{:}n-4{:}3)$.

Since no further multipliers occur, the full automorphism group of C_{2n} and D_{2n} is the dihedral group D_n with $2n$ elements. If we choose the vertex labels to be $V = \{1,\dots,2n\}$, the groups are given in the following permutation representation:

$$\mathrm{Aut(C_{2n})} = \langle (1,\dots,2n),(2,2n)(3,2n-1)\dots(n,n+2)\rangle = \mathrm{Aut(D_{2n})}\,.$$

□

Members of the series can be generated by using the functions `SCSeriesC2n(n)` and `SCSeriesD2n(n)` within the GAP-package simpcomp [40, 41, 42].

4.5.4 Overview and further series

In the following, we present a complete list of all combinatorial types of series of 3-dimensional bundles over the circle of order less or equal two where at least one member of the series has 15 vertices or less. The new dense series K^i_k, $1 \le k \le 14$ (cf. Table 4.3), and the series of order 2, namely L^i_k, $1 \le k \le 18$ (cf. Table 4.4), were found using an enumerative approach and simpcomp [40, 41, 42], starting with a classification of all cyclic triangulations. Note that, in order to find all combinatorial types of such series, we need to check all cyclic combinatorial 3-manifolds including possibly isomorphic complexes. The code of the algorithm is given in Appendix D. All series are integrated into the GAP-package simpcomp [40, 41, 42], all with the function prefix `SCSeries...` (`SCSeriesK` and `SCSeriesL`, cf. Table 4.6).

Table 4.3: List of all combinatorial types of dense series of 3-dimensional bundles over the circle starting with at most 15 vertices.

name	m	difference cycles	top. type (conj.)	reference
$\partial\mathbb{H}^3_{9+k}$	9	$(1{:}1{:}2{:}5+k)$, $(1{:}1{:}5+k{:}2)$, $(1{:}2{:}1{:}5+k)$	$\partial\mathbb{H}^3_{9+2k} \cong S^2 \tilde{\times} S^1$, $\partial\mathbb{H}^3_{10+2k} \cong S^2 \times S^{1**}$	[7, N^9_{51}], [83, II_9] ,[74, M^3], [85, M^3_2], [77, Theorem 5.5], [92, ${}^39^3_2$, ${}^310^3_2$, ${}^311^2_4$, ${}^312^{12}_2$,] [92, ${}^313^2_6$, ${}^314^3_6$, ${}^315^2_7$]
K^1_k	11	$(1{:}1{:}2{:}7+k)$, $(1{:}1{:}6+k{:}3)$, $(1{:}2{:}2{:}6+k)$, $(1{:}3{:}1{:}6+k)$	$K^1_{2k} \cong S^2 \tilde{\times} S^1$, $K^1_{2k+1} \cong S^2 \times S^1$	[92, ${}^311^1_2$, ${}^312^1_3$, ${}^313^1_7$,] [92, ${}^314^1_{16}$, ${}^315^1_{18}$], [83, 2_{11}]
K^2_k	13	$(1{:}1{:}2{:}9+k)$, $(1{:}1{:}7+k{:}4)$, $(1{:}2{:}2{:}8+k)$, $(1{:}3{:}2{:}7+k)$, $(1{:}4{:}1{:}7+k)$	$K^2_{2k} \cong S^2 \tilde{\times} S^1$, $K^2_{2k+1} \cong S^2 \times S^1$	[92, ${}^313^1_6$, ${}^314^1_{15}$, ${}^315^1_{17}$] [83, 5_{13}]
K^3_k	13	$(1{:}1{:}3{:}8+k)$, $(1{:}1{:}7+k{:}4)$, $(1{:}3{:}1{:}8+k)$, $(1{:}7+k{:}2{:}3)$, $(2{:}2{:}2{:}7+k)$	$K^3_{2k} \cong S^2 \tilde{\times} S^1$, $K^3_{2k+1} \cong S^2 \times S^1$	[92, ${}^313^1_2$, ${}^314^1_{22}$, ${}^315^1_8$] [83, 1_{13}]
K^{4*}_k	13	$(1{:}1{:}4{:}7+k)$, $(1{:}1{:}7+k{:}4)$, $(1{:}4{:}1{:}7+k)$, $(2{:}2{:}2{:}7+k)$	$K^4_{2k} \cong S^2 \tilde{\times} S^1$, $K^4_{2k+1} \cong S^2 \times S^1$	[92, ${}^313^2_3$, ${}^314^3_8$, ${}^315^2_3$]
K^{5*}_k	13	$(1{:}2{:}1{:}9+k)$, $(1{:}2{:}7+k{:}3)$, $(1{:}3{:}7+k{:}2)$, $(2{:}2{:}2{:}7+k)$	$K^5_{2k} \cong S^2 \tilde{\times} S^1$, $K^5_{2k+1} \cong S^2 \times S^1$	[92, ${}^313^2_4$, ${}^314^3_{10}$, ${}^315^2_4$]
K^6_k	13	$(1{:}2{:}3{:}7+k)$, $(1{:}2{:}8+k{:}2)$, $(1{:}3{:}2{:}7+k)$, $(1{:}3{:}7+k{:}2)$, $(2{:}2{:}2{:}7+k)$	$K^6_{2k} \cong S^2 \tilde{\times} S^1$, $K^6_{2k+1} \cong S^2 \times S^1$	[92, ${}^313^1_4$, ${}^314^1_{24}$, ${}^315^1_{11}$] [83, 3_{13}]
K^7_k	15	$(1{:}1{:}2{:}11+k)$, $(1{:}1{:}8+k{:}5)$, $(1{:}2{:}2{:}10+k)$, $(1{:}3{:}2{:}9+k)$, $(1{:}4{:}2{:}8+k)$, $(1{:}5{:}1{:}8+k)$	$K^7_{2k} \cong S^2 \tilde{\times} S^1$, $K^7_{2k+1} \cong S^2 \times S^1$	[92, ${}^315^1_{15}$], [83, 11_{15}]
K^8_k	15	$(1{:}1{:}4{:}9+k)$, $(1{:}1{:}8+k{:}5)$, $(1{:}2{:}2{:}10+k)$, $(1{:}2{:}4{:}8+k)$, $(1{:}5{:}1{:}8+k)$, $(2{:}2{:}8+k{:}3)$	$K^8_{2k} \cong S^2 \tilde{\times} S^1$, $K^8_{2k+1} \cong S^2 \times S^1$	[92, ${}^315^1_{20}$], [83, 14_{15}]
K^9_k	15	$(1{:}1{:}4{:}9+k)$, $(1{:}1{:}8+k{:}5)$, $(1{:}4{:}2{:}8+k)$, $(1{:}5{:}1{:}8+k)$, $(2{:}2{:}2{:}9+k)$	$K^9_{2k} \cong S^2 \tilde{\times} S^1$, $K^9_{2k+1} \cong S^2 \times S^1$	[92, ${}^315^1_7$]

continued on next page –

Table 4.3 – continued from previous page

name	m	difference cycles	top. type (conj.)	reference
K_k^{10}	15	$(1{:}2{:}2{:}10+k)$, $(1{:}2{:}8+k{:}4)$, $(1{:}3{:}1{:}10+k)$, $(1{:}3{:}8+k{:}3)$, $(2{:}2{:}8+k{:}3)$	$K_{2k}^{10} \cong S^2 \mathbin{\underline{\times}} S^1$, $K_{2k+1}^{10} \cong S^2 \times S^1$	[92, $^3 15^1_{21}$]
K_k^{11}	15	$(1{:}2{:}2{:}10+k)$, $(1{:}2{:}9+k{:}3)$,$(1{:}3{:}3{:}8+k)$, $(1{:}3{:}8+k{:}3)$, $(1{:}4{:}2{:}8+k)$, $(2{:}2{:}8+k{:}3)$	$K_{2k}^{11} \cong S^2 \mathbin{\underline{\times}} S^1$, $K_{2k+1}^{11} \cong S^2 \times S^1$	[92, $^3 15^1_{22}$], [83, 15_{15}]
K_k^{12}	15	$(1{:}2{:}4{:}8+k)$, $(1{:}2{:}8+k{:}4)$,$(1{:}3{:}1{:}10+k)$, $(1{:}3{:}8+k{:}3)$, $(1{:}4{:}2{:}8)$, $(2{:}2{:}2{:}9+k)$	$K_{2k}^{12} \cong S^2 \mathbin{\underline{\times}} S^1$, $K_{2k+1}^{12} \cong S^2 \times S^1$	[92, $^3 15^1_4$], [83, 4_{15}]
K_k^{13*}	15	$(1{:}2{:}4{:}8+k)$, $(1{:}2{:}8+k{:}4)$, $(1{:}4{:}2{:}8+k)$, $(1{:}4{:}8+k{:}2)$, $(1{:}8+k{:}2{:}4)$, $(1{:}8+k{:}4{:}2)$	$K_k^{13} \cong \mathbb{T}^{3**}$	[85, $M_1^3(15+k)$, p. 216], [82],[83, III_{15}],[84], [92, $^3 15^{11}_1$], [23]
K_k^{14}	15	$(1{:}2{:}4{:}8+k)$, $(1{:}2{:}9+k{:}3)$, $(1{:}3{:}3{:}8+k)$, $(1{:}3{:}8+k{:}3)$, $(2{:}2{:}2{:}9+k)$	$K_{2k}^{14} \cong S^2 \mathbin{\underline{\times}} S^1$, $K_{2k+1}^{14} \cong S^2 \times S^1$	[92, $^3 15^1_6$]

Table 4.4: List of all series of 3-dimensional bundles over the circle of order 2 starting with at most 15 vertices, as well as C_{2n} and D_{2n}.

name	m	difference cycles	top. type (conj.)	reference
L_k^{1*}	10	$(1{:}1{:}3+k{:}5+k)$, $(1{:}1{:}4+k{:}4+k)$, $(1{:}3+k{:}2{:}4+k)$, $(2{:}3+k{:}2{:}3+k)$	$L_k^1 \cong S^2 \times S^1$	[92, $^3 10^1_2$, $^3 12^{11}_1$, $^3 14^1_{33}$], [5, N^{10}_{3611}], [83, 2_{10}]
L_k^{2*}	10	$(1{:}1{:}3+k{:}5+k)$, $(1{:}1{:}5+k{:}3+k)$, $(1{:}3+k{:}1{:}5+k)$, $(2{:}3+k{:}2{:}3+k)$	$L_{2k}^2 \cong S^2 \mathbin{\underline{\times}} S^1, L_{2k+1}^2 \cong S^2 \times S^1$	[92, $^3 10^3_3$, $^3 12^{12}_3$, $^3 14^3_{11}$], [5, N^{10}_{3629}], [83, $\bar{\mathrm{II}}_{10}$]
L_k^{3*}	10	$(1{:}3+k{:}1{:}5+k)$, $(1{:}3+k{:}2{:}4+k)$, $(1{:}4+k{:}2{:}3+k)$, $(2{:}3+k{:}2{:}3+k)$	$L_{2k}^3 \cong S^2 \mathbin{\underline{\times}} S^1, L_{2k+1}^3 \cong S^2 \times S^1$	[92, $^3 10^3_4$, $^3 12^{12}_3$, $^3 14^3_{13}$], [5, N^{10}_{3631}], [83, II_{10}]

continued on next page –

Table 4.4 – continued from previous page

name	m	difference cycles	top. type (conj.)	reference
L^4_k	12	$(1{:}1{:}1{:}9+2k)$, $(1{:}2{:}4+k{:}5+k)$, $(1{:}4+k{:}2{:}5+k)$, $(1{:}4+k{:}5+k{:}2)$, $(2{:}4+k{:}2{:}4+k)$	$L^4_k \cong S^2 \times S^1$	[92, ${}^3 12^1_2$, ${}^3 14^1_{12}$] [83, 1_{12}]
$L^{5\star}_k$	12	$(1{:}1{:}1{:}9+2k)$, $(1{:}2{:}5+k{:}4+k)$, $(1{:}4+k{:}3{:}4+k)$, $(1{:}4+k{:}5+k{:}2)$	$L^5_k \cong S^2 \tilde{\times} S^1$	[92, ${}^3 12^{54}_1$, ${}^3 14^3_5$]
L^6_k	12	$(1{:}1{:}3+k{:}7+k)$, $(1{:}1{:}4+k{:}6+k)$, $(1{:}3+k{:}2{:}6+k)$, $(2{:}3+k{:}2{:}5+k)$,$(2{:}4+k{:}2{:}4+k)$	$L^6_k \cong S^2 \times S^1$	[92, ${}^3 12^1_2$, ${}^3 14^1_{29}$] [83, 1_{12}]
L^7_k	12	$(1{:}1{:}3+k{:}7+k)$, $(1{:}1{:}6+k{:}4+k)$, $(1{:}3+k{:}1{:}7+k)$, $(1{:}6+k{:}2{:}3+k)$, $(2{:}4+k{:}2{:}4+k)$	$L^7_{2k} \cong S^2 \times S^1, L^7_{2k+1} \cong S^2 \tilde{\times} S^1$	[92, ${}^3 12^1_4$, ${}^3 14^1_{32}$] [83, 2_{12}]
$L^{8\star}_k$	12	$(1{:}1{:}3+k{:}7+k)$, $(1{:}1{:}7+k{:}3+k)$, $(1{:}3+k{:}1{:}7+k)$, $(2{:}3+k{:}2{:}5+k)$	$L^8_{2k} \cong S^2 \tilde{\times} S^1, L^8_{2k+1} \cong S^2 \times S^1$	[92, ${}^3 12^{54}_1$, ${}^3 14^3_9$]
L^9_k	12	$(1{:}2{:}4+k{:}5+k)$, $(1{:}2{:}5+k{:}4+k)$, $(1{:}4+k{:}2{:}5+k)$, $(1{:}4+k{:}3{:}4+k)$,$(2{:}4+k{:}2{:}4+k)$	$L^9_{2k} \cong S^2 \times S^1, L^9_{2k+1} \cong S^2 \tilde{\times} S^1$	[92, ${}^3 12^1_4$, ${}^3 14^1_{36}$] [83, 2_{12}]
$L^{10\star}_k$	14	$(1{:}1{:}1{:}11+2k)$, $(1{:}2{:}1{:}10+2k)$, $(1{:}3{:}5+k{:}5+k)$, $(1{:}5+k{:}3{:}5+k)$, $(1{:}5+k{:}5+k{:}3)$	$L^{10}_k \cong S^2 \tilde{\times} S^1$	[92, ${}^3 14^3_3$]
L^{11}_k	14	$(1{:}1{:}1{:}11+2k)$, $(1{:}2{:}4+k{:}7+k)$, $(1{:}6+k{:}5+k{:}2)$, $(2{:}4+k{:}2{:}6+k)$, $(2{:}5+k{:}2{:}5+k)$, $(2{:}5+k{:}3{:}4+k)$	$L^{11}_k \cong S^2 \times S^1$	[92, ${}^3 14^1_9$], [83, 5_{14}]
L^{12}_k	14	$(1{:}1{:}3+k{:}9+k)$, $(1{:}1{:}4+k{:}8+k)$, $(1{:}3+k{:}2{:}8+k)$, $(2{:}3+k{:}2{:}7+k)$, $(2{:}4+k{:}2{:}6+k)$, $(2{:}5+k{:}2{:}5+k)$	$L^{12}_k \cong S^2 \times S^1$	[92, ${}^3 14^1_{19}$], [83, 12_{14}]
L^{13}_k	14	$(1{:}1{:}3+k{:}9+k)$, $(1{:}1{:}7+k{:}5+k)$, $(1{:}3+k{:}1{:}9+k)$, $(1{:}7+k{:}2{:}4+k)$, $(1{:}8+k{:}2{:}3+k)$, $(2{:}5+k{:}2{:}5+k)$	$L^{13}_{2k} \cong S^2 \tilde{\times} S^1, L^{13}_{2k+1} \cong S^2 \times S^1$	[92, ${}^3 14^1_{21}$], [83, 14_{14}]
L^{14}_k	14	$(1{:}1{:}3+k{:}9+k)$, $(1{:}1{:}8+k{:}4+k)$, $(1{:}3+k{:}1{:}9+k)$, $(1{:}8+k{:}2{:}3+k)$, $(2{:}4+k{:}2{:}6+k)$	$L^{14}_{2k} \cong S^2 \times S^1, L^{14}_{2k+1} \cong S^2 \tilde{\times} S^1$	[92, ${}^3 14^1_{23}$]
L^{15}_k	14	$(1{:}1{:}3+k{:}9+k)$, $(1{:}1{:}8+k{:}4+k)$, $(1{:}3+k{:}2{:}8+k)$, $(1{:}4+k{:}1{:}8+k)$, $(2{:}3+k{:}2{:}7+k)$, $(2{:}5+k{:}2{:}5+k)$	$L^{15}_{2k} \cong S^2 \tilde{\times} S^1, L^{15}_{2k+1} \cong S^2 \times S^1$	[92, ${}^3 14^1_{25}$], [83, 15_{14}]

continued on next page –

Table 4.4 – continued from previous page

name	m	difference cycles	top. type (conj.)	reference
$L_k^{16\star}$	14	$(1{:}1{:}3+k{:}9+k)$, $(1{:}1{:}9+k{:}3+k)$, $(1{:}3+k{:}1{:}9+k)$, $(2{:}3+k{:}2{:}7+k)$, $(2{:}5+k{:}2{:}5+k)$	$L_{2k}^{16} \cong S^2 \underline{\times} S^1, L_{2k+1}^{16} \cong S^2 \times S^1$	[92, $^3 14^3_7$]
L_k^{17}	14	$(1{:}4+k{:}1{:}8+k)$, $(1{:}4+k{:}2{:}7+k)$, $(1{:}5+k{:}3{:}5+k)$, $(1{:}6+k{:}3{:}4+k)$,$(2{:}5+k{:}2{:}5+k)$, $(2{:}5+k{:}3{:}4+k)$	$L_{2k}^{17} \cong S^2 \underline{\times} S^1, L_{2k+1}^{17} \cong S^2 \times S^1$	[92, $^3 14^1_{34}$], [83, 19_{14}]
L_k^{18}	14	$(1{:}4+k{:}1{:}8+k)$, $(1{:}4+k{:}7+k{:}2)$, $(1{:}5+k{:}3{:}5+k)$, $(1{:}6+k{:}3{:}4+k)$,$(1{:}6+k{:}5+k{:}2)$, $(2{:}5+k{:}2{:}5+k)$	$L_{2k}^{18} \cong S^2 \underline{\times} S^1, L_{2k+1}^{18} \cong S^2 \times S^1$	[92, $^3 14^1_{35}$], [83, 20_{14}]
Le_{2k}	14	$(1{:}1{:}1{:}2k-3)$, $(1{:}2{:}k-3{:}k)$, $(1{:}k-3{:}2{:}k)$, $(1{:}k-3{:}k{:}2)$, $(2{:}k-3{:}2{:}k-1)$, $(2{:}k-2{:}2{:}k-2)$	$\mathrm{Le}_{2k} \cong S^2 \times S^1$ for k odd **	[92, $^3 14^1_5$], [83, 2_{14}]
C_{2n}	16	$(1{:}1{:}n-5{:}n+3)$, $(1{:}1{:}n+3{:}n-5)$, $(1{:}n-5{:}1{:}n+3)$, $(2{:}n-5{:}2{:}n+1)$, $(2{:}n-3{:}2{:}n-1)$	$C_{4n+2} \cong S^2 \times S^1$, $C_{4n} \cong S^2 \underline{\times} S^{1**}$	
D_{2n}	20	$(1{:}1{:}1{:}2n-3)$, $(1{:}2{:}2n-5{:}2)$, $(3{:}n-4{:}5{:}n-4)$, $(2{:}3{:}n-4{:}n-1)$, $(2{:}n-1{:}n-4{:}3)$	$D_{2n} \cong S^2 \underline{\times} S^{1**}$	

* The members of these series have additional symmetry (i. e. multipliers of the set of difference cycles occur). For the (conjectured) full automorphism groups of these series see Table 4.5. Note that the members of L_k^5 and L_k^8 seem to have isomorphic automorphism groups for any $k \geq 0$ and are in fact combinatorially isomorphic whenever k is even. The series Ku_{4n} is of order 4 and thus not listed in Table 4.3 or 4.4.

** For members of these series the topological type has been verified.

Table 4.5: Full automorphism groups (conjectured) of K^i_k, $1 \le i \le 14$ and L^i_k, $1 \le i \le 18$, which have $\mathbb{Z}_n = \langle(1,\dots,n)\rangle$ as a proper subgroup.

name	generators of $G := \mathrm{Aut}(C)$ (conjectured)	structure of G (conjectured)
K^4_k	$\langle(1,\dots,13{+}k),(2,13{+}k)(3,12{+}k)\dots(\lceil\frac{13+k}{2}\rceil,\lfloor\frac{13+k}{2}\rfloor{+}2)\rangle$	D_{13+k}
K^5_k	$\langle(1,\dots,13{+}k),(2,13{+}k)(3,12{+}k)\dots(\lceil\frac{13+k}{2}\rceil,\lfloor\frac{13+k}{2}\rfloor{+}2)\rangle$	D_{13+k}
K^{13}_0	$\langle(2,15)(3,14)(4,13)(5,12)(6,11)(7,10)(8,9),(1,5,13,14)(2,7)(3,9,6,15)(4,11,10,8)\rangle$	$(C_5 \rtimes C_4) \times \sigma_3$
K^{13}_1	$\langle(1,\dots,16),(2,10)(4,12)(6,14)(8,16),(2,8)(3,15)(4,6)(5,13)(7,11)(10,16)(12,14)\rangle$	$(C_2 \times D_{16}) \rtimes C_2$
K^{13}_k	$\langle(1,\dots,15{+}k),(1,15{+}k)(2,14{+}k)\dots(\lfloor\frac{15+k}{2}\rfloor,\lceil\frac{15+k}{2}\rceil{+}1)\rangle$	D_{15+k}, $k \ge 2$
L^1_{2k}	$\langle(1,\dots,10{+}4k)\rangle$	$\mathbb{Z}_{10+k}$
L^1_{4k+1}	$\langle(1,\dots,12{+}8k),(1,2,7{+}4k,8{+}4k)$ $\Pi^k_{i=0}(3{+}2i,12{+}8k{-}2i,9{+}4k{+}2i,6{+}4k{-}2i)(4{+}2i,5{+}4k{-}2i,10{+}4k{+}2i,11{+}8k{-}2i)\rangle$	$C_4 \times D_{3+2k}$
L^1_{4k+3}	$\langle(1,\dots,16{+}8k),(1,2,7{+}4k,8{+}4k)(3,16{+}8k,11{+}4k,8{+}4k)$ $\Pi^k_{i=0}(4{+}2i,7{+}4k{-}2i,12{+}4k{+}2i,15{+}8k{-}2i)(5{+}2i,14{+}8k{-}2i,13{+}4k{+}2i,6{+}4k{-}2i)\rangle$	quasi-dihedral group $QD_{2^{l-1}}$ with $16+8k=2^l$ elements for some $l \in \mathbb{N}$, $C_{16+8k} \rtimes C_2$ otherwise
L^2_k	$\langle(1,\dots,10{+}2k),(2,10{+}2k)(3,9{+}2k)\dots(54{+}k,74{+}k)\rangle$	D_{10+k}
L^3_k	$\langle(1,\dots,10{+}2k),(2,10{+}2k)(3,9{+}2k)\dots(54{+}k,74{+}k)\rangle$	D_{10+k}
L^5_{3k+1}	$\langle(1,\dots,14{+}6k),(2,14{+}6k)(3,13{+}6k)\dots(7{+}3k,9{+}3k)\rangle$	D_{14+6k}
L^5_{3k+2}	$\langle(1,\dots,16{+}6k),(2,16{+}6k)(3,15{+}6k)\dots(8{+}3k,10{+}3k)\rangle$	D_{16+6k}
L^5_{3k}	$\langle(1,\dots,12{+}6k),(2,12{+}6k)(3,11{+}6k)\dots(6{+}3k,8{+}3k)$, $(1,10{+}3k)(4,13{+}3k)(7,16{+}3k)\dots(7{+}3k,16{+}6k)\rangle$	$(C_{4+2k} \times A_4) \rtimes C_2$, $k \bmod 6 \in \{0,2\}$ $C_2 \times ((C_{2+k} \times A_4) \rtimes C_2)$, $k \bmod 6 \in \{3,5\}$ $(C_2 \times (((C_2 \times C_2) \rtimes C_{3^q}) \rtimes C_2)) \rtimes C_2$, $\lvert G\rvert = 2^5 3^q$ $(C_{4l} \times ((C_2 \times C_2) \rtimes C_{3^q})) \rtimes C_2$, $\lvert G\rvert/(2^5 3^q) = l \in \mathbb{N}, l > 1$ $C_2 \times ((C_l \times ((C_2 \times C_2) \rtimes C_{3^q})) \rtimes C_2)$, $\lvert G\rvert = 2^4 3^q l$, $q > 1$, $2 \nmid l \wedge 3 \nmid l$
L^8_k	see L^5_k	
L^{10}_k	$\langle(1,\dots,14{+}2k),(2,14{+}2k)(3,13{+}2k)\dots(7{+}k,9{+}k)\rangle$	D_{14+2k}
L^{16}_k	$\langle(1,\dots,14{+}2k),(2,14{+}2k)(3,13{+}2k)\dots(7{+}k,9{+}k)\rangle$	D_{14+2k}

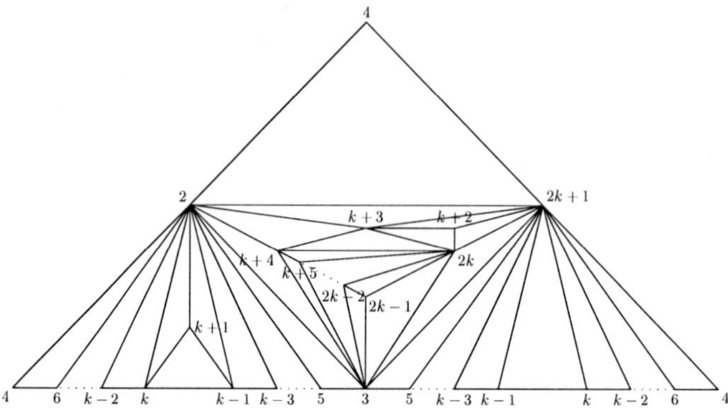

Figure 4.5.9: Link of vertex 1 in $\mathrm{NSB}_1(k)$ – a combinatorial 2-sphere with $2k$ vertices.

4.5.5 Three conjectured series of neighborly sphere bundles

All infinite series with cyclic symmetry presented above consist of a constant number of difference cycles and thus usually at most one member of each series is 2-neighborly. However, the 2-neighborly members of some of the series can be grouped together as the first members of three new infinite series of combinatorial 3-manifolds where all complexes contained in the infinite series are 2-neighborly. Namely we have:

$$\mathrm{NSB}_1(k) := \bigcup_{l=0}^{k-5} \{(1{:}l+2{:}2{:}2k-4-l)\} \cup \{(1{:}1{:}2{:}2k-3), (1{:}1{:}k+1{:}k-2), (1{:}k-2{:}1{:}k+1)\}, \quad (4.5.5)$$

with $2k+1$ vertices, $k \geq 4$. The first members of this series are:

1. $k = 4$: ∂B_9^3, ${}^3 9_2^3$,
2. $k = 5$: K_0^1, ${}^3 11_2^1$,
3. $k = 6$: K_0^2, ${}^3 13_6^1$ and
4. $k = 7$: K_0^7, ${}^3 15_{15}^1$.

All of the known complexes are 3-dimensional Klein bottles. The complexes are combinatorial manifolds for any $k \geq 4$. This can be deduced from Figure 4.5.9, which shows the vertex link of $\mathrm{NSB}_1(k)$. Note that the vertex link has exactly $2k$ vertices. Thus, $\mathrm{NSB}_1(k)$ is neighborly.

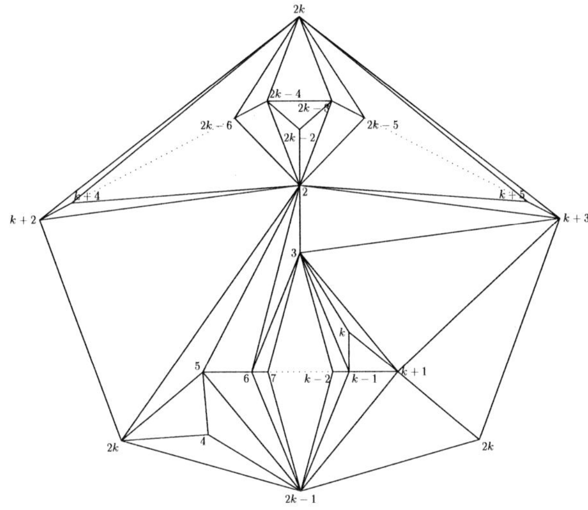

Figure 4.5.10: Link of vertex 1 in $\mathrm{NSB}_2(k)$ – a combinatorial 2-sphere with $2k-1$ vertices.

The complex

$$\mathrm{NSB}_2(k) := \bigcup_{l=0}^{k-6}\{(1{:}k{+}l{:}2{:}k{-}l{-}3)\}\cup\{(1{:}1{:}3{:}2k{-}5),(1{:}1{:}k{:}k{-}2),(1{:}3{:}1{:}2k{-}5),(2{:}k{-}2{:}2{:}k{-}2)\} \tag{4.5.6}$$

has $2k$ vertices, $k \geq 5$. The first members of this series are:

1. $k=5$: L_0^2, ${}^3 10_3^3$,
2. $k=6$: $L_0^7 \cong L_0^9$, ${}^3 12_4^1$ and
3. $k=7$: L_0^{13}, ${}^3 14_{21}^1$.

The first members of $\mathrm{NSB}_2(2m-1)$, $m \geq 3$, are 3-dimensional Klein bottles, $\mathrm{NSB}_2(2m)$, $m \geq 3$, seems to be PL homeomorphic to $S^2 \times S^1$. The vertex link of $\mathrm{NSB}_2(k)$ is shown in Figure 4.5.10. It is a combinatorial 2-sphere with $2k-1$ vertices.

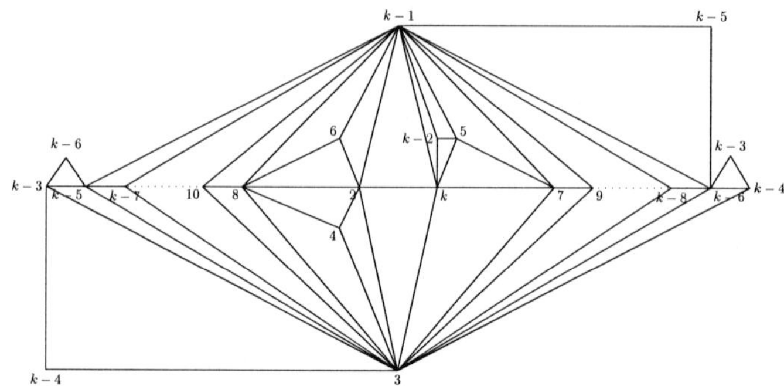

Figure 4.5.11: Link of vertex 1 in $\mathrm{NSB}_3(k)$ – a combinatorial 2-sphere with $k-1$ vertices.

Finally, the complex

$$\mathrm{NSB}_3(k) := \bigcup_{l=0}^{k-12} \{(2{:}l{+}4{:}2{:}k{-}l{-}8)\} \cup \{(1{:}1{:}1{:}k{-}3), (1{:}2{:}4{:}k{-}7), (1{:}4{:}2{:}k{-}7), (1{:}4{:}k{-}7{:}2)\} \tag{4.5.7}$$

is a 2-neighborly combinatorial manifold for any $k \geq 11$, as can be deduced from Figure 4.5.11. Its first members are

1. $k = 11$: ${}^{3}11^{2}_{3}$,
2. $k = 12$: $L^4_0 \cong L^6_0$, ${}^{3}12^{1}_{2}$,
3. $k = 13$: K^3_0, ${}^{3}13^{1}_{2}$,
4. $k = 14$: Le_{14}, ${}^{3}14^{1}_{5}$ and
5. $k = 15$: K^{12}_0, ${}^{3}15^{1}_{4}$.

It seems that the odd members of $\mathrm{NSB}_3(k)$ are 3-dimensional Klein bottles and that the even members are PL homeomorphic to $S^2 \times S^1$. The proof of the exact topological type of the series described above will be left open at this point. The series can be constructed using the simpcomp [40, 41, 42] functions `SCSeriesNSB1`, `SCSeriesNSB2` and `SCSeriesNSB3`, see also Table 4.6.

name	parameters	restrictions on parameters
`SCSeriesBdHandleBody`	d (dimension), n (# vert.)	$n \geq 2d+3$, $d \geq 2$
`SCSeriesC2n`	n ($\frac{1}{2}$ # vert.)	$n \geq 8$
`SCSeriesCSTSurface`	l, j, $2k$ (# vert.)	$l < j < 2k-l-j$, $l, j, l+j \neq k$
`SCSeriesCSTSurface`	l, $2k$ (# vert.)	$l \leq \frac{k-1}{2}$
`SCSeriesD2n`	n ($\frac{1}{2}$ # vert.)	$n \geq 8$, $n \neq 9$
`SCSeriesHandleBody`	d (dimension), n (# vert.)	$n \geq 2d+1$, $d \geq 3$
`SCSeriesK`	i, k	$1 \leq i \leq 14$, $k \geq 0$
`SCSeriesKu`	n ($\frac{1}{4}$ # vert.)	$n \geq 3$
`SCSeriesL`	i, k	$1 \leq i \leq 18$, $k \geq 0$
`SCSeriesLe`	n ($\frac{1}{2}$ # vert.)	$n \geq 7$
`SCSeriesNSB1`	n ($\frac{1}{2}$ (# vert. - 1))	$n \geq 4$
`SCSeriesNSB2`	n ($\frac{1}{2}$ # vert.)	$n \geq 5$
`SCSeriesNSB3`	n (# vert.)	$n \geq 11$
`SCSeriesPrimeTorus`	l, j, p (# vert.)	$l < j < p-l-j$, p pr., $\gcd(l, j) = 1$

Table 4.6: Overview over simpcomp's functionality regarding the construction of members of infinite transitive series of combinatorial manifolds.

Chapter 5

Doubly transitive combinatorial (pseudo-)manifolds

As of today, only very few examples of multiply transitive combinatorial manifolds and pseudomanifolds can be found in literature. In this chapter we concentrate on the search for multiply transitive simplicial complexes in dimension 4 and below: It is specified by a simple criterion, when exactly a sharply 2-transitive group generates a 2-transitive "pinch point" surface. Furthermore, a detailed topological description of a conjectured infinite sub-series of "pinch point" surfaces with affine symmetry is given. In addition, it is verified that there are no examples of 2-transitive 3-neighborly 4-manifolds with less than 64 vertices, apart from the 16-vertex $K3$ surface and the boundary of the 5-simplex. Finally, an infinite series of combinatorial 4-pseudomanifolds with 2-transitive affine symmetry is presented.

5.1 Introduction

One surprising aspect about symmetric simplicial complexes is that there exist very few examples of combinatorial manifolds with multiply transitive automorphism group, whereas the number of simply transitive complexes in comparison is very large (cf. Chapter 4). For triangulated closed surfaces, Kimmerle and Kouzoudi were able to prove that there exist exactly three complexes with multiply transitive automorphism group (see [67]). In dimension 3, it can be shown that the boundary of the 4-simplex is in fact the only multiply transitive combinatorial manifold (cf. Proposition 5.3.1). In dimension 4, the 16-vertex triangulation of the $K3$ surface (see [28] and Section 2.2) is the only known non-trivial, i. e. different to the boundary of the simplex, example of a multiply transitive combinatorial manifold so far.

However, when it comes to combinatorial pseudomanifolds the limitation on complexes with multiply transitive automorphism groups seems to be not as restrictive. This chapter focuses on the search for such combinatorial manifolds and pseudomanifolds up to dimension 4.

Remember that the automorphism group $\mathrm{Aut}(C)$ of a simplicial complex C is a permutation group isomorphic to a subgroup of the symmetric group σ_n where n is the number of vertices of C. Assuming that C is a combinatorial pseudomanifold further restrictions apply to $\mathrm{Aut}(C)$.

Lemma 5.1.1. *Let M be a strongly connected combinatorial d-pseudomanifold on n vertices and $\Delta = \langle v_1, \dots, v_{d+1} \rangle$ a facet of M. Then the pointwise stabilizer of Δ is trivial.*

Remark 5.1.2. Throughout this chapter we will write $\mathrm{stab}\langle\cdot\rangle$ for the pointwise stabilizer and $\mathrm{stab}(\cdot)$ for the stabilizer that preserves the argument as a set.

Proof. Let $g \in \mathrm{stab}_{\mathrm{Aut}(M)}\langle\Delta\rangle$. In particular, we have $g \in \mathrm{stab}_{\mathrm{Aut}(M)}\langle\delta\rangle$ for every $(d-1)$-face $\delta \subset \Delta$. As M is a combinatorial d-pseudomanifold, every $\delta \subset \Delta$ lies in exactly one other facet $\delta \subset \tilde{\Delta}$. Hence, every $g \in \mathrm{stab}_{\mathrm{Aut}(M)}\langle\delta\rangle$ either maps Δ to $\tilde{\Delta}$ or $g \in \mathrm{stab}_{\mathrm{Aut}(M)}\langle\tilde{\Delta}\rangle$. Since $g \in \mathrm{stab}_{\mathrm{Aut}(M)}\langle\Delta\rangle$ holds by assumption, the former case never occurs and since M is strongly connected we can iterate the process and it follows that $g = \mathbf{Id}$. □

Lemma 5.1.1 implies the following result.

Corollary 5.1.3. *Let M be a strongly connected combinatorial d-pseudomanifold with f_d facets and automorphism group $\mathrm{Aut}(M)$. Then the inequality*

$$|\mathrm{Aut}(M)| \leq (d+1)! f_d \tag{5.1.1}$$

holds.

Proof. Inequality (5.1.1) counts all possible images of a facet of M by an automorphism: Every facet can possibly be mapped on every other facet in any kind of vertex permutation of the $d+1$ vertices involved. Once the image of one facet is determined pointwise, the automorphism can uniquely be extended to all vertices of M by Lemma 5.1.1. This shows the statement. □

In particular, this means that if the symmetric group σ_n is the automorphism group of a strongly connected combinatorial pseudomanifold, then $d \in \{n-1, n-2\}$ and the only combinatorial pseudomanifold with this property is the boundary of the $(n-1)$-simplex. Further examples of non-trivial cases of equality in Inequality 5.1.1 are the boundary of the d-dimensional cross polytope and the 6-vertex triangulation of the real projective plane.

Inequality (5.1.1) is rather weak and admits automorphism groups of very high orders. In particular, equality is only attained in few cases. A more restrictive criterion for the order of an automorphism group is the following: The link of a face of codimension 2 of a combinatorial d-pseudomanifold, $d > 2$, has to be an n-gon, the link of a codimension-1-face consists of 2 points. Both, the n-gon as well as the two points, only admit small automorphism groups. This leads to the following observation.

Proposition 5.1.4. *Let G be the automorphism group of a strongly connected combinatorial d-pseudomanifold M, $\delta \in M$ a $(d-1)$-face and $\gamma \in M$ a $(d-2)$-face. Then*

$$\mathrm{stab}_G\langle\delta\rangle \leq C_2$$

if $d \geq 2$ and

$$\mathrm{stab}_G\langle\gamma\rangle \leq D_m$$

if $d \geq 3$, where D_m denotes the dihedral group with $2m$ elements and C_2 denotes the cyclic group with 2 elements. If the above condition holds, m is also an upper bound to the valence of the corresponding codimension-2-face.

Proof. If α is an l-face of M, $0 \leq l \leq d$, then every element in the pointwise stabilizer $g \in \mathrm{stab}_G\langle\alpha\rangle$ induces an automorphism of the link $h \in \mathrm{Aut}(\mathrm{lk}_M(\alpha))$. By Lemma 5.1.1 it follows that $h = \mathbf{Id}_{\mathrm{lk}_M(\alpha)}$ if and only if $g = \mathbf{Id}_M$. Hence, $\mathrm{stab}_G\langle\alpha\rangle$ is a subgroup of $\mathrm{Aut}(\mathrm{lk}_M(\alpha))$. If $l = d-2$, the link $\mathrm{lk}_M(\alpha)$ is a regular m-gon with dihedral symmetry D_m, if $l = d$ the link equals two vertices with C_2 symmetry. This proves the result. □

In principle, a statement analogous to Proposition 5.1.4 holds for higher dimensional links as well. However, in these cases little is known about the automorphism group of the links, which results in decisively weaker statements.

Remark 5.1.5. If we look at any d-dimensional, k-transitive simplicial complex C, $d \geq k-1$, we can observe that C has to be k-neighborly, since every $(k-1)$-face can be mapped on any other k-tuple of vertices by $\mathrm{Aut}(C)$ which thus have to span a $(k-1)$-face, too. As we will see, this property of a multiply transitive simplicial complex will be very helpful.

For example in dimension 1, any combinatorial pseudomanifold is a combinatorial manifold. Since an at least twofold transitive combinatorial manifold has to be at least 2-neighborly, the only candidate for such a complex is the boundary of a triangle which is in fact 2-transitive. All other combinatorial 1-manifolds are collections of polygons where the automorphism group is a direct sum of dihedral and symmetric groups and which are at most simply transitive.

5.2 Multiply transitive "pinch point" surfaces

In the case $d = 2$, the boundary of the tetrahedron $\partial\Delta^3$ is an example for a 3-transitive combinatorial triangulation of a surface. Since $\partial\Delta^3$ is the only 3-neighborly combinatorial 2-pseudomanifold, there is no further example of such a triangulation.

Surfaces with doubly transitive automorphism group were classified by Kimmerle and Kouzoudi in [67]. There are only three such complexes: the (sharply) 2-transitive 7-vertex Möbius torus $\mathbb{T}^2_7$, the unique 6-vertex real projective plane $\mathbb{R}P^2_6$ - and the boundary of the 3-simplex.

Combinatorial 2-pseudomanifolds, so-called *"pinch point" surfaces*, can be constructed from not necessarily connected triangulated surfaces $\hat{S}$ by identifying some of its vertices, such that the weak pseudomanifold property of the resulting complex S is still satisfied after the identification. The number of vertices $\hat{v}_i$ of $\hat{S}$ that are identified to one vertex v of S is called the *order* $o(v)$ of the vertex v. The surface $\hat{S}$ is called the *resolution of the "pinch point" surface* S.

In general, two different types of "pinch point" surfaces can be distinguished: Strongly connected "pinch point" surfaces S, where $\hat{S}$ is connected, and not strongly connected "pinch point" surfaces S, where $\hat{S}$ consists of multiple connected components. Of course, from a topological point of view, the first type of "pinch point" surface is the more interesting one. However, when searching for highly symmetric complexes in terms of orbit representatives it is hard to decide in general what type of "pinch point" surface is generated by the group action on the representatives in a particular example. In addition, note that in the case that S is not strongly connected, Lemma 5.1.1 does no longer hold and hence very large automorphism groups can occur.

A "pinch point" surface is a surface if and only if all of its vertices are of order 1. Hence, for the remainder of this Chapter we will consider surfaces to be a subclass in the class of "pinch point" surfaces. For doubly transitive "pinch point" surfaces and their automorphism groups we have the following necessary conditions.

Proposition 5.2.1. *Let S be a 2-transitive "pinch point" surface and G its automorphism group. Then the following statements hold.*

1. $f(S) = (n, \binom{n}{2}, \frac{2}{3}\binom{n}{2})$,
2. $\chi(S) = n - \frac{1}{3}\binom{n}{2}$,
3. $n \not\equiv 2\,(3)$,
4. $|\mathrm{stab}_G(u,v)| = \frac{2|G|}{n(n-1)}$,

Proof. The following proof is a straightforward generalization of the statements 2.5 and 2.6 from [67].

1. S is a 2-neighborly combinatorial pseudomanifold. By the weak pseudomanifold property the f-vector is thus determined by the number of vertices n.
2. Follows from *1.*
3. Follows from *1.* or the fact that a combinatorial 2-pseudomanifold can be seen as a simple twofold triple system for which Theorem 2.5 in [79] holds. This in turn states that $n \not\equiv 2\,(3)$ for the number of vertices n.
4. G is 2-transitive and S has $f_1 = \binom{n}{2}$ edges. Hence, the order of the stabilizer is given by $|G|/f_1$.

□

From Proposition 5.2.1 we can deduce the following corollary.

Corollary 5.2.2. *Let G be a group operating 2-transitively on $V := \{1,\dots,n\}$ and $\{a,b,c\} \subset V$. Then $G \cdot \langle a,b,c\rangle$ is a "pinch point" surface if and only if $|G \cdot \langle a,b,c\rangle| = \frac{1}{3}n(n-1)$.*

Proof. By Proposition 5.2.1, *(1.)*, every "pinch point" surface has $\frac{1}{3}n(n-1)$ triangles.

Now let $|G\cdot\langle a,b,c\rangle| = \frac{1}{3}n(n-1)$. In any case, the simplicial complex defined by $G\cdot\langle a,b,c\rangle$ is 2-transitive. Hence, it is 2-neighborly and every edge e is contained in the same number $\mathrm{val}_{G\cdot\langle a,b,c\rangle}(e)$ of triangles, that is $\mathrm{val}_{G\cdot\langle a,b,c\rangle}(e) = \frac{3f_2}{f_1} = \frac{3\frac{2}{3}\binom{n}{2}}{\binom{n}{2}} = 2$ and $G\cdot\langle a,b,c\rangle$ is a "pinch point" surface. □

Theorem 5.2.3 (Sharply 2-transitive "pinch point" surfaces). *Let G be a sharply 2-transitive permutation group. Then every "pinch point" surface generated by G is transitive on the set of triangles and $G\cdot\langle a,b,c\rangle$ is a "pinch point" surface if and only if* $\mathrm{stab}_G(a,b,c) \cong \mathbb{Z}_3$.

Proof. In order to prove that any sharply 2-transitive "pinch point" surface is transitive on the set of triangles, it suffices to show that any orbit of triangles of G must contain the edge $\langle 1,2\rangle$ at least twice. Since G operates 2-transitively on $\{1,\dots,n\}$, every orbit contains a triangle of the form $\langle 1,2,u\rangle$. Since S is sharply 2-transitive, the stabilizer of $\langle 1,2\rangle$ as a set has exactly two elements. Now let $g \in \mathrm{stab}_G(1,2) \setminus \mathbf{Id}$, then $g1 = 2$ and $g\langle 1,2,u\rangle = \langle 1,2,v\rangle$. If $u \neq v$, then $\langle 1,2\rangle$ is contained twice in any orbit. If $u = v$, then there exist a vertex w (the third vertex of the other triangle containing $\langle 1,2\rangle$) such that $g \in \mathrm{stab}_G\langle u,w\rangle$ and g fixes two vertices, contradiction to $g \neq \mathbf{Id}$ and G sharply 2-transitive.

Now let $G\cdot\langle a,b,c\rangle$ be a "pinch point" surface. Hence, $|G\cdot\langle a,b,c\rangle| = \frac{n(n-1)}{3}$ and since all stabilizers of triangles have to be isomorphic, we have $|\mathrm{stab}_G(a,b,c)| = 3$ and thus $\mathrm{stab}_G(a,b,c) \cong \mathbb{Z}_3$.

On the other hand, if $\mathrm{stab}_G(a,b,c) \cong \mathbb{Z}_3$, then the stabilizer of all triples $\langle d,e,f\rangle \in G\cdot\langle a,b,c\rangle$ is isomorphic to $\mathbb{Z}_3$. Hence, we have $|G\cdot\langle a,b,c\rangle| = \frac{n(n-1)}{3}$ and $G\cdot\langle a,b,c\rangle$ is a "pinch point" surface by Corollary 5.2.2. □

Remark 5.2.4. If all strongly connected components of a 2-transitive "pinch point" surface contain all vertices, Lemma 5.1.1 is valid and it can be shown that all such "pinch point" surfaces are transitive on the set of triangles. Moreover, they are either sharply 2-transitive (cf. Theorem 5.2.3) or they have an automorphism group of order $2n(n-1)$, where n denotes the number of vertices, and all triangle stabilizers are of type σ_3.

Table 5.1 lists all combinatorial types of 2-transitive "pinch point" surfaces up to 100 vertices. It is interesting to see that, apart from the 2-transitive non-singular surfaces, none of the complexes is strongly connected. The question, whether or not there exists an example of a strongly connected 2-transitive "pinch point" surface that is not a surface is interesting but has to be left open at this point. All calculations have been made using the GAP-package simpcomp (cf. Appendix A or [40, 41, 42]).

Table 5.1: All combinatorial types of 2-transitive "pinch point" surfaces with up to 100 vertices. The GAP-index (n,m) denotes the GAP-group PrimitiveGroup(n,m) from the library of all primitive permutation groups of order $n < 2500$ (see [49]) which is adopted from the classification given in [119].

GAP-index*	orbit	group	$f(S)$	$\chi(S)$	$H_\star(S)$	$o(v)$	$\hat{S}$
$(4,1)$	$(1,2,3)_{4}$	$A(4)$	$(4,6,4)$	2	$(\mathbb{Z},0,\mathbb{Z})$	1	S^2
$(6,1)$	$(1,2,3)_{10}$	$\mathrm{PSL}(2,5)$	$(6,15,10)$	1	$(\mathbb{Z},\mathbb{Z}_2,0)$	1	$\mathbb{R}P^2$
$(7,4)$	$(1,2,3)_{14}$	$\mathrm{AGL}(1,7)$	$(7,21,14)$	0	$(\mathbb{Z},\mathbb{Z}^2,\mathbb{Z})$	1	$\mathbb{T}^2$
$(13,6)$	$(1,2,4)_{52}$	$\mathrm{AGL}(1,13)$	$(13,78,52)$	−13	$(\mathbb{Z},\mathbb{Z}^{16},\mathbb{Z}^2)$	2	$\{1,2\}\times\mathbb{T}^2$
$(13,7)$	$(1,2,4)_{52}$	$\mathrm{L}(3,3)$	$(13,78,52)$	−13	$(\mathbb{Z},\mathbb{Z}^{27},\mathbb{Z}^{13})$	2	$\{1,\dots,13\}\times S^2$
$(16,12)$	$(1,2,7)_{80}$	$A\Gamma L(2,4)$	$(16,120,80)$	−24	$(\mathbb{Z},\mathbb{Z}^{45},\mathbb{Z}^{20})$	5	$\{1,\dots,20\}\times S^2$
$(19,6)$	$(1,2,5)_{114}$	$\mathrm{AGL}(1,19)$	$(19,171,114)$	−38	$(\mathbb{Z},\mathbb{Z}^{42},\mathbb{Z}^3)$	3	$\{1,2,3\}\times\mathbb{T}^2$
$(25,15)$	$(1,2,7)_{200}$	$C_5^2\rtimes(Q_8\rtimes C_3)$	$(25,300,200)$	−75	$(\mathbb{Z},\mathbb{Z}^{80},\mathbb{Z}^4)$	4	$\{1,\dots,4\}\times\mathbb{T}^2$
$(25,17)$	$(1,2,7)_{200}$	$A\Gamma L(1,25)$	$(25,300,200)$	−75	$(\mathbb{Z},\mathbb{Z}^{80},\mathbb{Z}^4)$	4	$\{1,\dots,4\}\times\mathbb{T}^2$
$(28,3)$	$(1,2,9)_{252}$	$P\Gamma L(2,8)$	$(28,378,252)$	−98	$(\mathbb{Z},\mathbb{Z}^{99}\times\mathbb{Z}_2^3,0)$	3	$\{1,2,3\}\times(\mathbb{K}^2)^{\#8}$
$(28,3)$	$(1,2,22)_{252}$	$P\Gamma L(2,8)$	$(28,378,252)$	−98	$(\mathbb{Z},\mathbb{Z}^{162},\mathbb{Z}^{63})$	9	$\{1,\dots,63\}\times S^2$
$(28,5)$	$(1,2,3)_{252}$	$P\Gamma U(3,3)$	$(28,378,252)$	−98	$(\mathbb{Z},\mathbb{Z}^{162},\mathbb{Z}^{63})$	9	$\{1,\dots,63\}\times S^2$
$(31,8)$	$(1,2,7)_{310}$	$\mathrm{AGL}(1,31)$	$(31,465,310)$	−124	$(\mathbb{Z},\mathbb{Z}^{130},\mathbb{Z}^5)$	5	$\{1,\dots,5\}\times\mathbb{T}^2$
$(37,9)$	$(1,2,8)_{444}$	$\mathrm{AGL}(1,37)$	$(37,666,444)$	−185	$(\mathbb{Z},\mathbb{Z}^{192},\mathbb{Z}^6)$	6	$\{1,\dots,6\}\times\mathbb{T}^2$
$(40,6)$	$(1,2,3)_{520}$	$\mathrm{PGL}(4,3)$	$(40,780,520)$	−220	$(\mathbb{Z},\mathbb{Z}^{351},\mathbb{Z}^{130})$	6	$\{1,\dots,130\}\times S^2$
$(43,8)$	$(1,2,9)_{602}$	$\mathrm{AGL}(1,43)$	$(43,903,602)$	−258	$(\mathbb{Z},\mathbb{Z}^{266},\mathbb{Z}^7)$	7	$\{1,\dots,7\}\times\mathbb{T}^2$
$(49,25)$	$(1,2,10)_{784}$	$C_7^2\rtimes(Q_8'D_6)$	$(49,1176,784)$	−343	$(\mathbb{Z},\mathbb{Z}^{352},\mathbb{Z}^8)$	8	$\{1,\dots,8\}\times\mathbb{T}^2$
$(49,28)$	$(1,2,3)_{784}$	$\mathrm{AGL}(2,7)$	$(49,1176,784)$	−343	$(\mathbb{Z},\mathbb{Z}^{400},\mathbb{Z}^{56})$	8	$\{1,\dots,56\}\times\mathbb{T}^2$
$(61,12)$	$(1,2,12)_{1220}$	$\mathrm{AGL}(1,61)$	$(61,1830,1220)$	−549	$(\mathbb{Z},\mathbb{Z}^{560},\mathbb{Z}^{10})$	10	$\{1,\dots,10\}\times\mathbb{T}^2$
$(64,43)$	$(1,2,33)_{1344}$	$A\Gamma L(3,4)$	$(64,2016,1344)$	−608	$(\mathbb{Z},\mathbb{Z}^{945},\mathbb{Z}^{336})$	21	$\{1,\dots,336\}\times S^2$
$(64,44)$	$(1,2,35)_{1344}$	$A\Sigma L(3,4)$	$(64,2016,1344)$	−608	$(\mathbb{Z},\mathbb{Z}^{945},\mathbb{Z}^{336})$	21	$\{1,\dots,336\}\times S^2$
$(67,8)$	$(1,2,13)_{1474}$	$\mathrm{AGL}(1,67)$	$(67,2211,1474)$	−670	$(\mathbb{Z},\mathbb{Z}^{682},\mathbb{Z}^{11})$	11	$\{1,\dots,11\}\times\mathbb{T}^2$
$(73,12)$	$(1,2,14)_{1752}$	$\mathrm{AGL}(1,73)$	$(73,2628,1752)$	−803	$(\mathbb{Z},\mathbb{Z}^{816},\mathbb{Z}^{12})$	12	$\{1,\dots,12\}\times\mathbb{T}^2$
$(79,8)$	$(1,2,15)_{2054}$	$\mathrm{AGL}(1,79)$	$(79,3081,2054)$	−948	$(\mathbb{Z},\mathbb{Z}^{962},\mathbb{Z}^{13})$	13	$\{1,\dots,13\}\times\mathbb{T}^2$
$(97,12)$	$(1,2,18)_{3104}$	$\mathrm{AGL}(1,97)$	$(97,4656,3104)$	−1455	$(\mathbb{Z},\mathbb{Z}^{1472},\mathbb{Z}^{16})$	16	$\{1,\dots,16\}\times\mathbb{T}^2$

Conjecture 5.2.5. *Let $p \equiv 1\,(6)$ be a prime, $G = \mathrm{AGL}(1,p)$ in the permutation representation given in the library of primitive groups in GAP and $k = \frac{p-1}{6} \in \mathbb{N}$. Then*

$$c_k = G \cdot \langle 1, 2, k+2 \rangle$$

is a "pinch point" surface with the following properties:

- *All singularities of c_k are of order k,*

- $f(c_k) = (p, \binom{p}{2}, \frac{2}{3}\binom{p}{2})$,
- $\chi(c_k) = p - \frac{1}{3}\binom{p}{2}$,
- $H_\star(c_k) = (\mathbb{Z}, \mathbb{Z}^{\beta_1}, \mathbb{Z}^k)$ *where* $\beta_1 = 6k^2 - 4k = \frac{1}{6}(p^2 - 6p + 5)$ *and*
- $\hat{c_k} \cong \{1, \dots, k\} \times \mathbb{T}^2_k$, *where* $\mathbb{T}^2_k$ *is a chiral triangulation of the the torus with* f*-vector* $f(\mathbb{T}^2_k) = (p, 3p, 2p)$ *and with an automorphism group of type* $(C_p \rtimes C_3) \rtimes C_2$.

Apart from the exact combinatorial type of the tori which build up the strongly connected components of the series c_k, Conjecture 5.2.5 was verified by computer for all primes $p \leq 1153$.

Remark 5.2.6. A classification of all simply transitive "pinch point" surfaces or even of all simply transitive surfaces seems hopeless for vertex numbers $n \geq 30$. However, it is of great importance for the classification of higher transitive complexes in higher dimensions that there are no transitive combinatorial 2-spheres apart from the boundaries of the three simplicial platonic solids.

5.3 Multiply transitive combinatorial 3-manifolds

The search for multiply transitive combinatorial 3-manifolds resolves itself by a fairly simple observation which is as follows.

Proposition 5.3.1 (cf. Knödler [72]). *The boundary of the* 4*-simplex is the only multiply transitive combinatorial* 3*-manifold.*

Proof. An n-vertex k-transitive combinatorial 3-manifold M, $k > 1$, has to be k-neighborly. Hence, its vertex links are $(k-1)$-transitive combinatorial 2-spheres with $n-1$ vertices. Thus, following Remark 5.2.6, M has $n \in \{5, 7, 13\}$ vertices. The case $n = 5$ leads to the (4-transitive) boundary of the 4-simplex. For $n \in \{7, 13\}$ we know from the classification of vertex transitive triangulations by Lutz (see [92, p. 47]) that no further examples of multiply transitive combinatorial 3-manifolds can exist. □

See [72] for an alternative proof avoiding the classification of transitive combinatorial 3-manifolds.

5.4 A 3-transitive combinatorial 3-pseudomanifold

We have seen that there are no non-trivial multiply transitive combinatorial 3-manifolds. However, there is an even 3-transitive combinatorial 3-pseudomanifold on 8 vertices (see the complex `pseudomanifold_3_8_43_1` on [91]) which we will denote by M_8^3. Since M_8^3 is 3-transitive, its vertex links must be 2-transitive surfaces. According to the classification of 2-transitive surfaces (see [67]) the vertex links of a 3-transitive combinatorial 3-pseudomanifold (different from the boundary of the 4-simplex) have to be isomorphic to either the 6-vertex real projective plane $\mathbb{R}P_6^2$ or the 7-vertex Möbius torus $\mathbb{T}_7^2$ and such a complex thus has to have either 7 or 8 vertices. However, no 3-transitive 3-pseudomanifold with 7 vertices exist and M_8^3 is the only complex which has vertex links isomorphic to $\mathbb{T}_7^2$. As a consequence, M_8^3 is the only pseudomanifold with a 3-transitive automorphism group.

M_8^3 is also referred to as *Emch's design* and is due to Emch, who first described the complex in terms of two complementary quadruple systems on 8 vertices and 14 quadruples in 1929 (see [43, Section III.4]). Moreover, in the language of modern design theory, M_8^3 is the unique Block design of type $S_2(3,4;8)$ without repeated blocks (see [79, Proposition 10.4], for more information about design theory see [29] and [11]).

The f-vector of M_8^3 is $f(M_8^3) = (8,28,56,28)$, its integral homology groups are $H_\star(M_8^3) = (\mathbb{Z},0,\mathbb{Z}^8,\mathbb{Z})$ and the automorphism group is isomorphic to the projective linear group $\mathrm{PGL}(2,7)$ of order 336 given by the generators $\langle(1,3)(2,5)(4,6)(7,8),(3,6,8,4,7,5)\rangle$. The complex is generated by the $\mathrm{PGL}(2,7)$-orbit $(1,2,3,5)_{28}$ of length 28.

From the geometric point of view, M_8^3 is a quotient of the tessellation $\{3,3,6\}$ of hyperbolic 3-space. Its fundamental domain is given by its vertex figure which is the 7-vertex Möbius torus or, equivalently, the quotient $\{3,6\}_{(2,1)}$ of the $\{3,6\}$-tessellation of the Euclidian plane (which in turn is the vertex figure of $\{3,3,6\}$). Figure 5.4.1 shows the edge star of $\{3,3,6\}$ which already gives a system of representatives of the 8 vertices of M_8^3. The vertices are labeled according to the system of representatives given by the 8 vertices k_1, $1 \leq k \leq 8$. This setting can be extended in a unique way to the complete fundamental domain of M_8^3. Starting from this, the complex M_8^3 can be constructed where the identification of the vertices is canonically up to one choice that has to be made:

If we extend the vertex link of 1_1 in Figure 5.4.1 by adding tetrahedra around it, the added vertices can be seen as vertices in the $\{3,6\}$-tessellation of the Euclidian plane. Thus, the $(2,1)$-identification of $\{3,6\}_{(2,1)}$ can be applied to the complex where one choice has to

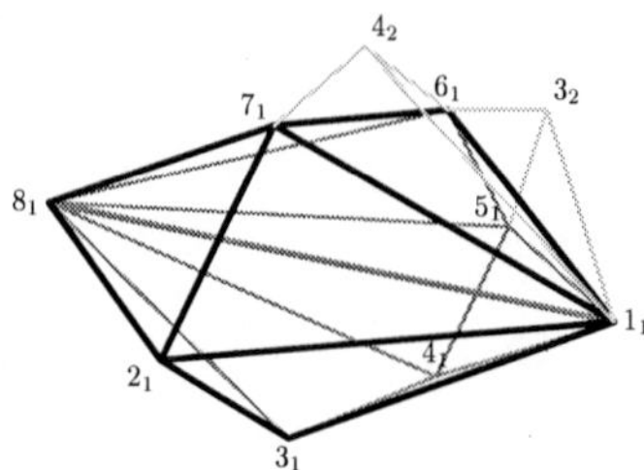

Figure 5.4.1: Edge star of the hyperbolic tessellation $\{3,3,6\}$ where two additional tetrahedra ($\langle 1_1, 4_2, 6_1, 7_1\rangle$ and $\langle 1_1, 3_2, 5_1, 6_1\rangle$) have already been added.

be made when identifying the first vertex (see Figure 5.4.2). Now we extend the vertex link of 8_1 in Figure 5.4.1. In order to be compatible with the extension of the vertex link of 1_1 as well as all other identifications, no further choice is possible and all vertex labels are already determined. It turns out that the identification of the vertices of $\{3,3,6\}$ according to this procedure results in a quotient isomorphic to M^3_8.

Since all vertex links of M^3_8 are tori, the complex is not only a combinatorial 3-pseudomanifold, but also an ideal triangulation (cf. [137]). Ideal triangulations are special kinds of cell decomposition in 3-manifold theory to describe the exterior of a knot or a link. In an ideal triangulation every vertex figure is either a torus or a Klein bottle, representing a connected component (a knot) of a link. Ideal triangulations can be understood by the normal surface software Regina [26] by Burton. With the help of Regina we were able to compute various fillings of the knot exterior described by M^3_8 and to verify that M^3_8 admits a hyperbolic metric.

5.5 2-transitive combinatorial 3-pseudomanifolds

The automorphism group G of a strongly connected combinatorial 3-pseudomanifold M must have pointwise stabilizers of edges that are all subgroups of the dihedral group D_m, where $m \leq n-2$ is a lower bound to the edge valence of M.

Furthermore, by the weak pseudomanifold property the complex must contain $f_3 \leq \frac{1}{2}\binom{n}{3}$ tetrahedra. The 2-neighborliness implies $f_1 = \binom{n}{2}$ for the number of edges. Altogether we

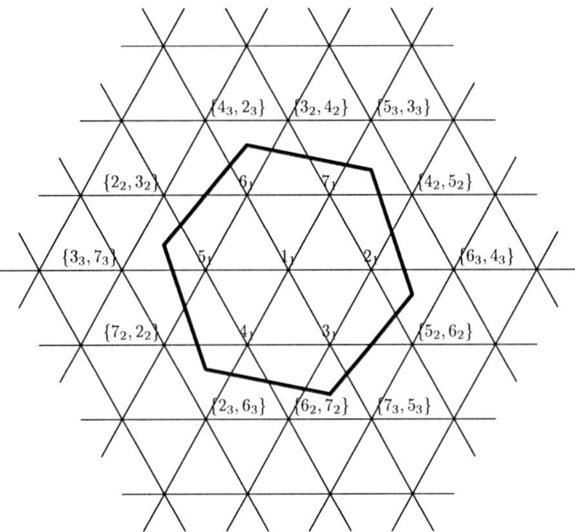

Figure 5.4.2: Part of $\mathrm{lk}_{\{3,3,6\}}(8_1)$. Both choices for identifying the vertices are indicated by the two labels at each vertex.

have for the valence of an edge $\langle u, v \rangle$ in M:

$$m = \mathrm{val}(\langle u, v \rangle) = \frac{6 f_3}{f_1} \leq (n-2).$$

A check of all such 2-transitive permutation groups of degree less or equal 30 using the GAP-package simpcomp [40, 41, 42] resulted in a list of 43 doubly transitive combinatorial 3-pseudomanifolds of 29 different combinatorial types shown in Table 5.2.

Proposition 5.5.1. *Let M be a transitive combinatorial 3-pseudomanifold M with f_0 vertices and let g be the genus of $\mathrm{lk}_M(v)$ for any vertex v of M. Then*

$$g = \frac{\chi(M)}{f_0}$$

holds.

Proof. To see this, write the f-vector $f(M) = (f_0, f_1, f_2.f_3)$ of M in terms of the f-vectors

$$f(\text{lk}_{\text{M}}(\text{v})) = (n, e, t) = (n, 3(n - \chi(\text{lk}_{\text{M}}(\text{v})), 2(n - \chi(\text{lk}_{\text{M}}(\text{v})))$$

of the links $\text{lk}_{\text{M}}(\text{v})$ which are all the same due to the fact that M is transitive. Namely we have

$$\begin{aligned}
\chi(M) &= f_0 - f_1 + f_2 - f_3 \\
&= f_0 - \frac{f_0 n}{2} + \frac{f_0 e}{3} - \frac{f_0 t}{4} \\
&= f_0(1 - \frac{n}{2} + \frac{3(n - \chi(\text{lk}_{\text{M}}(\text{v}))}{3} - \frac{2(n - \chi(\text{lk}_{\text{M}}(\text{v}))}{4}) \\
&= f_0(1 - \frac{1}{2}\chi(\text{lk}_{\text{M}}(\text{v})) \\
&= f_0 g(\text{lk}_{\text{M}}(\text{v})).
\end{aligned}$$

□

Table 5.2: All combinatorial types of strongly connected 2-transitive combinatorial 3-pseudomanifolds with up to 30 vertices.

GAP-index	orbits	group	f-vector	$\chi(S)$	$H_\star(S)$	$\text{lk}(v)$
$(8,5)$	$(1,2,3,5)_{28}$	PGL(2,7)	$(8,28,56,28)$	8	$(\mathbb{Z},0,\mathbb{Z}^8,\mathbb{Z})$	$^{**}\mathbb{T}^2$
$(9,5)$	$(1,2,4,5)_{36}$	$\text{A}\Gamma\text{L}(1,9)$	$(9,36,72,36)$	9	$(\mathbb{Z},0,\mathbb{Z}^9,\mathbb{Z})$	$\mathbb{T}^2$
$(11,4)$	$(1,2,4,6)_{55}$	AGL(1,11)	$(11,55,110,55)$	11	$(\mathbb{Z},0,\mathbb{Z}^{10}\times\mathbb{Z}_2,0)$	$\mathbb{K}^2$
$(14,1)$	$(1,2,3,5)_{91}$	PSL(2,13)	$(14,91,182,91)$	14	$(\mathbb{Z},0\times\mathbb{Z}_2,\mathbb{Z}^{14},\mathbb{Z})$	$\mathbb{T}^2$
$(17,5)$	$(1,2,3,9)_{272}$	AGL(1,17)	$(17,136,544,272)$	153	$(\mathbb{Z},0,\mathbb{Z}^{152}\times\mathbb{Z}_2,0)$	$(\mathbb{K}^2)^{\#9}$
$(17,5)$	$(1,2,3,4)_{272},(1,2,6,11)_{68}$	AGL(1,17)	$(17,136,680,340)$	221	$(\mathbb{Z},0,\mathbb{Z}^{220}\times\mathbb{Z}_2,0)$	$^{**}(\mathbb{K}^2)^{\#13}$
$(19,6)$	$(1,2,4,6)_{171}$	AGL(1,19)	$(19,171,342,171)$	19	$(\mathbb{Z},0\times\mathbb{Z}_2^2,\mathbb{Z}^{18}\times\mathbb{Z}_2,0)$	$\mathbb{K}^2$
$(19,6)$	$(1,2,4,9)_{342}$	AGL(1,19)	$(19,171,684,342)$	190	$(\mathbb{Z},0,\mathbb{Z}^{189}\times\mathbb{Z}_2,0)$	$(\mathbb{K}^2)^{\#10}$
$(23,4)$	$(1,2,3,10)_{506}$	AGL(1,23)	$(23,253,1012,506)$	276	$(\mathbb{Z},0,\mathbb{Z}^{276},\mathbb{Z})$	$(\mathbb{T}^2)^{\#12}$
$(23,4)$	$(1,2,6,11)_{506}$	AGL(1,23)	$(23,253,1012,506)$	276	$(\mathbb{Z},0,\mathbb{Z}^{275}\times\mathbb{Z}_2,0)$	$(\mathbb{K}^2)^{\#12}$
$(23,4)$	$(1,2,3,8)_{506},(1,2,6,18)_{253}$	AGL(1,23)	$(23,253,1518,759)$	529	$(\mathbb{Z},0,\mathbb{Z}^{528}\times\mathbb{Z}_2,0)$	$(\mathbb{K}^2)^{\#23}$
$(23,4)$	$(1,2,3,17)_{253},(1,2,3,19)_{506}$	AGL(1,23)	$(23,253,1518,759)$	529	$(\mathbb{Z},0,\mathbb{Z}^{528}\times\mathbb{Z}_2,0)$	$(\mathbb{K}^2)^{\#23}$
$(25,12)$	$(1,2,6,10)_{600},(1,2,9,18)_{300}$	$C_5^2 \rtimes C_3 \rtimes C_8$	$(25,300,1800,900)$	625	$(\mathbb{Z},0,\mathbb{Z}^{625},\mathbb{Z})$	$(\mathbb{T}^2)^{\#25}$
$(25,14)$	$(1,2,6,10)_{600}$	AGL(1,25)	$(25,300,1200,600)$	325	$(\mathbb{Z},0,\mathbb{Z}^{324}\times\mathbb{Z}_2,0)$	$(\mathbb{K}^2)^{\#13}$
$(27,6)$	$(1,2,4,12)_{702}$	AGL(1,27)	$(27,351,1404,702)$	378	$(\mathbb{Z},0,\mathbb{Z}^{377}\times\mathbb{Z}_2,0)$	$(\mathbb{K}^2)^{\#14}$
$(27,6)$	$(1,2,4,5)_{351},(1,2,4,13)_{702}$	AGL(1,27)	$(27,351,2106,1053)$	729	$(\mathbb{Z},0,\mathbb{Z}^{728}\times\mathbb{Z}_2,0)$	$(\mathbb{K}^2)^{\#27}$
$(27,6)$	$(1,2,4,6)_{351},(1,2,4,15)_{702}$	AGL(1,27)	$(27,351,2106,1053)$	729	$(\mathbb{Z},0,\mathbb{Z}^{728}\times\mathbb{Z}_2,0)$	$(\mathbb{K}^2)^{\#27}$
$(27,9)$	$(1,2,4,19)_{702},(1,2,4,26)_{702}$	$\text{A}\Gamma\text{L}(1,27)$	$(27,351,2808,1404)$	729	$(\mathbb{Z},0,\mathbb{Z}^{1080},\mathbb{Z})$	$(\mathbb{T}^2)^{\#40}$
$(29,6)$	$(1,2,3,10)_{812},(1,2,9,20)_{203}$	AGL(1,29)	$(29,406,2030,1015)$	638	$(\mathbb{Z},0,\mathbb{Z}^{638},\mathbb{Z})$	$(\mathbb{T}^2)^{\#22}$
$(29,6)$	$(1,2,3,12)_{812},(1,2,9,20)_{203}$	AGL(1,29)	$(29,406,2030,1015)$	638	$(\mathbb{Z},0,\mathbb{Z}^{638},\mathbb{Z})$	$(\mathbb{T}^2)^{\#22}$
$(29,6)$	$(1,2,4,15)_{812},(1,2,4,21)_{406}$	AGL(1,29)	$(29,406,2436,1218)$	841	$(\mathbb{Z},0,\mathbb{Z}^{841},\mathbb{Z})$	$(\mathbb{T}^2)^{\#29}$

continuation on next last page –

Table5.2 – continuation from last page

GAP-index*	orbits	group	f-vector	$\chi(S)$	$H_*(S)$	$\mathrm{lk}(v)$
(29,6)	$(1,2,4,19)_{812}, (1,2,6,23)_{406}$	AGL(1,29)	(29,406,2436,1218)	841	$(\mathbb{Z},0,\mathbb{Z}^{841},\mathbb{Z})$	$(\mathbb{T}^2)^{\#29}$
(29,6)	$(1,2,4,7)_{406}, (1,2,4,10)_{812}$	AGL(1,29)	(29,406,2436,1218)	841	$(\mathbb{Z},0,\mathbb{Z}^{840}\times\mathbb{Z}_2,0)$	$(\mathbb{K}^2)^{\#29}$
(29,6)	$(1,2,4,22)_{812}, (1,2,4,24)_{406}$	AGL(1,29)	(29,406,2436,1218	841	$(\mathbb{Z},0,\mathbb{Z}^{840}\times\mathbb{Z}_2,0)$	$(\mathbb{K}^2)^{\#29}$
(29,6)	$(1,2,4,12)_{812}, (1,2,6,10)_{812}$	AGL(1,29)	(29,406,3248,1624)	1247	$(\mathbb{Z},0,\mathbb{Z}^{1246}\times\mathbb{Z}_2,0)$	$(\mathbb{K}^2)^{\#43}$
(29,6)	$(1,2,4,19)_{812}, (1,2,6,14)_{812}$	AGL(1,29)	(29,406,3248,1624)	1247	$(\mathbb{Z},0,\mathbb{Z}^{1246}\times\mathbb{Z}_2,0)$	$(\mathbb{K}^2)^{\#43}$
(29,6)	$(1,2,3,8)_{812}, (1,2,4,19)_{812}, (1,2,9,20)_{203}$	AGL(1,29)	(29,406,3654,1827)	1450	$(\mathbb{Z},0,\mathbb{Z}^{1450},\mathbb{Z})$	$^{**}(\mathbb{T}^2)^{\#50}$
(29,6)	$(1,2,3,5)_{812}, (1,2,6,12)_{812}, (1,2,9,20)_{203}$	AGL(1,29)	(29,406,3654,1827)	1450	$(\mathbb{Z},0,\mathbb{Z}^{1449}\times\mathbb{Z}_2,0)$	$^{**}(\mathbb{K}^2)^{\#50}$
(29,6)	$(1,2,3,15)_{812}, (1,2,4,6)_{812}, (1,2,9,20)_{203}$	AGL(1,29)	(29,406,3654,1827)	1450	$(\mathbb{Z},0,\mathbb{Z}^{1449}\times\mathbb{Z}_2,0)$	$^{**}(\mathbb{K}^2)^{\#50}$

** These links are transitive triangulations of surfaces with the minimum number of vertices. Hence, due to Proposition 5.5.1, the corresponding combinatorial 3-pseudomanifold has the largest Euler characteristic possible for the given vertex number. Furthermore, note that the triangulations of $(\mathbb{T}^2)^{\#50}$ and $(\mathbb{K}^2)^{\#50}$ are a pair of topologically distinct combinatorial manifolds with the same f-vector which both realize a triangulation with the minimum number of vertices (cf. Problem 1 of Chapter 2).

5.6 The 4-dimensional case

As of today, precisely two multiply transitive combinatorial 4-manifolds are known, namely the boundary of the 5-simplex and the 16-vertex triangulation of a $K3$ surface $(K3)_{16}$ due to Casella and Kühnel (see [28]).

Thus, further examples of multiply 4-dimensional combinatorial manifolds would be very interesting. Such examples would have to be at most 2-transitive by the consideration below.

Proposition 5.6.1. *Except for the boundary of the* 5*-simplex there are no combinatorial* 4*-pseudomanifolds with a* 3*-transitive automorphism group.*

Proof. First recall that a 3-transitive combinatorial 4-pseudomanifold M on n vertices has to be 3-neighborly. Hence, its edge links have to be transitive 2-spheres on $n-2$ vertices. Since the boundary of the 3-simplex (Schläfli-symbol $\{3,3\}$), the boundary of the octahedron ($\{3,4\}$) and the boundary of the icosahedron ($\{3,5\}$) are the only transitive triangulated 2-spheres, M has to have 6, 10 or 14 vertices. If M has 6 vertices, it must be the boundary of the 5-simplex. For $n \in \{10,14\}$, the statement follows directly from the classification of vertex transitive triangulations by Lutz (see [92, p. 52]). □

5.7 2-transitive 3-neighborly 4-manifolds

Since the search for arbitrary 2-transitive combinatorial 4-manifolds needs a lot of computational power, we first want to make some reasonable additional assumptions. Under all combinatorial 4-manifolds, those having the minimum number of vertices possible (given a certain topological type) are of particular interest. In general, it is very hard to determine this number. However, in the case of dimension 4 we have Inequality (1.3.8) from Section 1.3:

$$\binom{n-4}{3} \geq 10(\chi(M)-2),$$

with equality if and only if the triangulation is 3-neighborly and thus simply connected and tight.

Now recall that by Freedman's theorem (see [47]) it follows that the topological type of simply connected combinatorial 4-manifolds is completely determined by the intersection form (note that M admits a PL structure and thus the Kirby-Siebenmann-invariant always vanishes). Such complexes have been found for the cases $n \in \{6, 9, 16\}$, or equivalently $\chi(M) \in \{2, 3, 24\}$. The topological types are S^4 (the boundary of the 5-simplex), $\mathbb{C}P^2$ (see [80] for the minimal 9-vertex triangulation of $\mathbb{C}P^2$ due to Kühnel and Banchoff) and $K3$ (see [28] for the minimal 16-vertex triangulation of $K3$ due to Casella and Kühnel), where the first and the third complex are the only 2-transitive combinatorial 4-manifolds known so far (cf. Section 5.6). Finding further complexes satisfying these properties or even a pair with the same number of vertices but distinct intersection forms would be an interesting endeavor (cf. Chapter 2). Hence, we will focus on the search for 2-transitive 3-neighborly 4-manifolds throughout this section.

Proposition 5.7.1. *Let M be a 3-neighborly combinatorial 4-manifold. Then M has the following properties:*

1. *M is simply connected (by the 3-neighborliness) and thus orientable (Freedman's theorem [47]),*
2. *$n \geq 6$, with equality if and only if M is the boundary of the 5-simplex,*
3. *$n \equiv 0, 1, 4, 5, 6, 9, 10, 14, 16 \mod 20$ (see [77, p. 69]),*
4. *$f(M) = (n, \binom{n}{2}, \binom{n}{3}, \frac{n^3}{4} - \frac{5n^2}{4} + n, \frac{n^3}{10} - \frac{n^2}{2} + \frac{2n}{5})$ (by virtue of the Dehn-Sommerville equations for 4-manifolds, see Section 1.3 or [77, p. 67]),*

5. $\chi(M) = \frac{n^3}{60} - \frac{n^2}{4} + \frac{37n}{30}$,

6. $H_*(M) = (\mathbb{Z}, 0, \mathbb{Z}^{\chi(M)-2}, 0, \mathbb{Z})$ *(by Poincaré duality),*

7. $M \cong (\mathbb{CP}^2)^{\#m_1} \#(-\mathbb{CP}^2)^{\#m_2} \#(S^2 \times S^2)^{\#m_3} \#(K3)^{\#m_4} \#(-K3)^{\#m_5}$ *where* $m_1 + m_2 + 2m_3 + 22(m_4 + m_5) = \chi(M) - 2$ *(by Freedman's theorem [47]) and*

8. *for the order of the automorphism group we have*

$$n(n-1) \leq |G| \leq n(n-1)\,\mathrm{val}(\langle v_1, v_2, v_3\rangle) \cdot n_{\langle v_1, v_2, v_3\rangle}, \tag{5.7.1}$$

where $\mathrm{val}(\langle v_1, v_2, v_3\rangle)$ *is the valence of an arbitrary triangle of* M *and* $n_{\langle v_1,v_2,v_3\rangle}$ *denotes the number of triangles with isomorphic links in an arbitrary edge star of* M*. In the worst case, Inequality (5.7.1) transforms into* $|G| \leq 60\binom{n}{3}$.

Proof. The proofs are either direct conclusions from the theorems given or their proof can easily be carried out.

For *8.* note that the order of every 2-transitive permutation group G of degree n is $|G| = n(n-1)|\mathrm{stab}\langle v_1, v_2\rangle|$. The pointwise stabilizer $\mathrm{stab}\langle v_1, v_2\rangle$ is a subgroup of the automorphism group of the edge link S of M which is a combinatorial 2-sphere. Since M is not 3-transitive, $\mathrm{stab}\langle v_1, v_2\rangle$ does not operate transitively on the vertices of S. As a consequence, the upper bound in Inequality (5.1.1) cannot be attained. Altogether G has at most half of the elements as stated in Inequality (5.1.1). □

Group candidates

Propositions 5.7.1 and 5.1.4 together with the 3-neighborliness lead to the following list of 2-transitive permutation groups of degree $6 \leq n \leq 100$, which are candidates for automorphism groups of 3-neighborly 4-manifolds.

Table 5.3: 2-transitive permutation groups of degree $6 \leq n \leq 100$. See Table 5.4 for an explanation of the symbols used in the following.

GAP-index*	group	order	triangle stabilizer	remark
(6, 1)	PSL(2, 5)	60	$\mathrm{stab}(\Delta) \in \{1\}$	checked
(9, 3)	$M(9)$	72	$\mathrm{stab}(\Delta) \in \{1\}$	checked

continuation on next page –

Table 5.3 – continuation from last page

GAP-index*	group	order	triangle stabilizer	remark
(9,4)	AGL(1,9)	72	stab(Δ) ∈ {1}	**checked**
(9,5)	AΓL(1,9)	144	stab(Δ) ∈ $\{1, C_2\}$	**checked**
(9,6)	$(C_3^2 \rtimes Q_8) \rtimes C_3$	216	stab(Δ) ∈ $\{1, C_3\}$	**checked**
(9,7)	AGL(2,3)	432	stab(Δ) ∈ $\{1, \sigma_3\}$	**checked**
(10,3)	PSL(2,9)	360	stab(Δ) ∈ {1}	**checked**
(10,5)	σ_5	720	stab(Δ) ∈ $\{C_2\}$	**checked**
(14,1)	PSL(2,13)	1092	stab(Δ) ∈ {1}	**checked**
(16,3)	AGL(1,16)	240	stab(Δ) ∈ {1}	**checked**
(16,6)	AGL(1,16) $\rtimes C_2$	480	stab(Δ) ∈ $\{1, C_2\}$	**checked**
(16,9)	AΓL(1,16)	960	stab(Δ) ∈ $\{1, C_2\}$	**checked**
(16,12)	AΓL(2,4)	5760	stab(Δ) = A_4	too large
(16,13)	ASL(2,4) $\rtimes C_2$	1920	stab(Δ) ∈ $\{1, C_2, C_2 \times C_2\}$	**checked**
(16,14)	AGL(2,4)	2880	stab(Δ) = A_4	too large
(16,15)	ASL(2,4)	960	stab(Δ) ∈ $\{1, C_2 \times C_2\}$	**checked**
(16,16)	$C_2^4.\sigma_5$	11520	stab(Δ) = $C_2 \times C_2 \times C_2$	too large
(16,17)	$C_2^4.A_6$	5760	stab(Δ) ∈ $\{C_2 \times C_2, C_3\}$	**checked**
(20,1)	PSL(2,19)	3420	stab(Δ) ∈ {1}	**checked**
(21,4)	$M(21)$	20160	stab(Δ) = C_2^4	too large
(21,5)	PΣL(3,4)	40320	stab(Δ) = $C_2^4 \rtimes C_2$	too large
(21,6)	$PGL(3,4)$	60480	stab(Δ) = $C_3 \times C_3$	too large
(21,7)	PΓL(3,4)	120960		too large
(24,2)	PSL(2,23)	6072	stab(Δ) ∈ {1}	**checked**
(25,12)	$C_5^2 \rtimes C_3 \rtimes C_8$	600	stab(Δ) ∈ {1}	**checked**
(25,14)	AGL(1,25)	600	stab(Δ) ∈ {1}	**checked**
(25,15)	$C_5^2 \rtimes (Q_8 \rtimes C_3)$	600	stab(Δ) ∈ {1}	**checked**
(25,17)	AΓL(1,25)	1200	stab(Δ) ∈ $\{1, C_2\}$	**checked**
(25,18)	$C_5^2 \rtimes ((Q_8 \rtimes C_3) \rtimes C_2)$	1200	stab(Δ) ∈ $\{1, C_2\}$	**checked**
(25,19)	$C_5^2 \rtimes ((Q_8 \rtimes C_3) \rtimes C_4)$	2400	stab(Δ) ∈ $\{1, C_4\}$	**checked**
(25,20)	ASL(2,5)	3000	stab(Δ) ∈ $\{1, C_5\}$	**checked**
(25,21)	ASL(2,5) $\rtimes C_2$	6000	stab(Δ) ∈ $\{1, D_{10}\}$	**checked**
(25,22)	AGL(2,5)	12000	stab(Δ) = $C_5 \rtimes C_4$	too large
(26,1)	PSL(2,25)	7800	stab(Δ) ∈ {1}	**checked**
(26,3)	PΣL(2,25)	15600	stab(Δ) ∈ $\{C_2\}$	**checked**
(29,6)	AGL(1,29)	812	stab(Δ) ∈ {1}	**checked**
(30,1)	PSL(2,29)	12180	stab(Δ) ∈ {1}	**checked**
(36,10)	PSp(6,2)	1451520		too large
(40,5)	PSL(4,3)	6065280		too large
(40,6)	$PGL(4,3)$	12130560		too large
(41,8)	AGL(1,41)	1640	stab(Δ) ∈ {1}	**checked**
(44,1)	PSL(2,43)	39732	stab(Δ) ∈ {1}	**checked**
(49,22)	$C_7^2 \rtimes (C_3 \times Q_{16})$	2352	stab(Δ) ∈ {1}	**checked**
(49,23)	AGL(1,49)	2352	stab(Δ) ∈ {1}	**checked**
(49,25)	$C_7^2 \rtimes C_2.\sigma_4$	2352	stab(Δ) ∈ {1}	**checked**
(49,28)	AΓL(1,49)	4704	stab(Δ) ∈ $\{1, C_2\}$	**checked**
(49,29)	$(C_7^2 \rtimes C_2.\sigma_4) \rtimes C_3)$	7056	stab(Δ) ∈ $\{1, C_3\}$	**checked**

continuation on next page –

Table 5.3 – continuation from last page

GAP-index*	group	order	triangle stabilizer	remark
(49, 30)	ASL(2, 7)	16464	stab(Δ) ∈ $\{1, C_7\}$	**checked**
(49, 31)	ASL(2, 7) ⋊ C_2	32928	stab(Δ) ∈ $\{1, D_{14}\}$	**checked**
(49, 32)	ASL(2, 7) ⋊ C_3	49392	stab(Δ) = $C_7 \rtimes C_3$	too large
(49, 33)	AGL(2, 7)	98784	stab(Δ) = $(C_7 \rtimes C_3) \rtimes C_2$	too large
(50, 3)	PSL(2, 49)	58800	stab(Δ) ∈ $\{1\}$	**checked**
(50, 5)	PΣL(2, 49)	117600	stab(Δ) ∈ $\{C_2\}$	**checked**
(54, 1)	PSL(2, 53)	74412	stab(Δ) ∈ $\{1\}$	**checked**
(60, 6)	PSL(2, 59)	102660	stab(Δ) ∈ $\{1\}$	**checked**
(61, 12)	AGL(1, 61)	3660	stab(Δ) ∈ $\{1\}$	**checked**
(64, 12)	AGL(1, 2^6)	4032	stab(Δ) ∈ $\{1\}$	
(64, 14)	$C_2^6 \rtimes C_7 \rtimes C_9$	4032	stab(Δ) ∈ $\{1\}$	
(64, 20)	$C_2^6 \rtimes (C_7 \times D_{18})$	8064	stab(Δ) ∈ $\{1, C_2\}$	
(64, 24)	$C_2^6 \rtimes C_7 \rtimes C_9 \rtimes C_3$	12096	stab(Δ) ∈ $\{1, C_3\}$	
(64, 32)	$C_2^6 \rtimes C_7 \rtimes C_9 \rtimes C_6$	24192	stab(Δ) ∈ $\{1, C_2, C_3\}$	
(64, 43)	AΓL(3, 4)	23224320		too large
(64, 44)	AΣL(3, 4)	7741440		too large
(64, 45)	AGL(3, 4)	11612160		too large
(64, 46)	ASL(3, 4)	3870720		too large
(64, 49)	AΓL(2, 8)	677376	stab(Δ) = $C_2^3 \rtimes C_7$	too large
(64, 50)	AΣL(2, 8)	96768	stab(Δ) = C_2^3	too large
(64, 51)	AGL(2, 8)	225792	stab(Δ) = $C_2^3 \rtimes C_7$	too large
(64, 52)	$C_2^6 \rtimes$ PSL(2, 8)	32256	stab(Δ) = C_2^3	too large
(64, 55)	$C_2^6 \rtimes$ Sp(6, 2)	92897280		too large
(64, 62)	$C_2^6 \rtimes$ ΣU(3, 3)	774144	stab(Δ) = $C_2 \times C_2 \times C_2$	too large
(64, 63)	$C_2^6 \rtimes SU(3, 3)$	387072	stab(Δ) = $(C_4 \times C_2) \rtimes C_2$	too large
(65, 3)	PSU(3, 4)	62400	stab(Δ) ∈ $\{1, C_5\}$	
(65, 4)	PSU(3, 4).2	124800	stab(Δ) ∈ $\{1, D_{10}\}$	
(65, 5)	PΓU(3, 4)	249600	stab(Δ) = $C_5 \rtimes C_4$	too large
(65, 6)	PSz(8)	29120	stab(Δ) ∈ $\{1\}$	
(65, 7)	PSz(8).C_3	87360	stab(Δ) ∈ $\{1, C_3\}$	
(74, 1)	PSL(2, 73)	194472	stab(Δ) ∈ $\{1\}$	
(80, 1)	PSL(2, 79)	246480	stab(Δ) ∈ $\{1\}$	
(81, 44)	$C_3^4 \rtimes C_5 \rtimes C_{16}$	6480	stab(Δ) ∈ $\{1\}$	
(81, 47)	$C_3^4 \rtimes C_{80}$	6480	stab(Δ) ∈ $\{1\}$	
(81, 69)	$C_3^4 \rtimes C_5 \rtimes C_2^5$	12960	stab(Δ) ∈ $\{1, C_2\}$	
(81, 70)	$C_3^4 \rtimes C_{16} \rtimes D_{10}$	12960	stab(Δ) ∈ $\{1, C_2\}$	
(81, 71)	$C_3^4 \rtimes ((C_2 \times Q_8) \rtimes C_2) \rtimes C_5$	12960	stab(Δ) ∈ $\{1, C_2\}$	
(81, 89)	$C_3^4 \rtimes C_{16} \rtimes C_5 \rtimes C_4$	25920	stab(Δ) ∈ $\{1, C_2, C_4\}$	
(81, 90)	$C_3^4 \rtimes ((C_2 \times Q_8) \rtimes C_2) \rtimes D_{10}$	25920	stab(Δ) ∈ $\{1, C_4\}$	
(81, 99)	$C_3^4 \rtimes ((C_2 \times Q_8) \rtimes C_2) \rtimes C_5 \rtimes C_4$	51840	stab(Δ) ∈ $\{1, C_4, C_8\}$	
(81, 109)	AGL(4, 3)	1965150720		too large
(81, 110)	ASL(4, 3)	982575360		too large
(81, 111)	$C_3^4 \rtimes C_2.A_6$	58320	stab(Δ) = $C_3 \times C_3$	too large
(81, 112)	$C_3^4 \rtimes C_4.A_6$	116640	stab(Δ) = $(C_3 \times C_3) \rtimes C_2$	too large
(81, 113)	$C_3^4 \rtimes C_2.\sigma_5$	116640	stab(Δ) = $C_3 \times \sigma_3$	too large

continuation on next page –

Table 5.3 – continuation from last page

GAP-index*	group	order	triangle stabilizer	remark
(81, 114)	$C_3^4 \rtimes C_8.A_6$	233280	$\mathrm{stab}(\Delta) = (C_3 \times C_3) \rtimes C_4$	too large
(81, 115)	$C_3^4 \rtimes \mathrm{SL}(2,9) \rtimes C_2^2$	233280	$\mathrm{stab}(\Delta) = \sigma_3 \times \sigma_3$	too large
(81, 116)	$C_3^4 \rtimes C_4.A_6.C_2$	233280	$\mathrm{stab}(\Delta) = (C_3 \times C_3) \rtimes C_4$	too large
(81, 117)	$C_3^4 \rtimes C_2.A_6 \rtimes D_8$	466560	$\mathrm{stab}(\Delta) = (\sigma_3 \times \sigma_3) \rtimes C_2$	too large
(81, 118)	$\mathrm{AGL}(2,9)$	466560	$\mathrm{stab}(\Delta) = (C_3 \times C_3) \rtimes C_8$	too large
(81, 119)	$C_3^4 \rtimes C_2.A_6 \rtimes Q_8$	466560	$\mathrm{stab}(\Delta) = (C_3 \times C_3) \rtimes Q8$	too large
(81, 120)	$\mathrm{A\Gamma L}(2,9)$	933120	$\mathrm{stab}(\Delta) = ((C_3 \times C_3) \rtimes C_8) \rtimes C_2$	too large
(81, 124)	$C_3^4 \rtimes C_2.A_5.C_2$	19440	$\mathrm{stab}(\Delta) \in \{1, C_3\}$	
(81, 126)	$C_3^4 \rtimes C_8.A_5$	38880	$\mathrm{stab}(\Delta) \in \{1, \sigma_3\}$	
(81, 127)	$C_3^4 \rtimes C_4.\sigma_5$	38880	$\mathrm{stab}(\Delta) \in \{1, \sigma_3\}$	
(81, 128)	$C_3^4 \rtimes C_8.\sigma_5$	77760	$\mathrm{stab}(\Delta) \in \{1, C_2, D_{12}, \sigma_3\}$	
(81, 129)	$C_3^4 \rtimes C_2^5.A_5 \rtimes C_2$	311040	$\mathrm{stab}(\Delta) = \mathrm{GL}(2,3)$	too large
(81, 130)	$C_3^4 \rtimes C_2^5.A_5$	155520	$\mathrm{stab}(\Delta) = \mathrm{SL}(2,3)$	too large
(81, 131)	$C_3^4 \rtimes \mathrm{Sp}(4,3) \rtimes C_2$	8398080		too large
(81, 132)	$C_3^4 \rtimes \mathrm{Sp}(4,3)$	4199040	$\mathrm{stab}(\Delta) = ((C_3^2 \rtimes C_3) \rtimes Q8) \rtimes C_3$	too large
(84, 3)	$\mathrm{PSL}(2,83)$	285852	$\mathrm{stab}(\Delta) \in \{1\}$	
(85, 3)	$\mathrm{PSL}(4,4)$	987033600		too large
(85, 4)	$\mathrm{P\Sigma L}(4,4)$	1974067200		too large
(89, 8)	$\mathrm{AGL}(1,89)$	7832	$\mathrm{stab}(\Delta) \in \{1\}$	
(90, 1)	$\mathrm{PSL}(2,89)$	352440	$\mathrm{stab}(\Delta) \in \{1\}$	

Table 5.4: Explanation of the notations used in Table 5.3

notation	meaning
$\mathrm{GL}(n,q)$	Linear group of dimension n over $\mathbb{F}_q$
$\mathrm{SL}(n,q)$	Special linear group of dimension n over $\mathbb{F}_q$
$\mathrm{Sp}(n,q)$	Symplectic group of dimension n over $\mathbb{F}_q$
$\mathrm{AGL}(n,q)$	Affine linear group of dimension n over $\mathbb{F}_q$
$\mathrm{ASL}(n,q)$	Affine special linear group of dimension n over $\mathbb{F}_q$
$\mathrm{PGL}(n,q)$	Projective linear group of dimension n over $\mathbb{F}_q$
$\mathrm{PSL}(n,q)$	Projective special linear group of dimension n over $\mathbb{F}_q$
$\mathrm{P\Sigma L}(n,q^2)$	Automorphism group of the projective special linear group of dimension n over $\mathbb{F}_q$
$\mathrm{A\Gamma L}(n,q^2)$	Automorphism group of the affine linear group of dimension n over $\mathbb{F}_q$
$\mathrm{SU}(n,q)$	Special unitary group of dimension n over $\mathbb{F}_q$
$\mathrm{PSU}(n,q)$	Projective special unitary group of dimension n over $\mathbb{F}_q$
$\mathrm{\Sigma U}(n,q)$	Automorphism group of the unitary group of dimension n over $\mathbb{F}_q$
$\mathrm{A\Gamma U}(n,q^2)$	Automorphism group of the affine unitary group of dimension n over $\mathbb{F}_q$
$\mathrm{PSz}(n)$	Suzuki group
C_n	Cyclic group of order n
D_n	Dihedral group of order n
Q_n	(Generalized) quaternionic group of order n
A_n	Alternating group
σ_n	Symmetric group
$M(n)$	Matthieu group of degree n
$\times$, $\rtimes$	Direct product, semi direct product (not unique)
.	Non-split extension

Using the GAP-package simpcomp [40, 41, 42] we were able to proof the following theorem.

Theorem 5.7.2. *$\partial\Delta^5$ and $(K3)_{16}$ are the only 2-transitive 3-neighborly combinatorial 4-manifolds on $n < 64$ vertices.*

Checking the remaining 2-transitive groups with degree $n \geq 64$ needs a lot of computational resources. However, given the size of the resulting manifolds, the number of orbits involved, as well as the topological type of the expected complexes, we believe the following statement to be true.

Conjecture 5.7.3. *$\partial\Delta^5$ and $(K3)_{16}$ are the only 2-transitive 3-neighborly combinatorial 4-manifolds.*

In order to prove Conjecture 5.7.3, it could be useful to take a closer look at the pointwise stabilizers of various 3-tuples of a group candidate G. It seems that there is an upper bound for the size of an automorphism group of a combinatorial manifold G which is much more restrictive than 5.7.1. In most cases, group orders of $|G| \geq 20n(n-1)$ lead to stabilizers of triangles which are not compatible with Proposition 5.1.4.

In a second step one could try to exclude the remaining series of smaller 2-transitive groups.

5.8 Sharply 2-transitive combinatorial 4-pseudomanifolds

As of today, there are very few examples of 2-transitive combinatorial 4-pseudomanifolds. So far, only a few sporadic examples of such complexes are known (cf. [37]). However, their topological properties as well as the question of existence for a given number of vertices do not seem to have any obvious regularities at the first glance.

Nonetheless, one can observe that most of these sporadic 2-transitive complexes are generated by only one orbit under the action of the affine linear group $\mathrm{AGL}(1,p)$ of dimension 1 over the field with p elements, where p denotes a prime number. As a matter of fact, most of these complexes are part of an infinite series of combinatorial 4-pseudomanifolds as will be shown in the following.

Theorem 5.8.1. *Let p be a prime, $p > 14$, $\mathrm{AGL}(1,p) := \langle(0,1,\ldots,p-1),(x,x^2,x^{p-1}=1)\rangle$ where $x \in \mathbb{F}_p^\star$, $o(x) = p-1$, and $1 \le l \le p-2$ such that*

$$x^{3l} + x^{2l} \equiv 1(p). \tag{5.8.1}$$

Then the orbit

$$C_p^{x^l} := \mathrm{AGL}(1,p) \cdot \langle 0,1,x^l,x^{2l},x^{3l}\rangle$$

is a combinatorial pseudomanifold (necessarily with 2-transitive automorphism group).

Proof. Since $C_p^{x^l}$ is 2-transitive, it suffices to show that $\mathrm{lk}_{C_p^{x^l}}(\langle 0,1\rangle)$ is a triangulated 2-sphere. As we will prove below, we have $o(x^l) > 4$. Thus, the orbit of $\langle 0,1,x^l,x^{2l},x^{3l}\rangle$ is of full length and $\mathrm{lk}_{C_p^{x^l}}(\langle 0,1\rangle)$ contains exactly 20 triangles. Assuming that $C_p^{x^l}$ is in fact a combinatorial pseudomanifold, it follows by the 2-transitivity and the Dehn-Sommerville equations that

$$f(C_p^{x^l}) = (p,\binom{p}{2},4\binom{p}{2},5\binom{p}{2},2\binom{p}{2}).$$

Thus, the edge links satisfy

$$f(\mathrm{lk}_{C_p^{x^l}}(\langle u,v\rangle)) = (12,30,20).$$

Note that we have $x^{3l} + x^{2l} \equiv 1(p)$, what results in the following equalities that will be of help in the following.

$$\begin{aligned}
x^l &= (x^l+1)(x^{2l}+x^l+1)^{-1} && (5.8.2)\\
x^{2l} &= 1-x^{3l} = (1+x^l)^{-1} = (x^{2l}+x^l)(x^{2l}+x^l+1)^{-1} && (5.8.3)\\
x^{3l} &= 1-x^{2l} = (x^{2l}+x^l+1)^{-1} = x^l(x^l+1)^{-1} = (1+x^{-l})^{-1} && (5.8.4)\\
x^{5l} &= 1-x^l = (1+x^{-l}+x^{-2l})^{-1} && (5.8.5)\\
-x^l &= 1-x^{-2l} = -x^{-l}(x^l+1)^{-1} = x^{3l}(x^{3l}-1)^{-1} = (1-x^{-3l})^{-1} && (5.8.6)\\
-x^{4l} &= 1-x^{-l} = (-x^{-2l}-x^{-l})^{-1} && (5.8.7)
\end{aligned}$$

In the following we will calculate explicitly the 20 simplices of $\mathrm{star}_{C_p^{x^l}}(\langle 0,1\rangle)$.

Multiplication of the generator $\langle 0, 1, x^l, x^{2l}, x^{3l}\rangle$:

$$\begin{aligned}
\langle 0, 1, x^l, x^{2l}, x^{3l}\rangle &= \langle 0, 1, x^l, x^{2l}, x^{3l}\rangle && (5.8.8)\\
x^{-l}\langle 0, 1, x^l, x^{2l}, x^{3l}\rangle &= \langle 0, x^{-l}, 1, x^l, x^{2l}\rangle && (5.8.9)\\
x^{-2l}\langle 0, 1, x^l, x^{2l}, x^{3l}\rangle &= \langle 0, x^{-2l}, x^{-l}, 1, x^l\rangle && (5.8.10)\\
x^{-3l}\langle 0, 1, x^l, x^{2l}, x^{3l}\rangle &= \langle 0, x^{-3l}, x^{-2l}, x^{-l}, 1\rangle && (5.8.11)
\end{aligned}$$

Translation of the generator $\langle 0, 1, x^l, x^{2l}, x^{3l}\rangle$ by -1, followed by a suitable multiplication:

$$\begin{aligned}
(-1)\langle -1, 0, x^l-1, x^{2l}-1, x^{3l}-1\rangle &= \langle 1, 0, 1-x^l, 1-x^{2l}, 1-x^{3l}\rangle \\
&= \langle 1, 0, x^{5l}, x^{3l}, x^{2l}\rangle && (5.8.12)\\
(x^l-1)^{-1}\langle -1, 0, x^l-1, x^{2l}-1, x^{3l}-1\rangle &= \langle (1-x^l)^{-1}, 0, 1, x^l+1, x^{2l}+x^l+1\rangle \\
&= \langle x^{-5l}, 0, 1, x^{-2l}, x^{-3l}\rangle && (5.8.13)\\
(x^{2l}-1)^{-1}\langle -1, 0, x^l-1, x^{2l}-1, x^{3l}-1\rangle &= \langle (1-x^{2l})^{-1}, 0, (x^l+1)^{-1}, 1, (x^l+1)^{-1}(x^{2l}+x^l+1)\rangle \\
&= \langle x^{-3l}, 0, x^{2l}, 1, x^{-l}\rangle && (5.8.14)\\
(x^{3l}-1)^{-1}\langle -1, 0, x^l-1, x^{2l}-1, x^{3l}-1\rangle &= \langle (1-x^{3l})^{-1}, 0, (x^{2l}+x^l+1)^{-1}, (x^l+1)(x^{2l}+x^l+1)^{-1}, 1\rangle \\
&= \langle x^{-2l}, 0, x^{3l}, x^l, 1\rangle && (5.8.15)
\end{aligned}$$

Translation of the generator $\langle 0, 1, x^l, x^{2l}, x^{3l}\rangle$ by $-x^l$, followed by a suitable multiplication:

$$\begin{aligned}
(-x^{-l})\langle -x^l, 1-x^l, 0, x^{2l}-x^l, x^{3l}-x^l\rangle &= \langle 1, 1-x^{-l}, 0, 1-x^l, 1-x^{2l}\rangle \\
&= \langle 1, -x^{4l}, 0, x^{5l}, x^{3l}\rangle && (5.8.16)\\
(1-x^l)^{-1}\langle -x^l, 1-x^l, 0, x^{2l}-x^l, x^{3l}-x^l\rangle &= \langle x^l(x^l-1)^{-1}, 1, 0, -x^l, -x^l(x^l+1)\rangle \\
&= \langle -x^{-4l}, 1, 0, -x^l, -x^{-l}\rangle && (5.8.17)\\
(x^{2l}-x^l)^{-1}\langle -x^l, 1-x^l, 0, x^{2l}-x^l, x^{3l}-x^l\rangle &= \langle (1-x^l)^{-1}, -x^{-l}, 0, 1, x^l+1\rangle \\
&= \langle x^{-5l}, -x^{-l}, 0, 1, x^{-2l}\rangle && (5.8.18)\\
(x^{3l}-x^l)^{-1}\langle -x^l, 1-x^l, 0, x^{2l}-x^l, x^{3l}-x^l\rangle &= \langle (1-x^{2l})^{-1}, -x^{-l}(x^l+1)^{-1}, 0, (x^l+1)^{-1}, 1\rangle \\
&= \langle x^{-3l}, -x^l, 0, x^{2l}, 1\rangle && (5.8.19)
\end{aligned}$$

Translation of the generator $\langle 0,1,x^l,x^{2l},x^{3l}\rangle$ by $-x^{2l}$, followed by a suitable multiplication:

$$\begin{aligned}
(-x^{-2l})\langle -x^{2l},1-x^{2l},x^l-x^{2l},0,x^{3l}-x^{2l}\rangle &= \langle 1,1-x^{-2l},1-x^{-l},0,1-x^l\rangle \\
&= \langle 1,-x^l,-x^{4l},0,x^{5l}\rangle && (5.8.20)\\
(1-x^{2l})^{-1}\langle -x^{2l},1-x^{2l},x^l-x^{2l},0,x^{3l}-x^{2l}\rangle &= \langle (1-x^{-2l})^{-1},1,(1+x^{-l})^{-1},0,-x^l(x^{-l}+1)^{-1}\rangle \\
&= \langle -x^{-l},1,x^{3l},0,-x^{4l}\rangle && (5.8.21)\\
(x^l-x^{2l})^{-1}\langle -x^{2l},1-x^{2l},x^l-x^{2l},0,x^{3l}-x^{2l}\rangle &= \langle x^l(x^l-1)^{-1},x^{-l}+1,1,0,-x^l\rangle \\
&= \langle -x^{-4l},x^{-3l},1,0,-x^l\rangle && (5.8.22)\\
(x^{3l}-x^{2l})^{-1}\langle -x^{2l},1-x^{2l},x^l-x^{2l},0,x^{3l}-x^{2l}\rangle &= \langle (1-x^l)^{-1},-(x^{-2l}+x^{-l}),-x^{-l},0,1\rangle \\
&= \langle x^{-5l},-x^{-4l},-x^{-l},0,1\rangle && (5.8.23)
\end{aligned}$$

Translation of the generator $\langle 0,1,x^l,x^{2l},x^{3l}\rangle$ by $-x^{3l}$, followed by a suitable multiplication:

$$\begin{aligned}
(-x^{-3l})\langle -x^{3l},1-x^{3l},x^l-x^{3l},x^{2l}-x^{3l},0\rangle &= \langle 1,1-x^{-3l},1-x^{-2l},1-x^{-l},0\rangle \\
&= \langle 1,-x^{-l},-x^l,-x^{4l},0\rangle && (5.8.24)\\
(1-x^{3l})^{-1}\langle -x^{3l},1-x^{3l},x^l-x^{3l},x^{2l}-x^{3l},0\rangle &= \langle x^{3l}(x^{3l}-1)^{-1},1,(x^{2l}+x^l)(x^{2l}+x^l+1)^{-1},(1+x^{-l}+x^{-2l})^{-1},0\rangle \\
&= \langle -x^l,1,x^{2l},x^{5l},0\rangle && (5.8.25)\\
(x^l-x^{3l})^{-1}\langle -x^{3l},1-x^{3l},x^l-x^{3l},x^{2l}-x^{3l},0\rangle &= \langle (1-x^{-2l})^{-1},(x^{2l}+x^l+1)(x^{2l}+x^l)^{-1},1,x^l(x^l+1)^{-1},0\rangle \\
&= \langle -x^{-l},x^{-2l},1,x^{3l},0\rangle && (5.8.26)\\
(x^{2l}-x^{3l})^{-1}\langle -x^{3l},1-x^{3l},x^l-x^{3l},x^{2l}-x^{3l},0\rangle &= \langle x^l(x^l-1)^{-1},1+x^{-l}+x^{-2l},1+x^{-l},1,0\rangle \\
&= \langle -x^{-4l},x^{-5l},x^{-3l},1,0\rangle && (5.8.27)
\end{aligned}$$

Altogether, we get the following triangles for the edge link of $\langle 0,1\rangle$:

$$\begin{aligned}
\mathrm{lk}_{C_p^{x^l}}(\langle 0,1\rangle) = \langle &\langle x^l,x^{2l},x^{3l}\rangle,\langle x^{-2l},x^{-l},x^l\rangle,\langle x^{-l},x^l,x^{2l}\rangle,\langle x^{-3l},x^{-2l},x^{-l}\rangle,\\
&\langle x^{5l},x^{3l},x^{2l}\rangle,\langle x^{-5l},x^{-2l},x^{-3l}\rangle,\langle x^{-3l},x^{2l},x^{-l}\rangle,\langle x^{-2l},x^{3l},x^l\rangle,\\
&\langle -x^{4l},x^{5l},x^{3l}\rangle,\langle -x^{-4l},-x^l,-x^{-l}\rangle,\langle x^{-5l},-x^{-l},x^{-2l}\rangle,\langle x^{-3l},-x^l,x^{2l}\rangle,\\
&\langle -x^l,-x^{4l},x^{5l}\rangle,\langle -x^{-l},x^{3l},-x^{4l}\rangle,\langle -x^{-4l},x^{-3l},-x^l\rangle,\langle x^{-5l},-x^{-4l},-x^{-l}\rangle,\\
&\langle -x^{-l},-x^l,-x^{4l}\rangle,\langle -x^l,x^{2l},x^{5l}\rangle,\langle -x^{-l},x^{-2l},x^{3l}\rangle,\langle -x^{-4l},x^{-5l},x^{-3l}\rangle\rangle.
\end{aligned}$$

Looking at Figure 5.8.1, we can see that $\mathrm{lk}_{C_p^{x^l}}(\langle 0,1\rangle)$ is a triangulated 2-sphere if all 12 vertices are distinct elements in the vertex set $V=\{2,\dots,p-1\}$, $p>14$. This will be shown in the following.

If the order of the element $o(x^l)$ is an even number, we have

$$x^{\frac{o(x^l)}{2}l} = -1$$

and thus the twelve vertices can be written as

$$\begin{array}{llll} x^l, & x^{2l}, & x^{3l}, & x^{5l}, \\ x^{(\frac{o(x^l)}{2}-4)l}, & x^{(\frac{o(x^l)}{2}-1)l}, & x^{(\frac{o(x^l)}{2}+1)l}, & x^{(\frac{o(x^l)}{2}+4)l}, \\ x^{-5l}, & x^{-3l}, & x^{-2l}, & x^{-l}. \end{array}$$

This means that all vertices are distinct if and only if $o(x^l) \in \{16 + 2k \,|\, k \in \mathbb{N}_0\} \setminus \{18\}$, $k \in \mathbb{N}$.

If $o(x^l)$ is odd, we have

$$x^{ml} \neq -1$$

for all $m \in \{0, \dots p-1\}$. Thus, the set of vertices can be split into two disjoint subsets

$$\begin{array}{llll} x^l, & x^{2l}, & x^{3l}, & x^{5l}, \\ x^{-5l}, & x^{-3l}, & x^{-2l}, & x^{-l} \end{array}$$

and

$$-x^l, \quad -x^{4l}, \quad -x^{-4l}, \quad -x^{-l}.$$

Hence, all twelve vertices are distinct for $o(x^l) \in \{9 + 2k \,|\, k \in \mathbb{N}_0\}$.

To keep things as simple as possible we will write $y = x^l$ for the remainder of the proof.

It now remains to show that for $o(y) \in \{1, \dots, 8, 10, 12, 14, 18\}$, Equation (5.8.1) either already implies that $p < 14$ or leads to a contradiction. By Equation (5.8.1) it follows immediately that $o(y) > 3$ and we will start with the case that $o(y) = 4$.

$o(y) = 4$: $y^2 = -1$

$$\begin{array}{rcl} y^2(1+y) & \equiv & 1\,(p) \\ -1-y & \equiv & 1\,(p) \\ y & \equiv & -2\,(p) \\ -4\,(p) & = & 1\,(p) \\ \Leftrightarrow \qquad p & = & 5 \end{array}$$

$o(y) = 5$:

$$\begin{aligned} y^{-1}(1+y)^2 &\equiv 1\,(p) \\ y^{-1}+2+y &\equiv 1\,(p) \\ 1+y+y^2 &\equiv 0\,(p) \\ (1-y)(1+y+y^2) = 1-y^3 &\equiv 0\,(p) \\ \Leftrightarrow \qquad o(y) &= 3 \end{aligned}$$

$o(y) = 6$:

$$\begin{aligned} y^2(1+y) &\equiv 1\,(p) \\ -y^{-1}(1+y) &\equiv 1\,(p) \\ y^{-1} &\equiv -2\,(p) \\ 8\,(p) &= 1\,(p) \\ \Leftrightarrow \qquad p &= 7 \end{aligned}$$

$o(y) = 7$:

$$\begin{aligned} y^{-1}(1+y)^3 &\equiv 1\,(p) \\ 1+2y+3y^2+y^3 &\equiv 0\,(p) \\ 2+2y+2y^2 &\equiv 0\,(p) \\ \Leftrightarrow \qquad o(y) &= 3 \end{aligned}$$

$o(y) = 8$:

$$\begin{aligned} y^2(1+y) &\equiv 1\,(p) \\ y^{-2}+y^{-1} &\equiv 1\,(p) \\ 1+y+y^2 &\equiv 0\,(p) \\ \Leftrightarrow \qquad o(y) &= 3 \end{aligned}$$

$o(y) = 10$:

$$\begin{aligned} y^6(1+y)^3 &\equiv 1\,(p) \\ -y(1+y)^3 &\equiv 1\,(p) \\ 1+y+3y^2+3y^3+y^4 &\equiv 0\,(p) \\ 1+2y+3y^2+2y^3 &\equiv 0\,(p) \\ 1+2y+y^2 &\equiv -2\,(p) \\ (y+1)^2 &\equiv -2\,(p) \\ \Leftrightarrow \qquad y &\equiv 2\,(p) \\ 12\,(p) &= 1\,(p) \\ \Leftrightarrow \qquad p &= 11 \end{aligned}$$

$o(y) = 12$:

$$\begin{aligned} (1+y)^3 &\equiv -1\,(p) \\ 2+3y+3y^2+y^3 &\equiv 0\,(p) \\ 3+3y+2y^2 &\equiv 0\,(p) \\ 2+3y+y^2 &\equiv 0\,(p) \\ 1+2y+2y^2 &\equiv 0\,(p) \\ y &\equiv -2\,(p) \\ \Leftrightarrow \qquad p &= 5 \end{aligned}$$

$o(y) = 14$:

$$\begin{aligned} y(1+y)^4 &\equiv -1\,(p) \\ 1+y+4y^2+6y^3+4y^4+y^5 &\equiv 0\,(p) \\ 1+y+5y^2+6y^3+3y^4 &\equiv 0\,(p) \\ 1+4y+5y^2+3y^3 &\equiv 0\,(p) \\ 2+2y+y^2 &\equiv 0\,(p) \\ 1+2y+y^2 &\equiv 0\,(p) \\ \Leftrightarrow \qquad 1+y &\equiv 0\,(p) \end{aligned}$$

$o(y) = 18$:

$$\begin{aligned}
y(1+y)^5 &\equiv -1\,(p)\\
1+y+5y^2+10y^3+10y^4+5y^5+y^6 &\equiv 0\,(p)\\
1+y+5y^2+11y^3+10y^4+4y^5 &\equiv 0\,(p)\\
1+y+9y^2+11y^3+6y^4 &\equiv 0\,(p)\\
1+7y+9y^2+5y^3 &\equiv 0\,(p)\\
6+7y+4y^2 &\equiv 0\,(p)\\
4+6y+3y^2 &\equiv 0\,(p)\\
3+4y+3y^2 &\equiv 0\,(p)\\
3+3y+y^2 &\equiv 0\,(p)\\
1+3y+2y^2 &\equiv 0\,(p)\\
(y+1)(2y+1) &\equiv 0\,(p)\\
\Leftrightarrow \qquad 2y &\equiv -1\,(p)\\
4y^2\,(p) &= y^2+y^3\,(p)\\
\Leftrightarrow \qquad y &\equiv 3\,(p)\\
36\,(p) &= 1\,(p)\\
\Leftrightarrow \qquad p &\in \{5,7\}
\end{aligned}$$

This states the result. The smallest pairs (p, x^l) to satisfy Equation (5.8.1) are $(17,7)$, $(19,16)$, $(23,8)$ and $(23,7)$. □

Moreover, we can prove the following.

Proposition 5.8.2. *There are infinitely many pairs (p, x^l) satisfying Equation (5.8.1).*

Proof. Finding primes satisfying the condition

$$y^2 + y^3 \equiv 1(p)$$

for some $y \in \mathbb{N}$ is equivalent to finding prime divisors of the polynomial

$$\alpha : \mathbb{N} \to \mathbb{N}; y \mapsto y^2 + y^3 - 1.$$

We will show by iteration that there are infinitely many of such prime divisors greater 14. As a start, we compute $\alpha(7) = 7^2 + 7^3 - 1 = 391 = 17 \cdot 23$. Now let $p_1, \ldots, p_n$ be all known

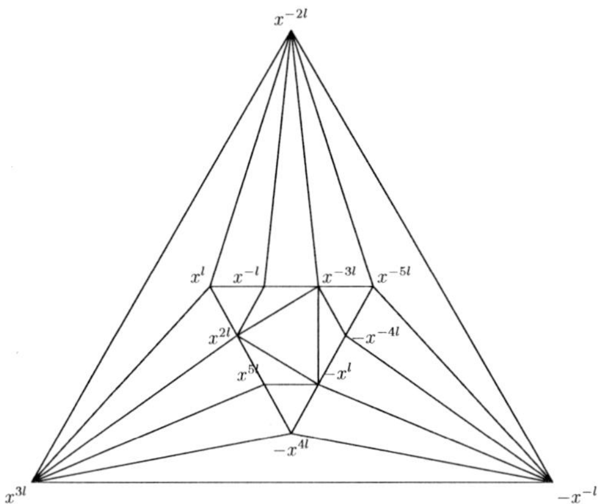

Figure 5.8.1: Edge link of $C_p^{x^l}$, a 12-vertex 2-sphere with dihedral automorphism group of order 6.

prime divisors of α, then clearly $\gcd(\alpha(p_1 \cdot \ldots \cdot p_n), p_i) = 1$, $1 \leq i \leq n$. Thus, $\alpha(p_1 \cdot \ldots \cdot p_n)$ contains a new prime divisor p_{n+1}. □

A list of all generators for $p < 300$ is shown in Table 5.5. Representatives for large values of p can be computed using the the function `SCSeriesAGL(p)` from the GAP-package simpcomp [40, 41, 42]. Here the vertex labels are $1, 2, \ldots, p$ instead of $0, 1, \ldots, p-1$.

Table 5.5: List of $C_p^{x^l}$ for $p < 300$.

p	x	l	x^l	$o(x^l)$	$C_p^{x^l}$	$\langle 0, 1, x^l, x^{2l}, x^{3l} \rangle$
17	3	11	7	16	$C_{17}^{3^{11}}$	$\langle 0, 1, 7, 15, 3 \rangle$
19	13	8	16	9	$C_{19}^{13^{8}}$	$\langle 0, 1, 16, 9, 11 \rangle$
23	5	6	8	11	$C_{23}^{5^{6}}$	$\langle 0, 1, 8, 18, 6 \rangle$
		19	7	22	$C_{23}^{5^{19}}$	$\langle 0, 1, 7, 3, 21 \rangle$
37	5	23	20	36	$C_{37}^{5^{23}}$	$\langle 0, 1, 20, 30, 8 \rangle$
43	5	8	13	21	$C_{43}^{5^{8}}$	$\langle 0, 1, 13, 40, 4 \rangle$
53	3	38	43	26	$C_{53}^{3^{38}}$	$\langle 0, 1, 43, 47, 7 \rangle$
59	11	33	50	58	$C_{59}^{11^{33}}$	$\langle 0, 1, 50, 22, 38 \rangle$
		39	52	58	$C_{59}^{11^{39}}$	$\langle 0, 1, 52, 49, 11 \rangle$

Continuation on next page –

Table 5.5 – Continuation from last page

p	x	l	x^l	$o(x^l)$	$C_p^{x^l}$	$\langle 0,1,x^l,x^{2l},x^{3l}\rangle$
		44	15	29	$C_{59}^{11^{44}}$	$\langle 0,1,15,48,12\rangle$
61	7	52	15	15	$C_{61}^{7^{52}}$	$\langle 0,1,15,42,20\rangle$
67	7	65	48	66	$C_{67}^{7^{65}}$	$\langle 0,1,48,26,42\rangle$
79	7	34	4	39	$C_{79}^{7^{34}}$	$\langle 0,1,4,16,64\rangle$
83	5	45	39	82	$C_{83}^{5^{45}}$	$\langle 0,1,39,27,57\rangle$
89	3	82	21	44	$C_{89}^{3^{82}}$	$\langle 0,1,21,85,5\rangle$
97	5	33	78	32	$C_{97}^{5^{33}}$	$\langle 0,1,78,70,28\rangle$
101	3	46	96	50	$C_{101}^{3^{46}}$	$\langle 0,1,96,25,77\rangle$
		63	63	100	$C_{101}^{3^{63}}$	$\langle 0,1,63,30,72\rangle$
		91	42	100	$C_{101}^{3^{91}}$	$\langle 0,1,42,47,55\rangle$
103	5	75	80	34	$C_{103}^{5^{75}}$	$\langle 0,1,80,14,90\rangle$
107	5	58	85	53	$C_{107}^{5^{58}}$	$\langle 0,1,85,56,52\rangle$
109	11	100	21	27	$C_{109}^{11^{100}}$	$\langle 0,1,21,5,105\rangle$
113	3	97	90	112	$C_{113}^{3^{97}}$	$\langle 0,1,90,77,37\rangle$
137	3	8	122	17	$C_{137}^{3^{8}}$	$\langle 0,1,122,88,50\rangle$
149	3	8	5	37	$C_{149}^{3^{8}}$	$\langle 0,1,5,25,125\rangle$
157	5	142	10	78	$C_{157}^{5^{142}}$	$\langle 0,1,10,100,58\rangle$
167	5	77	151	166	$C_{167}^{5^{77}}$	$\langle 0,1,151,89,79\rangle$
		127	86	166	$C_{167}^{5^{127}}$	$\langle 0,1,86,48,120\rangle$
		128	96	83	$C_{167}^{5^{128}}$	$\langle 0,1,96,31,137\rangle$
173	3	21	66	172	$C_{173}^{3^{21}}$	$\langle 0,1,66,31,143\rangle$
		55	162	172	$C_{173}^{3^{55}}$	$\langle 0,1,162,121,53\rangle$
		96	117	43	$C_{173}^{3^{96}}$	$\langle 0,1,117,22,152\rangle$
181	23	159	175	60	$C_{181}^{23^{159}}$	$\langle 0,1,175,36,146\rangle$
191	21	135	31	38	$C_{191}^{21^{135}}$	$\langle 0,1,31,6,186\rangle$
199	41	61	129	198	$C_{199}^{41^{61}}$	$\langle 0,1,129,124,76\rangle$
211	17	109	35	210	$C_{211}^{17^{109}}$	$\langle 0,1,35,170,42\rangle$
		123	124	70	$C_{211}^{17^{123}}$	$\langle 0,1,124,184,28\rangle$
		188	51	105	$C_{211}^{17^{188}}$	$\langle 0,1,51,69,143\rangle$
223	5	46	72	111	$C_{223}^{5^{46}}$	$\langle 0,1,72,55,169\rangle$
		50	177	111	$C_{223}^{5^{50}}$	$\langle 0,1,177,109,115\rangle$
		126	196	37	$C_{223}^{5^{126}}$	$\langle 0,1,196,60,164\rangle$
227	5	145	20	226	$C_{227}^{5^{145}}$	$\langle 0,1,20,173,55\rangle$
229	7	216	16	19	$C_{229}^{7^{216}}$	$\langle 0,1,16,27,203\rangle$
241	7	157	66	240	$C_{241}^{7^{157}}$	$\langle 0,1,66,18,224\rangle$
251	11	141	6	250	$C_{251}^{11^{141}}$	$\langle 0,1,6,36,216\rangle$
263	5	188	221	131	$C_{263}^{5^{188}}$	$\langle 0,1,221,186,78\rangle$

Continuation on next page –

Table 5.5 – Continuation from last page

p	x	l	x^l	$o(x^l)$	$C_p^{x^l}$	$\langle 0,1,x^l,x^{2l},x^{3l}\rangle$
271	43	5	157	54	$C_{271}^{43^5}$	$\langle 0,1,157,259,13\rangle$
		66	166	45	$C_{271}^{43^{66}}$	$\langle 0,1,166,185,87\rangle$
		199	218	270	$C_{271}^{43^{199}}$	$\langle 0,1,218,99,173\rangle$
281	3	147	61	40	$C_{281}^{3^{147}}$	$\langle 0,1,61,68,214\rangle$
283	5	175	104	282	$C_{283}^{5^{175}}$	$\langle 0,1,104,62,222\rangle$
293	3	222	83	146	$C_{293}^{3^{222}}$	$\langle 0,1,83,150,144\rangle$

Remark 5.8.3. It is worthwhile mentioning that the set of primes allowing more than one solution seems to be equal to the set of primes of type $p = x^2+23y^2$ for any pair of non-negative integers $(x,y) \in \mathbb{N}_0 \times \mathbb{N}_0$. This seems to be interesting to investigate in more detail.

Remark 5.8.4. The series $C_p^{x^l}$ probably expands to a very similar series $C_{p^r}^{x^l}$ on the set of prime powers p^r as there exist corresponding complexes on 25 and 49 vertices.

The vertex links of $C_p^{x^l}$

Let $S := \mathrm{stab}_{\mathrm{AGL}(1,p)}(0) = \mathrm{AGL}(1,p)/\langle(0,1,\dots,p-1)\rangle = \langle(x,x^2,\dots,x^{p-1}=1)\rangle \cong \mathbb{Z}_{p-1}$. Since S fixes vertex 0, $\mathrm{star}_{C_p^{x^l}}(0)$ equals the union of the S-orbits of (5.8.8), (5.8.12), (5.8.16), (5.8.20) and (5.8.24). For the link of vertex 0 we have

$$\begin{aligned}\mathrm{lk}_{C_p^{x^l}}(0) &= \{S\cdot\langle 1,x^l,x^{2l},x^{3l}\rangle, S\cdot\langle 1,x^{5l},x^{3l},x^{2l}\rangle,\\ &\quad S\cdot\langle 1,-x^{4l},x^{5l},x^{3l}\rangle, S\cdot\langle 1,-x^l,-x4l,x^{5l}\rangle,\\ &\quad S\cdot\langle 1,-x^{-l},-x^l,-x^{4l}\rangle\}.\end{aligned}$$

We were able to verify for small values of p that the vertex links of $C_p^{x^l}$ are PL homeomorphic to $S^1\times S^2$, $S^1 \mathbin{\underline{\times}} S^2$ or a union of disjoint copies of $S^1\times S^2$ and $S^1 \mathbin{\underline{\times}} S^2$. Moreover, the connected components of the links seem to be combinatorially isomorphic to members of the series C_{2n} and D_{2n} from Section 4.5.3. An explanation for this seems to be the fact that the edge links of $C_p^{x^l}$ coincide with the vertex links of C_{2n} and D_{2n} (cf. Figure 5.8.1 and Figure 4.5.5).

5.9 Summary

Table 5.6 and Table 5.7 summarize the results about k-transitive combinatorial d-manifolds and d-pseudomanifolds for $d \leq 4$ presented in this chapter.

Generally, one can say that transitivity in dimension 1 is trivial and thus this case is completely understood.

In dimension 2, all multiply transitive surfaces and all sharply 2-transitive "pinch point" surfaces can now be constructed, whereas the simply transitive case seems hopeless to classify.

In dimension 3, no non-trivial multiply transitive combinatorial manifolds exist. For the unique 3-transitive combinatorial 3-pseudomanifold, a geometric interpretation is presented. The 2-transitive case is solved for vertex numbers less or equal to 30.

In dimension 4, an infinite series of 2-transitive pseudomanifolds is now known which contains most of the previously known sporadic examples of such complexes for small vertex numbers. On the other hand, no further examples for multiply transitive manifolds are presented. In the 3-neighborly case it is shown that no further multiply transitive examples exist with less than 64 vertices.

k / d	1	2	3	4	5
1	all n-gons, $n \geq 4$	$\partial\Delta^2$			
2	$\partial\beta^3$, the boundary of the icosahedron, surfaces different from S^2 classified up to $n = 15$ vertices in [92]	$\mathbb{T}^2_7$, $\mathbb{R}P^2_6$, see [67]	$\partial\Delta^3$		
3	classified up to $n = 15$ vertices in [92]	$\varnothing$ by Section 5.3	$\varnothing$	$\partial\Delta^4$	
4	classified up to $n = 15$ vertices in [92]	$(K3)_{16}$	$\varnothing$ by Section 5.6.	$\varnothing$	$\partial\Delta^5$

Table 5.6: k-transitive combinatorial d-manifolds for $d \leq 4$, $k \leq 4$.

k / d	2	3
2	see Theorem 5.2.3 and Table 5.1	$\varnothing$ by Section 5.2
3	see Table 5.2	M^3_8 from Section 5.4
4	series $C^{x^l}_p$ 5.8	$\varnothing$ by Section 5.6.

Table 5.7: k-transitive combinatorial d-pseudomanifolds for $d \leq 4$, $k \leq 4$ (without combinatorial manifolds).

APPENDIX A

simpcomp – A GAP toolbox for simplicial complexes

simpcomp[1] is an extension (a so-called *package*) to GAP [49], the well known system for computational discrete algebra. The package enables the user to compute numerous properties of (abstract) simplicial complexes, provides functions to construct new complexes from existing ones and has an extensive library of triangulations of manifolds and pseudomanifolds.

The development of the software package started roughly in spring 2009. The first publicly available version was released in October 2009. simpcomp was awarded the "Best Software Presentation Award" by the Fachgruppe Computeralgebra for a presentation given at ISSAC 2010 in Munich. The current version (1.5.1) was released on August 16th, 2011. The package is accepted as an official GAP share package.

The package, as well as its documentation is available from [41].

[1]The development of simpcomp is joint work with Effenberger and has partly been developed within the DFG projects Ku 1203/5-2 and Ku 1203/5-3.

A.1 Why simpcomp

simpcomp is a package for working with simplicial complexes in the GAP system. In contrast to the package homology [34] which focuses on simplicial homology computation, simpcomp claims to provide the user with a broader spectrum of functionality regarding simplicial constructions.

simpcomp allows the user to interactively construct complexes and to compute their properties in the GAP shell. Furthermore, it makes use of GAP's expertise in groups and group operations. For example, automorphism groups and fundamental groups of complexes can be computed and examined further within the GAP system. Apart from supplying a facet list, the user can as well construct simplicial complexes from a set of generators and a prescribed automorphism group – the latter form being the common in which a complex is presented in a publication. This feature is to our knowledge unique to simpcomp.

Furthermore, simpcomp has an extensive library of known triangulations of manifolds. This is the first time that they are easily accessible without having to look them up in the literature [86], [28], or online [91]. This allows the user to work with many different known triangulations without having to construct them first. As of Version 1.4.0 the library contains triangulations of roughly 650 manifolds and roughly 7000 pseudomanifolds, including all vertex transitive triangulations from [91]. Most properties that simpcomp can handle are precomputed for complexes in the library. Searching in the library is possible by the complexes' names as well as some of their properties (such as f-, g- and h-vectors and their homology).

A.2 simpcomp benefits

simpcomp is written entirely in the GAP scripting language, thus giving the user the possibility to see behind the scenes and to customize or alter simpcomp functions if needed.

The main benefit when working with simpcomp over implementing the needed functions from scratch is that simpcomp encapsulates all methods and properties of a simplicial complex in a new GAP object type (as an abstract data type). This way, among other things, simpcomp can transparently cache properties already calculated, thus preventing unnecessary double calculations. It also takes care of the error-prone vertex labeling of a complex.

simpcomp provides the user with functions to save and load the simplicial complexes to and from files and to import and export a complex in various formats (e. g. from and to `polymake/TOPAZ` [50], `Macaulay2` [52], LaTeX, etc.).

In contrast to the software package `polymake` [50], providing the most efficient algorithms for each task in form of a heterogeneous package (where algorithms are implemented in various languages), the primary goal when developing simpcomp was not efficiency (this is already limited by the GAP scripting language), but rather ease of use and ease of extensibility by the user in the GAP language with all its mathematical and algebraic capabilities.

The package includes an extensive manual (see [41]) in which all functionality of simpcomp is documented and also makes use of GAP's built in help system such that all the documentation is available directly from the GAP prompt in an interactive way.

A.3 Some operations and constructions that simpcomp supports

simpcomp implements many standard and often needed functions for working with simplicial complexes. These functions can be roughly divided into three groups: (i) functions generating simplicial complexes, (ii) functions to construct new complexes from old and (iii) functions calculating properties of complexes – for a full list of supported features see the documentation [41].

simpcomp also implements a variety of functions connected to bistellar moves (cf. Section 1.7) on simplicial complexes.

Furthermore, the package supports slicings of 3-manifolds (see Section 1.6 or [71], [55]) and related constructions.

As of Version 1.3.0, simpcomp supports the construction of simplicial complexes of prescribed dimension, vertex number and transitive automorphism group as described in [92], [28] (see also Section A.4.2).

With the release of Version 1.4.0, support for simplicial blowups was added (cf. Chapter 2), i. e. the resolutions of ordinary double points in combinatorial 4-pseudomanifolds. This functionality is to our knowledge not provided by any other software package so far. In addition, a lot of infinite series of combinatorial manifolds and pseudomanifolds presented in this work were implemented.

A.4 Examples

This section contains a small demonstration of the capabilities of simpcomp in form of some example constructions.

A.4.1 4-manifolds

Casella and Kühnel constructed a triangulated $K3$ surface with the minimum number of 16 vertices in [28]. They presented it in terms of the complex obtained by the automorphism group $G \cong AGL(1, \mathbb{F}_{16})$ given (for example) by the two generators

$$G = \langle (1,3,8,4,9,16,15,2,14,12,6,7,13,5,10), \\ (1,11,16)(2,10,14)(3,12,13)(4,9,15)(5,7,8) \rangle$$

acting on the two generating simplices $\Delta_1 = \langle 2,3,4,5,9 \rangle$ and $\Delta_2 = \langle 2,5,7,10,11 \rangle$ (cf. Section 2.2 for more information about this complex). It turned out to be a non-trivial problem to show that the complex obtained in this way is homeomorphic to the $K3$ surface. This turns out to be a rather easy task using simpcomp, as will be shown below:

After starting GAP, load simpcomp and then construct the complex from its representation given above.

```
gap> LoadPackage("simpcomp");; #load the package
Loading simpcomp 1.5.1
by F.Effenberger and J.Spreer
http://www.igt.uni-stuttgart.de/LstDiffgeo/simpcomp
gap> SCInfoLevel(0);; #suppress simpcomp info messages
http://www.igt.uni-stuttgart.de/LstDiffgeo/simpcomp
gap> G:=Group((1,3,8,4,9,16,15,2,14,12,6,7,13,5,10),
              (1,11,16)(2,10,14)(3,12,13)(4,9,15)(5,7,8));;
gap> K3:=SCFromGenerators(G,[[2,3,4,5,9],[2,5,7,10,11]]);
[SimplicialComplex

 Properties known: Dim, FacetsEx, Name, Vertices.

 Name="complex from generators under unknown group"
 Dim=4

/SimplicialComplex]
```

We first compute the f-vector, the Euler characteristic and the homology groups of K3:

```
gap> K3.F;
[ 16, 120, 560, 720, 288 ]
gap> K3.Chi;
24
gap> K3.Homology;
[ [ 0, [  ] ], [ 0, [  ] ], [ 22, [  ] ], [ 0, [  ] ], [ 1, [  ] ] ]
```

Now we verify that the complex K3 is a combinatorial manifold using a heuristic algorithm based on bistellar moves:

```
gap> K3.IsManifold;
true
```

In a next step, we compute the parity and the signature of the intersection form of the complex K3:

```
gap> K3.IntersectionFormParity;
0
gap> K3.IntersectionFormSignature;
[ 22, 3, 19 ]
```

This means that the intersection form of the complex K3 is even. It has dimension 22 and signature 19−3 = 16. Furthermore, K3 is simply connected as can either be verified by showing that the fundamental group is trivial or by checking that the complex is 3-neighborly:

```
gap> K3.FundamentalGroup;
<fp group with 105 generators>
gap> Size(last);
1
gap> K3.Neighborliness;
3
```

It now follows from Freedman's theorem [47] that the complex is in fact homeomorphic to the $K3$ surface because it has the same (even) intersection form. Furthermore, K3 is a tight triangulation [86, 77] as can be verified as follows:

```
gap> K3.IsTight;
true
```

We can also have a look at the multiplicity vectors of some rsl-functions:

```
gap> SCMorseIsPerfect(K3,[1..16]);
true
gap> SCMorseMultiplicityVector(K3,[1..16]);
[ [ 1, 0, 0, 0, 0 ], [ 0, 0, 0, 0, 0 ], [ 0, 0, 0, 0, 0 ], [ 0, 0, 1, 0, 0 ],
  [ 0, 0, 2, 0, 0 ], [ 0, 0, 1, 0, 0 ], [ 0, 0, 4, 0, 0 ], [ 0, 0, 3, 0, 0 ],
  [ 0, 0, 3, 0, 0 ], [ 0, 0, 4, 0, 0 ], [ 0, 0, 1, 0, 0 ], [ 0, 0, 2, 0, 0 ],
  [ 0, 0, 1, 0, 0 ], [ 0, 0, 0, 0, 0 ], [ 0, 0, 0, 0, 0 ], [ 0, 0, 0, 0, 1 ] ]
```

Finally, instead of constructing the triangulation K3 from scratch, we could have also loaded it directly from the library, saving us a lot of work. We now load the complex from the library and verify that the complex from the library is combinatorially isomorphic to the complex K3 that we constructed – note below the two different searching methods provided by the library, one using complex names and the other using complex properties:

```
gap> SCLib.SearchByName("K3");
[ [ 7494, "K3 surface" ] ]
gap> SCLib.SearchByAttribute("Dim=4 and F[3]=Binomial(F[1],3)");
[ [ 16, "CP^2 (VT)" ], [ 7494, "K3 surface" ] ]
gap> M:=SCLib.Load(7494);;
gap> M.IsIsomorphic(K3);
true
```

A.4.2 The classification of transitive combinatorial manifolds

The search for highly symmetric, transitive combinatorial manifolds has been one of the major topics in combinatorial topology over the past decades. As discussed in more detail in Chapter 5, there is an algorithm classifying all transitive combinatorial manifolds with a prescribed number of vertices n and dimension d which are invariant under the automorphisms provided by a permutation group G.

The idea is to partition the set of all $(d+1)$-tuples and the set of all d-tuples on n vertices into orbits of G. Then it is checked how often the d-tuple orbits occur in a certain combination of $(d+1)$-tuple orbits. If this number is always 0 or 2, this is equivalent to the fact that the underlying complex fulfills the weak pseudomanifold property and thus is a candidate for a transitive combinatorial manifold. This method has been used for the first time by Kühnel and Lassmann in [83]. In 2003, a classification of all transitive combinatorial manifolds with up to 15 was given by Lutz in [92] using his GAP-programm MANIFOLD_VT. In the meanwhile the program has been integrated to simpcomp together with a number of tools allowing to build own classifications of transitive combinatorial manifolds on the fly:

Let us assume we want to classify all simply transitive combinatorial 2-manifolds with 17 or 19 vertices. We do not want to compute whether the full automorphism group coincides with the one from the construction process and we do not want to remove isomorphic entries.

```
gap> SCsFromGroupByTransitivity;
function( n, d, k, maniflag, computeAutGroup, removeDoubleEntries ) ... end
gap> list:=SCsFromGroupByTransitivity([17,19],2,1,true,false,false);;
#I  SCsFromGroupByTransitivity: Building list of groups...
```

```
#I  SCsFromGroupByTransitivity: ...3 groups found.
#I  degree 17: [ C(17)=17, L(17)=PSL(2,16) ]
#I  degree 19: [ C(19)=19 ]
#I  SCsFromGroupByTransitivity: Processing dimension 2.
#I  SCsFromGroupByTransitivity: Processing degree 17.
#I  SCsFromGroupByTransitivity: 1 / 2 groups calculated, found 336 complexes.
#I  SCsFromGroupByTransitivity: 2 / 2 groups calculated, found 0 complexes.
#I  SCsFromGroupByTransitivity: ...done dim = 2, deg =  17, 80 manifolds, 256
pseudomanifolds, 0 candidates found.
#I  SCsFromGroupByTransitivity: Processing degree 19.
#I  SCsFromGroupByTransitivity: 1 / 1 groups calculated, found 3617 complexes.
#I  SCsFromGroupByTransitivity: ...done dim = 2, deg =  19, 1437 manifolds, 2180
pseudomanifolds, 0 candidates found.
#I  SCsFromGroupByTransitivity: ...done dim = 2.
gap> Sum(List(list,x->Size(x)));
3953
gap> List(list,x->Size(x));
[ 1517, 2436, 0 ]
gap> Set(List(list[1],x->[x.Chi,x.IsOrientable]));
[ [ -38, true ], [ -38, false ], [ -19, false ], [ -17, false ], [ 0, true ] ]
```

This means that the algorithm found 3953 complexes, 1517 of which turned out to be combinatorial manifolds, 2436 to be combinatorial pseudomanifolds and no complexes where the attempt to verify via bistellar moves whether or not the complex is a combinatorial pseudomanifold or manifold failed. The last line computes the Euler characteristic together with the orientability and thus the topological type of all 1517 triangulated surfaces which were found by the algorithm.

A.4.3 Bistellar moves

If we quickly want to produce example manifolds of a certain topological type, simpcomp offers a variety of functionalities to ease the construction process. Let us say we want to construct the 3-manifold $(S^2 \times S^1)^{\#3}$:

```
gap> SCBistellarOptions.MaxRounds:=10000;
10000
gap> s2s1:=SCCartesianProduct(SCBdSimplex(2),SCBdSimplex(3));;
gap> s2s13:=SCConnectedProduct(s2s1,3);;
gap> s2s13.F;
[ 28, 132, 208, 104 ]
gap> res:=SCReduceComplex(s2s13);;
gap> res[1];
fail
gap> res[2].F;
[ 15, 80, 130, 65 ]
```

This means, we constructed a version of $(S^2 \times S^1)^{\#3}$ with $((3\cdot4)\cdot3) - 2\cdot4 = 28$ vertices using standard techniques and reduced it via bistellar moves to a version with only 16 vertices (note that this cannot be expected to be the minimal number of vertices needed to triangulate $(S^2 \times S^1)^{\#3}$, hence `res[1]:=fail`).

A.4.4 Simplicial blowups

The abstract singular Kummer variety K^4 of dimension 4 leads to the $K3$ surface by a series of 16 blowups (see [128] or Chapter 2). K^4 is available from the simpcomp library:

```
gap> SCLib.SearchByName("Kummer");
[ [ 7493, "4-dimensional Kummer variety (VT)" ] ]
gap> k4:=SCLib.Load(7493);;
gap> k4.F;
[ 16, 120, 400, 480, 192 ]
gap> lks:=SCLinks(k4,0);;
gap> Set(List(lks,x->SCHomology(x)));
[ [ [ 0, [ ] ], [ 0, [ 2 ] ], [ 0, [ ] ], [ 1, [ ] ] ] ]
gap> SCHomology(k4);
[ [ 0, [ ] ], [ 0, [ ] ], [ 6, [ 2, 2, 2, 2, 2 ] ], [ 0, [ ] ], [ 1, [ ] ] ]
```

We want to blowup the first singularity:

```
gap> c:=SCBlowup(k4,1);; time;
#I  SCBlowup: checking if singularity is a combinatorial manifold...
#I  SCBlowup: ...true
#I  SCBlowup: checking type of singularity...
#I  SCReduceComplexEx: complexes are bistellarly equivalent.
#I  SCBlowup: ...ordinary double point (supported type).
#I  SCBlowup: starting blowup...
#I  SCBlowup: map boundaries...
#I  SCBlowup: boundaries not isomorphic, initializing bistellar moves...
#I  SCBlowup: found complex with smaller boundary: f = [ 14, 71, 114, 57 ].
#I  SCBlowup: found complex with smaller boundary: f = [ 14, 69, 110, 55 ].
#I  SCBlowup: found complex with smaller boundary: f = [ 14, 68, 108, 54 ].
#I  SCBlowup: found complex with smaller boundary: f = [ 14, 67, 106, 53 ].
#I  SCBlowup: found complex with smaller boundary: f = [ 13, 63, 100, 50 ].
#I  SCBlowup: found complex with smaller boundary: f = [ 13, 62, 98, 49 ].
#I  SCBlowup: found complex with smaller boundary: f = [ 13, 61, 96, 48 ].
#I  SCBlowup: found complex with smaller boundary: f = [ 12, 57, 90, 45 ].
#I  SCBlowup: found complex with smaller boundary: f = [ 12, 56, 88, 44 ].
#I  SCBlowup: found complex with smaller boundary: f = [ 12, 55, 86, 43 ].
#I  SCBlowup: found complex with smaller boundary: f = [ 11, 51, 80, 40 ].
#I  SCBlowup: found complex with isomorphic boundaries.
#I  SCBlowup: ...boundaries mapped succesfully.
#I  SCBlowup: build complex...
```

```
#I  SCBlowup: ...done.
#I  SCBlowup: ...blowup completed.
#I  SCBlowup: You may now want to reduce the complex via 'SCReduceComplex'.
64561
```

The resulting complex has 15 singularities. Its second Betti number has increased by one:

```
gap> c.F;
[ 38, 351, 1114, 1320, 528 ]
gap> c.Homology;
[ [ 0, [ ] ], [ 0, [ ] ], [ 7, [ 2, 2, 2, 2 ] ], [ 0, [ ] ], [ 1, [ ] ] ]
gap>
```

Repeating this process for the remaining singularities yields a $K3$ surface, as described in [128] or Chapter 2.

A.4.5 Handlebodies and collapsing

There is a series of bounded d-manifolds with $2d+2$ vertices given as the difference cycle $\{(1:1:\ldots:1:d+3)\}$ (cf. Chapter 4, Definition 4.2.3). In dimension $d=7$ we have

```
gap> c:=SCFromDifferenceCycles([[1,1,1,1,1,1,1,9]]);;
gap> SCHomology(c);
[[0,[]],[1,[]],[0,[]],[0,[]],[0,[]],[0,[]],[0,[]],[0,[]]]
gap> coll:=SCCollapseGreedy(c);;
gap>  SCFacets(coll);
[ [ 7, 14 ], [ 7, 16 ], [ 14, 16 ] ]
```

and thus a handlebody. Its boundary is the sphere product $S^5 \times S^1$:

```
gap> bd:=SCBoundary(c);;
gap> bd.Homology;
[[0,[]],[1,[]],[0,[]],[0,[]],[0,[]],[1,[]],[1,[]]]
```

which follows by the simplicial homology groups and a theorem due to Kreck (see [73]).

A.4.6 Slicings

simpcomp supports slicings (cf. Chapter 3). In the case of 3-manifolds, slicings are returned using another abstract data type `SCNormalSurface`.

The following calculation yields a slicing of a lens space of type $L(3,1)$ inducing a genus 1 handlebody decomposition of $L(3,1)$, the surface is shown in Figure 4.5.1.

```
gap> SCLib.SearchByName("L(3,1)");
[ [ 245, "Mapping cylinder L(3,1)" ], [ 595, "L(3,1) (VT)" ],
  [ 758, "L(3,1) (VT)" ], [ 762, "L(3,1) (VT)" ], [ 768, "L(3,1) (VT)" ],
  [ 773, "L(3,1) (VT)" ] ]
gap> c:=SCLib.Load(595);;
gap> s:=SCSlicing(c,[[1,3,5,7,9,11,13],[2,4,6,8,10,12,14]]);;
gap> SCTopologicalType(s);
"T^2"
gap> SCFVector(s);
[ 35, 84, 28, 21 ]
```

A.4.7 Fundamental group and double suspension

We can use simpcomp to construct non-combinatorial triangulations of manifolds as well: The double suspension of Poincaré's homology sphere, also called "Edward's sphere", is a well-known example for such a complex:

```
gap> SCLib.SearchByName("Poincare_sphere");
[ [ 7469, "Poincare_sphere" ] ]
gap> c:=SCLib.Load(7469);;
gap> SCHomology(c);
[ [ 0, [  ] ], [ 0, [  ] ], [ 0, [  ] ], [ 1, [  ] ] ]
gap> SCFundamentalGroup(c);
<fp group with 91 generators>
gap> Size(last);
120
gap> es:=SCSuspension(SCSuspension(c));; # is homeomorphic to S^5
gap> lk:=SCLink(es,[[1,17],[2,1]]);; # should be a sphere...
gap> SCIsIsomorphic(lk,c); # ...but is a Poincare-sphere
true
```

Appendix B

The blowup process, step by step

In the following, all 17 complexes from the minimal resolution of the 4-dimensional Kummer variety $(K^4)_{16}$ to the 17-vertex $K3$ surface $(K3)_{17}$ are listed.

In every step the signature of the intersection form and the second Betti number and thus also the Euler characteristic increases by one and the number of singularities decreases by one. Also, the torsion part of $H_2(K^4)$ gradually declines. It is not known if for every step there exists a 16-vertex version of the corresponding pseudomanifold. However, at least for the first 8 complexes this is true.

The complexes are labeled C_n^m, where m indicates the number of blowups already performed and n denotes the number of vertices of the complex. In every step the smallest complex (with respect to the f-vector) found by the bistellar flip process is listed. In particular, most complexes are not 2-neighborly.

Table B.1: The 4-dimensional Kummer variety $(K^4)_{16}$ with 16 isolated singularities.

$(K^4)_{16}$ =	⟨1 2 3 6 11⟩,	⟨1 2 3 6 14⟩,	⟨1 2 3 7 10⟩,	⟨1 2 3 7 15⟩,	⟨1 2 3 10 14⟩,
	⟨1 2 3 11 15⟩,	⟨1 2 4 5 12⟩,	⟨1 2 4 5 13⟩,	⟨1 2 4 8 9⟩,	⟨1 2 4 8 16⟩,
	⟨1 2 4 9 13⟩,	⟨1 2 4 12 16⟩,	⟨1 2 5 7 10⟩,	⟨1 2 5 7 15⟩,	⟨1 2 5 10 12⟩,
	⟨1 2 5 13 15⟩,	⟨1 2 6 8 9⟩,	⟨1 2 6 8 16⟩,	⟨1 2 6 9 11⟩,	⟨1 2 6 14 16⟩,
	⟨1 2 9 11 15⟩,	⟨1 2 9 13 15⟩,	⟨1 2 10 12 16⟩,	⟨1 2 10 14 16⟩,	⟨1 3 4 5 12⟩,
	⟨1 3 4 5 13⟩,	⟨1 3 4 8 9⟩,	⟨1 3 4 8 16⟩,	⟨1 3 4 9 13⟩,	⟨1 3 4 12 16⟩,
	⟨1 3 5 6 11⟩,	⟨1 3 5 6 14⟩,	⟨1 3 5 11 12⟩,	⟨1 3 5 13 14⟩,	⟨1 3 7 8 9⟩,
	⟨1 3 7 8 16⟩,	⟨1 3 7 9 10⟩,	⟨1 3 7 15 16⟩,	⟨1 3 9 10 14⟩,	⟨1 3 9 13 14⟩,
	⟨1 3 11 12 16⟩,	⟨1 3 11 15 16⟩,	⟨1 5 6 8 9⟩,	⟨1 5 6 8 16⟩,	⟨1 5 6 9 11⟩,

Continuation on next page –

Table B.1 – Continuation from last page

⟨1 5 6 14 16⟩,	⟨1 5 7 8 9⟩,	⟨1 5 7 8 16⟩,	⟨1 5 7 9 10⟩,	⟨1 5 7 15 16⟩,
⟨1 5 9 10 12⟩,	⟨1 5 9 11 12⟩,	⟨1 5 13 14 16⟩,	⟨1 5 13 15 16⟩,	⟨1 9 10 12 16⟩,
⟨1 9 10 14 16⟩,	⟨1 9 11 12 16⟩,	⟨1 9 11 15 16⟩,	⟨1 9 13 14 16⟩,	⟨1 9 13 15 16⟩,
⟨2 3 4 6 11⟩,	⟨2 3 4 6 14⟩,	⟨2 3 4 7 10⟩,	⟨2 3 4 7 15⟩,	⟨2 3 4 10 14⟩,
⟨2 3 4 11 15⟩,	⟨2 4 5 6 12⟩,	⟨2 4 5 6 13⟩,	⟨2 4 6 11 12⟩,	⟨2 4 6 13 14⟩,
⟨2 4 7 8 10⟩,	⟨2 4 7 8 15⟩,	⟨2 4 8 9 10⟩,	⟨2 4 8 15 16⟩,	⟨2 4 9 10 13⟩,
⟨2 4 10 13 14⟩,	⟨2 4 11 12 15⟩,	⟨2 4 12 15 16⟩,	⟨2 5 6 7 10⟩,	⟨2 5 6 7 15⟩,
⟨2 5 6 10 12⟩,	⟨2 5 6 13 15⟩,	⟨2 6 7 8 10⟩,	⟨2 6 7 8 15⟩,	⟨2 6 8 9 10⟩,
⟨2 6 8 15 16⟩,	⟨2 6 9 10 11⟩,	⟨2 6 10 11 12⟩,	⟨2 6 13 14 15⟩,	⟨2 6 14 15 16⟩,
⟨2 9 10 11 15⟩,	⟨2 9 10 13 15⟩,	⟨2 10 11 12 15⟩,	⟨2 10 12 15 16⟩,	⟨2 10 13 14 15⟩,
⟨2 10 14 15 16⟩,	⟨3 4 5 7 12⟩,	⟨3 4 5 7 13⟩,	⟨3 4 6 8 11⟩,	⟨3 4 6 8 14⟩,
⟨3 4 7 10 12⟩,	⟨3 4 7 13 15⟩,	⟨3 4 8 9 11⟩,	⟨3 4 8 14 16⟩,	⟨3 4 9 11 13⟩,
⟨3 4 10 12 14⟩,	⟨3 4 11 13 15⟩,	⟨3 4 12 14 16⟩,	⟨3 5 6 7 11⟩,	⟨3 5 6 7 14⟩,
⟨3 5 7 11 12⟩,	⟨3 5 7 13 14⟩,	⟨3 6 7 8 11⟩,	⟨3 6 7 8 14⟩,	⟨3 7 8 9 11⟩,
⟨3 7 8 14 16⟩,	⟨3 7 9 10 11⟩,	⟨3 7 10 11 12⟩,	⟨3 7 13 14 15⟩,	⟨3 7 14 15 16⟩,
⟨3 9 10 11 14⟩,	⟨3 9 11 13 14⟩,	⟨3 10 11 12 14⟩,	⟨3 11 12 14 16⟩,	⟨3 11 13 14 15⟩,
⟨3 11 14 15 16⟩,	⟨4 5 6 8 12⟩,	⟨4 5 6 8 13⟩,	⟨4 5 7 8 12⟩,	⟨4 5 7 8 13⟩,
⟨4 6 8 11 12⟩,	⟨4 6 8 13 14⟩,	⟨4 7 8 10 12⟩,	⟨4 7 8 13 15⟩,	⟨4 8 9 10 12⟩,
⟨4 8 9 11 12⟩,	⟨4 8 13 14 16⟩,	⟨4 8 13 15 16⟩,	⟨4 9 10 12 13⟩,	⟨4 9 11 12 13⟩,
⟨4 10 12 13 14⟩,	⟨4 11 12 13 15⟩,	⟨4 12 13 14 16⟩,	⟨4 12 13 15 16⟩,	⟨5 6 7 10 14⟩,
⟨5 6 7 11 15⟩,	⟨5 6 8 9 13⟩,	⟨5 6 8 12 16⟩,	⟨5 6 9 11 13⟩,	⟨5 6 10 12 14⟩,
⟨5 6 11 13 15⟩,	⟨5 6 12 14 16⟩,	⟨5 7 8 9 13⟩,	⟨5 7 8 12 16⟩,	⟨5 7 9 10 13⟩,
⟨5 7 10 13 14⟩,	⟨5 7 11 12 15⟩,	⟨5 7 12 15 16⟩,	⟨5 9 10 12 13⟩,	⟨5 9 11 12 13⟩,
⟨5 10 12 13 14⟩,	⟨5 11 12 13 15⟩,	⟨5 12 13 14 16⟩,	⟨5 12 13 15 16⟩,	⟨6 7 8 10 14⟩,
⟨6 7 8 11 15⟩,	⟨6 8 9 10 14⟩,	⟨6 8 9 13 14⟩,	⟨6 8 11 12 16⟩,	⟨6 8 11 15 16⟩,
⟨6 9 10 11 14⟩,	⟨6 9 11 13 14⟩,	⟨6 10 11 12 14⟩,	⟨6 11 12 14 16⟩,	⟨6 11 13 14 15⟩,
⟨6 11 14 15 16⟩,	⟨7 8 9 11 15⟩,	⟨7 8 9 13 15⟩,	⟨7 8 10 12 16⟩,	⟨7 8 10 14 16⟩,
⟨7 9 10 11 15⟩,	⟨7 9 10 13 15⟩,	⟨7 10 11 12 15⟩,	⟨7 10 12 15 16⟩,	⟨7 10 13 14 15⟩,
⟨7 10 14 15 16⟩,	⟨8 9 10 12 16⟩,	⟨8 9 10 14 16⟩,	⟨8 9 11 12 16⟩,	⟨8 9 11 15 16⟩,
⟨8 9 13 14 16⟩,	⟨8 9 13 15 16⟩			

f-vector:	$(16, 120, 400, 480, 192)$
# singularities:	16
Simplicial homology:	$(\mathbb{Z}, 0, \mathbb{Z}^6 \times \mathbb{Z}_2^5, 0, \mathbb{Z})$
Euler characteristic:	8

Table B.2: 1st blowup: $(K^4)_1 6$ after the resolution of the first singularity. The torsion part of the second homology group already decreased by one factor.

$C^1_{16} =$ ⟨1 2 3 4 8⟩, ⟨1 2 3 4 11⟩, ⟨1 2 3 7 8⟩, ⟨1 2 3 7 13⟩, ⟨1 2 3 11 13⟩,
⟨1 2 4 8 14⟩, ⟨1 2 4 10 11⟩, ⟨1 2 4 10 14⟩, ⟨1 2 5 8 14⟩, ⟨1 2 5 8 16⟩,
⟨1 2 5 14 16⟩, ⟨1 2 7 8 16⟩, ⟨1 2 7 13 15⟩, ⟨1 2 7 15 16⟩, ⟨1 2 10 11 13⟩,
⟨1 2 10 13 15⟩, ⟨1 2 10 14 16⟩, ⟨1 2 10 15 16⟩, ⟨1 3 4 5 11⟩, ⟨1 3 4 5 12⟩,
⟨1 3 4 8 12⟩, ⟨1 3 5 11 14⟩, ⟨1 3 5 12 16⟩, ⟨1 3 5 14 16⟩, ⟨1 3 6 7 9⟩,
⟨1 3 6 7 14⟩, ⟨1 3 6 9 16⟩, ⟨1 3 6 14 16⟩, ⟨1 3 7 8 9⟩, ⟨1 3 7 13 14⟩,
⟨1 3 8 9 12⟩, ⟨1 3 9 12 16⟩, ⟨1 3 11 13 14⟩, ⟨1 4 5 7 9⟩, ⟨1 4 5 7 12⟩,
⟨1 4 5 9 11⟩, ⟨1 4 6 7 9⟩, ⟨1 4 6 7 14⟩, ⟨1 4 6 9 16⟩, ⟨1 4 6 14 16⟩,
⟨1 4 7 12 14⟩, ⟨1 4 8 12 14⟩, ⟨1 4 9 10 11⟩, ⟨1 4 9 10 15⟩, ⟨1 4 9 15 16⟩,
⟨1 4 10 14 16⟩, ⟨1 4 10 15 16⟩, ⟨1 5 7 8 9⟩, ⟨1 5 7 8 16⟩, ⟨1 5 7 12 15⟩,
⟨1 5 7 15 16⟩, ⟨1 5 8 9 14⟩, ⟨1 5 9 11 14⟩, ⟨1 5 12 15 16⟩, ⟨1 7 12 14 15⟩,
⟨1 7 13 14 15⟩, ⟨1 8 9 12 14⟩, ⟨1 9 10 11 15⟩, ⟨1 9 11 14 15⟩, ⟨1 9 12 14 15⟩,
⟨1 9 12 15 16⟩, ⟨1 10 11 13 15⟩, ⟨1 11 13 14 15⟩, ⟨2 3 4 6 11⟩, ⟨2 3 4 6 12⟩,
⟨2 3 4 8 12⟩, ⟨2 3 5 7 10⟩, ⟨2 3 5 7 13⟩, ⟨2 3 5 10 14⟩, ⟨2 3 5 13 16⟩,
⟨2 3 5 14 16⟩, ⟨2 3 6 11 16⟩, ⟨2 3 6 12 14⟩, ⟨2 3 6 14 16⟩, ⟨2 3 7 8 10⟩,
⟨2 3 8 10 12⟩, ⟨2 3 10 12 14⟩, ⟨2 3 11 13 16⟩, ⟨2 4 5 7 10⟩, ⟨2 4 5 7 13⟩,
⟨2 4 5 8 14⟩, ⟨2 4 5 8 16⟩, ⟨2 4 5 10 14⟩, ⟨2 4 5 13 16⟩, ⟨2 4 6 7 9⟩,
⟨2 4 6 7 11⟩, ⟨2 4 6 9 12⟩, ⟨2 4 7 9 13⟩, ⟨2 4 7 10 11⟩, ⟨2 4 8 12 13⟩,
⟨2 4 8 13 16⟩, ⟨2 4 9 12 13⟩, ⟨2 6 7 8 11⟩, ⟨2 6 7 8 16⟩, ⟨2 6 7 9 15⟩,
⟨2 6 7 15 16⟩, ⟨2 6 8 11 16⟩, ⟨2 6 9 12 15⟩, ⟨2 6 12 14 15⟩, ⟨2 6 14 15 16⟩,
⟨2 7 8 10 11⟩, ⟨2 7 9 13 15⟩, ⟨2 8 10 11 13⟩, ⟨2 8 10 12 13⟩, ⟨2 8 11 13 16⟩,
⟨2 9 12 13 15⟩, ⟨2 10 12 13 15⟩, ⟨2 10 12 14 15⟩, ⟨2 10 14 15 16⟩, ⟨3 4 5 6 12⟩,
⟨3 4 5 6 15⟩, ⟨3 4 5 11 15⟩, ⟨3 4 6 11 15⟩, ⟨3 5 6 7 13⟩, ⟨3 5 6 7 15⟩,
⟨3 5 6 12 13⟩, ⟨3 5 7 10 15⟩, ⟨3 5 10 11 14⟩, ⟨3 5 10 11 15⟩, ⟨3 5 12 13 16⟩,
⟨3 6 7 9 15⟩, ⟨3 6 7 13 14⟩, ⟨3 6 9 11 15⟩, ⟨3 6 9 11 16⟩, ⟨3 6 12 13 14⟩,
⟨3 7 8 9 15⟩, ⟨3 7 8 10 15⟩, ⟨3 8 9 10 12⟩, ⟨3 8 9 10 15⟩, ⟨3 9 10 11 12⟩,
⟨3 9 10 11 15⟩, ⟨3 9 11 12 13⟩, ⟨3 9 11 13 16⟩, ⟨3 9 12 13 16⟩, ⟨3 10 11 12 14⟩,
⟨3 11 12 13 14⟩, ⟨4 5 6 8 10⟩, ⟨4 5 6 8 16⟩, ⟨4 5 6 10 12⟩, ⟨4 5 6 15 16⟩,
⟨4 5 7 9 13⟩, ⟨4 5 7 10 12⟩, ⟨4 5 8 10 14⟩, ⟨4 5 9 11 13⟩, ⟨4 5 11 13 15⟩,
⟨4 5 13 15 16⟩, ⟨4 6 7 11 14⟩, ⟨4 6 8 9 10⟩, ⟨4 6 8 9 16⟩, ⟨4 6 9 10 12⟩,
⟨4 6 11 14 15⟩, ⟨4 6 14 15 16⟩, ⟨4 7 10 11 12⟩, ⟨4 7 11 12 14⟩, ⟨4 8 9 10 15⟩,
⟨4 8 9 15 16⟩, ⟨4 8 10 14 15⟩, ⟨4 8 12 13 14⟩, ⟨4 8 13 14 15⟩, ⟨4 8 13 15 16⟩,
⟨4 9 10 11 12⟩, ⟨4 9 11 12 13⟩, ⟨4 10 14 15 16⟩, ⟨4 11 12 13 14⟩, ⟨4 11 13 14 15⟩,
⟨5 6 7 8 13⟩, ⟨5 6 7 8 16⟩, ⟨5 6 7 15 16⟩, ⟨5 6 8 10 13⟩, ⟨5 6 10 12 13⟩,
⟨5 7 8 9 13⟩, ⟨5 7 10 12 15⟩, ⟨5 8 9 11 13⟩, ⟨5 8 9 11 14⟩, ⟨5 8 10 11 13⟩,
⟨5 8 10 11 14⟩, ⟨5 10 11 13 15⟩, ⟨5 10 12 13 15⟩, ⟨5 12 13 15 16⟩, ⟨6 7 8 11 14⟩,

Continuation on next page –

Table B.2 – Continuation from last page

⟨6 7 8 13 14⟩,	⟨6 8 9 10 12⟩,	⟨6 8 9 11 14⟩,	⟨6 8 9 11 16⟩,	⟨6 8 9 12 14⟩,
⟨6 8 10 12 13⟩,	⟨6 8 12 13 14⟩,	⟨6 9 11 14 15⟩,	⟨6 9 12 14 15⟩,	⟨7 8 9 13 15⟩,
⟨7 8 10 11 14⟩,	⟨7 8 10 14 15⟩,	⟨7 8 13 14 15⟩,	⟨7 10 11 12 14⟩,	⟨7 10 12 14 15⟩,
⟨8 9 11 13 16⟩,	⟨8 9 13 15 16⟩,	⟨9 12 13 15 16⟩		

f-vector:	$(16, 120, 410, 495, 198)$
# singularities:	15
Simplicial homology:	$(\mathbb{Z}, 0, \mathbb{Z}^7 \times \mathbb{Z}_2^4, 0, \mathbb{Z})$
Euler characteristic:	9

Table B.3: 2nd blowup

$C_{16}^2 =$	⟨1 2 3 4 8⟩,	⟨1 2 3 4 11⟩,	⟨1 2 3 6 11⟩,	⟨1 2 3 6 12⟩,	⟨1 2 3 8 12⟩,
	⟨1 2 4 7 8⟩,	⟨1 2 4 7 13⟩,	⟨1 2 4 11 13⟩,	⟨1 2 5 7 10⟩,	⟨1 2 5 7 13⟩,
	⟨1 2 5 10 14⟩,	⟨1 2 5 13 16⟩,	⟨1 2 5 14 16⟩,	⟨1 2 6 11 16⟩,	⟨1 2 6 12 14⟩,
	⟨1 2 6 14 16⟩,	⟨1 2 7 8 10⟩,	⟨1 2 8 10 12⟩,	⟨1 2 10 12 14⟩,	⟨1 2 11 13 16⟩,
	⟨1 3 4 8 16⟩,	⟨1 3 4 10 11⟩,	⟨1 3 4 10 16⟩,	⟨1 3 5 7 10⟩,	⟨1 3 5 7 13⟩,
	⟨1 3 5 10 14⟩,	⟨1 3 5 13 16⟩,	⟨1 3 5 14 16⟩,	⟨1 3 6 7 9⟩,	⟨1 3 6 7 11⟩,
	⟨1 3 6 9 12⟩,	⟨1 3 7 9 13⟩,	⟨1 3 7 10 11⟩,	⟨1 3 8 12 13⟩,	⟨1 3 8 13 16⟩,
	⟨1 3 9 12 13⟩,	⟨1 3 10 14 16⟩,	⟨1 4 7 8 16⟩,	⟨1 4 7 13 15⟩,	⟨1 4 7 15 16⟩,
	⟨1 4 10 11 13⟩,	⟨1 4 10 13 15⟩,	⟨1 4 10 15 16⟩,	⟨1 6 7 9 15⟩,	⟨1 6 7 11 16⟩,
	⟨1 6 7 15 16⟩,	⟨1 6 9 12 15⟩,	⟨1 6 10 14 15⟩,	⟨1 6 10 14 16⟩,	⟨1 6 10 15 16⟩,
	⟨1 6 12 14 15⟩,	⟨1 7 8 10 11⟩,	⟨1 7 8 11 16⟩,	⟨1 7 9 13 15⟩,	⟨1 8 10 11 13⟩,
	⟨1 8 10 12 13⟩,	⟨1 8 11 13 16⟩,	⟨1 9 12 13 15⟩,	⟨1 10 12 13 15⟩,	⟨1 10 12 14 15⟩,
	⟨2 3 4 5 11⟩,	⟨2 3 4 5 12⟩,	⟨2 3 4 8 12⟩,	⟨2 3 5 6 11⟩,	⟨2 3 5 6 12⟩,
	⟨2 4 5 11 14⟩,	⟨2 4 5 12 16⟩,	⟨2 4 5 14 16⟩,	⟨2 4 7 8 9⟩,	⟨2 4 7 9 14⟩,
	⟨2 4 7 13 14⟩,	⟨2 4 8 9 12⟩,	⟨2 4 9 12 16⟩,	⟨2 4 9 14 16⟩,	⟨2 4 11 13 14⟩,
	⟨2 5 6 7 13⟩,	⟨2 5 6 7 15⟩,	⟨2 5 6 11 15⟩,	⟨2 5 6 12 13⟩,	⟨2 5 7 10 15⟩,
	⟨2 5 10 11 14⟩,	⟨2 5 10 11 15⟩,	⟨2 5 12 13 16⟩,	⟨2 6 7 9 14⟩,	⟨2 6 7 9 15⟩,
	⟨2 6 7 13 14⟩,	⟨2 6 9 11 15⟩,	⟨2 6 9 11 16⟩,	⟨2 6 9 14 16⟩,	⟨2 6 12 13 14⟩,
	⟨2 7 8 9 15⟩,	⟨2 7 8 10 15⟩,	⟨2 8 9 10 12⟩,	⟨2 8 9 10 15⟩,	⟨2 9 10 11 12⟩,
	⟨2 9 10 11 15⟩,	⟨2 9 11 12 16⟩,	⟨2 10 11 12 14⟩,	⟨2 11 12 13 14⟩,	⟨2 11 12 13 16⟩,
	⟨3 4 5 7 9⟩,	⟨3 4 5 7 12⟩,	⟨3 4 5 9 11⟩,	⟨3 4 7 8 12⟩,	⟨3 4 7 8 16⟩,
	⟨3 4 7 9 14⟩,	⟨3 4 7 14 16⟩,	⟨3 4 9 10 11⟩,	⟨3 4 9 10 14⟩,	⟨3 4 10 14 16⟩,
	⟨3 5 6 8 10⟩,	⟨3 5 6 8 16⟩,	⟨3 5 6 10 12⟩,	⟨3 5 6 11 15⟩,	⟨3 5 6 15 16⟩,
	⟨3 5 7 9 13⟩,	⟨3 5 7 10 12⟩,	⟨3 5 8 10 14⟩,	⟨3 5 8 14 16⟩,	⟨3 5 9 11 13⟩,

Continuation on next page –

Table B.3 – Continuation from last page

	⟨3 5 11 13 15⟩,	⟨3 5 13 15 16⟩,	⟨3 6 7 9 14⟩,	⟨3 6 7 11 14⟩,	⟨3 6 8 10 15⟩,
	⟨3 6 8 15 16⟩,	⟨3 6 9 10 12⟩,	⟨3 6 9 10 14⟩,	⟨3 6 10 14 15⟩,	⟨3 6 11 14 15⟩,
	⟨3 7 8 12 16⟩,	⟨3 7 10 11 12⟩,	⟨3 7 11 12 14⟩,	⟨3 7 12 14 16⟩,	⟨3 8 10 14 15⟩,
	⟨3 8 12 13 14⟩,	⟨3 8 12 14 16⟩,	⟨3 8 13 14 15⟩,	⟨3 8 13 15 16⟩,	⟨3 9 10 11 12⟩,
	⟨3 9 11 12 13⟩,	⟨3 11 12 13 14⟩,	⟨3 11 13 14 15⟩,	⟨4 5 7 8 9⟩,	⟨4 5 7 8 12⟩,
	⟨4 5 8 9 11⟩,	⟨4 5 8 11 14⟩,	⟨4 5 8 12 14⟩,	⟨4 5 12 14 16⟩,	⟨4 6 8 9 11⟩,
	⟨4 6 8 9 12⟩,	⟨4 6 8 11 14⟩,	⟨4 6 8 12 14⟩,	⟨4 6 9 11 15⟩,	⟨4 6 9 12 15⟩,
	⟨4 6 11 14 15⟩,	⟨4 6 12 14 15⟩,	⟨4 7 13 14 15⟩,	⟨4 7 14 15 16⟩,	⟨4 9 10 11 15⟩,
	⟨4 9 10 14 16⟩,	⟨4 9 10 15 16⟩,	⟨4 9 12 15 16⟩,	⟨4 10 11 13 15⟩,	⟨4 11 13 14 15⟩,
	⟨4 12 14 15 16⟩,	⟨5 6 7 8 13⟩,	⟨5 6 7 8 16⟩,	⟨5 6 7 15 16⟩,	⟨5 6 8 10 13⟩,
	⟨5 6 10 12 13⟩,	⟨5 7 8 9 13⟩,	⟨5 7 8 12 16⟩,	⟨5 7 10 12 15⟩,	⟨5 7 12 15 16⟩,
	⟨5 8 9 11 13⟩,	⟨5 8 10 11 13⟩,	⟨5 8 10 11 14⟩,	⟨5 8 12 14 16⟩,	⟨5 10 11 13 15⟩,
	⟨5 10 12 13 15⟩,	⟨5 12 13 15 16⟩,	⟨6 7 8 11 14⟩,	⟨6 7 8 11 16⟩,	⟨6 7 8 13 14⟩,
	⟨6 8 9 10 12⟩,	⟨6 8 9 10 16⟩,	⟨6 8 9 11 16⟩,	⟨6 8 10 12 13⟩,	⟨6 8 10 15 16⟩,
	⟨6 8 12 13 14⟩,	⟨6 9 10 14 16⟩,	⟨7 8 9 13 15⟩,	⟨7 8 10 11 14⟩,	⟨7 8 10 14 15⟩,
	⟨7 8 13 14 15⟩,	⟨7 10 11 12 14⟩,	⟨7 10 12 14 15⟩,	⟨7 12 14 15 16⟩,	⟨8 9 10 15 16⟩,
	⟨8 9 11 13 16⟩,	⟨8 9 13 15 16⟩,	⟨9 11 12 13 16⟩,	⟨9 12 13 15 16⟩	

f-vector:	$(16, 120, 420, 510, 204)$
# singularities:	14
Simplicial homology:	$(\mathbb{Z}, 0, \mathbb{Z}^8 \times \mathbb{Z}_2^3, 0, \mathbb{Z})$
Euler characteristic:	10

Table B.4: 3rd blowup: Note that the complex is not 2-neighborly anymore.

$C_{16}^3 =$	⟨1 2 3 4 9⟩,	⟨1 2 3 4 14⟩,	⟨1 2 3 5 9⟩,	⟨1 2 3 5 11⟩,	⟨1 2 3 11 16⟩,
	⟨1 2 3 14 16⟩,	⟨1 2 4 5 9⟩,	⟨1 2 4 5 14⟩,	⟨1 2 5 11 14⟩,	⟨1 2 7 13 14⟩,
	⟨1 2 7 13 16⟩,	⟨1 2 7 14 16⟩,	⟨1 2 11 13 14⟩,	⟨1 2 11 13 16⟩,	⟨1 3 4 6 9⟩,
	⟨1 3 4 6 10⟩,	⟨1 3 4 10 14⟩,	⟨1 3 5 9 16⟩,	⟨1 3 5 11 16⟩,	⟨1 3 6 7 9⟩,
	⟨1 3 6 7 14⟩,	⟨1 3 6 10 14⟩,	⟨1 3 7 9 16⟩,	⟨1 3 7 14 16⟩,	⟨1 4 5 7 10⟩,
	⟨1 4 5 7 13⟩,	⟨1 4 5 9 12⟩,	⟨1 4 5 10 14⟩,	⟨1 4 5 12 13⟩,	⟨1 4 6 9 11⟩,
	⟨1 4 6 10 16⟩,	⟨1 4 6 11 16⟩,	⟨1 4 7 10 16⟩,	⟨1 4 7 13 16⟩,	⟨1 4 9 11 12⟩,
	⟨1 4 11 12 13⟩,	⟨1 4 11 13 16⟩,	⟨1 5 6 7 8⟩,	⟨1 5 6 7 15⟩,	⟨1 5 6 8 12⟩,
	⟨1 5 6 11 15⟩,	⟨1 5 6 11 16⟩,	⟨1 5 6 12 16⟩,	⟨1 5 7 8 13⟩,	⟨1 5 7 10 15⟩,
	⟨1 5 8 12 13⟩,	⟨1 5 9 12 16⟩,	⟨1 5 10 11 14⟩,	⟨1 5 10 11 15⟩,	⟨1 6 7 8 14⟩,
	⟨1 6 7 9 15⟩,	⟨1 6 8 12 14⟩,	⟨1 6 9 11 15⟩,	⟨1 6 10 12 14⟩,	⟨1 6 10 12 16⟩,

Continuation on next page –

Table B.4 – Continuation from last page

⟨1 7 8 13 14⟩,	⟨1 7 9 10 15⟩,	⟨1 7 9 10 16⟩,	⟨1 8 12 13 14⟩,	⟨1 9 10 11 12⟩,
⟨1 9 10 11 15⟩,	⟨1 9 10 12 16⟩,	⟨1 10 11 12 14⟩,	⟨1 11 12 13 14⟩,	⟨2 3 4 7 10⟩,
⟨2 3 4 7 13⟩,	⟨2 3 4 9 10⟩,	⟨2 3 4 13 16⟩,	⟨2 3 4 14 16⟩,	⟨2 3 5 9 11⟩,
⟨2 3 6 7 11⟩,	⟨2 3 6 7 13⟩,	⟨2 3 6 11 16⟩,	⟨2 3 6 13 16⟩,	⟨2 3 7 10 11⟩,
⟨2 3 9 10 11⟩,	⟨2 4 5 8 15⟩,	⟨2 4 5 8 16⟩,	⟨2 4 5 9 15⟩,	⟨2 4 5 14 16⟩,
⟨2 4 6 8 9⟩,	⟨2 4 6 8 11⟩,	⟨2 4 6 9 10⟩,	⟨2 4 6 10 11⟩,	⟨2 4 7 8 11⟩,
⟨2 4 7 8 16⟩,	⟨2 4 7 10 11⟩,	⟨2 4 7 13 16⟩,	⟨2 4 8 9 15⟩,	⟨2 5 6 7 8⟩,
⟨2 5 6 7 15⟩,	⟨2 5 6 8 15⟩,	⟨2 5 7 8 16⟩,	⟨2 5 7 14 15⟩,	⟨2 5 7 14 16⟩,
⟨2 5 9 11 14⟩,	⟨2 5 9 14 15⟩,	⟨2 6 7 8 11⟩,	⟨2 6 7 13 15⟩,	⟨2 6 8 9 10⟩,
⟨2 6 8 10 15⟩,	⟨2 6 10 11 16⟩,	⟨2 6 10 13 15⟩,	⟨2 6 10 13 16⟩,	⟨2 7 13 14 15⟩,
⟨2 8 9 10 15⟩,	⟨2 9 10 11 15⟩,	⟨2 9 11 14 15⟩,	⟨2 10 11 13 15⟩,	⟨2 10 11 13 16⟩,
⟨2 11 13 14 15⟩,	⟨3 4 5 7 10⟩,	⟨3 4 5 7 13⟩,	⟨3 4 5 8 10⟩,	⟨3 4 5 8 15⟩,
⟨3 4 5 13 15⟩,	⟨3 4 6 9 10⟩,	⟨3 4 8 10 14⟩,	⟨3 4 8 13 15⟩,	⟨3 4 8 13 16⟩,
⟨3 4 8 14 16⟩,	⟨3 5 6 8 12⟩,	⟨3 5 6 8 15⟩,	⟨3 5 6 11 15⟩,	⟨3 5 6 11 16⟩,
⟨3 5 6 12 16⟩,	⟨3 5 7 9 13⟩,	⟨3 5 7 9 16⟩,	⟨3 5 7 10 12⟩,	⟨3 5 7 12 16⟩,
⟨3 5 8 10 12⟩,	⟨3 5 9 11 13⟩,	⟨3 5 11 13 15⟩,	⟨3 6 7 9 13⟩,	⟨3 6 7 11 14⟩,
⟨3 6 8 9 10⟩,	⟨3 6 8 9 12⟩,	⟨3 6 8 10 15⟩,	⟨3 6 9 12 13⟩,	⟨3 6 10 14 15⟩,
⟨3 6 11 14 15⟩,	⟨3 6 12 13 16⟩,	⟨3 7 10 11 12⟩,	⟨3 7 11 12 14⟩,	⟨3 7 12 14 16⟩,
⟨3 8 9 10 12⟩,	⟨3 8 10 14 15⟩,	⟨3 8 13 14 15⟩,	⟨3 8 13 14 16⟩,	⟨3 9 10 11 12⟩,
⟨3 9 11 12 13⟩,	⟨3 11 12 13 14⟩,	⟨3 11 13 14 15⟩,	⟨3 12 13 14 16⟩,	⟨4 5 8 10 14⟩,
⟨4 5 8 14 16⟩,	⟨4 5 9 12 15⟩,	⟨4 5 12 13 15⟩,	⟨4 6 8 9 11⟩,	⟨4 6 10 11 16⟩,
⟨4 7 8 11 16⟩,	⟨4 7 10 11 16⟩,	⟨4 8 9 11 13⟩,	⟨4 8 9 13 15⟩,	⟨4 8 11 13 16⟩,
⟨4 9 11 12 13⟩,	⟨4 9 12 13 15⟩,	⟨5 7 8 9 13⟩,	⟨5 7 8 9 16⟩,	⟨5 7 10 12 15⟩,
⟨5 7 12 14 15⟩,	⟨5 7 12 14 16⟩,	⟨5 8 9 11 13⟩,	⟨5 8 9 11 14⟩,	⟨5 8 9 14 16⟩,
⟨5 8 10 11 13⟩,	⟨5 8 10 11 14⟩,	⟨5 8 10 12 13⟩,	⟨5 9 12 14 15⟩,	⟨5 9 12 14 16⟩,
⟨5 10 11 13 15⟩,	⟨5 10 12 13 15⟩,	⟨6 7 8 11 14⟩,	⟨6 7 9 13 15⟩,	⟨6 8 9 11 14⟩,
⟨6 8 9 12 14⟩,	⟨6 9 11 14 15⟩,	⟨6 9 12 13 15⟩,	⟨6 9 12 14 15⟩,	⟨6 10 12 13 15⟩,
⟨6 10 12 13 16⟩,	⟨6 10 12 14 15⟩,	⟨7 8 9 10 15⟩,	⟨7 8 9 10 16⟩,	⟨7 8 9 13 15⟩,
⟨7 8 10 11 14⟩,	⟨7 8 10 11 16⟩,	⟨7 8 10 14 15⟩,	⟨7 8 13 14 15⟩,	⟨7 10 11 12 14⟩,
⟨7 10 12 14 15⟩,	⟨8 9 10 12 16⟩,	⟨8 9 12 14 16⟩,	⟨8 10 11 13 16⟩,	⟨8 10 12 13 16⟩,
⟨8 12 13 14 16⟩				

f-vector:	$(16, 118, 422, 515, 206)$
# singularities:	13
Simplicial homology:	$(\mathbb{Z}, 0, \mathbb{Z}^9 \times \mathbb{Z}_2^2, 0, \mathbb{Z})$
Euler characteristic:	11

Table B.5: 4th blowup

$C^4_{16} =$ ⟨1 2 3 5 9⟩, ⟨1 2 3 5 10⟩, ⟨1 2 3 6 10⟩, ⟨1 2 3 6 15⟩, ⟨1 2 3 9 14⟩,
⟨1 2 3 14 15⟩, ⟨1 2 4 5 7⟩, ⟨1 2 4 5 10⟩, ⟨1 2 4 7 11⟩, ⟨1 2 4 8 10⟩,
⟨1 2 4 8 16⟩, ⟨1 2 4 11 16⟩, ⟨1 2 5 7 13⟩, ⟨1 2 5 9 13⟩, ⟨1 2 6 8 10⟩,
⟨1 2 6 8 12⟩, ⟨1 2 6 12 15⟩, ⟨1 2 7 9 11⟩, ⟨1 2 7 9 13⟩, ⟨1 2 8 12 15⟩,
⟨1 2 8 15 16⟩, ⟨1 2 9 11 14⟩, ⟨1 2 11 14 16⟩, ⟨1 2 14 15 16⟩, ⟨1 3 4 7 11⟩,
⟨1 3 4 7 13⟩, ⟨1 3 4 9 12⟩, ⟨1 3 4 9 16⟩, ⟨1 3 4 11 16⟩, ⟨1 3 4 12 13⟩,
⟨1 3 5 9 16⟩, ⟨1 3 5 10 11⟩, ⟨1 3 5 11 16⟩, ⟨1 3 6 7 8⟩, ⟨1 3 6 7 10⟩,
⟨1 3 6 8 14⟩, ⟨1 3 6 14 15⟩, ⟨1 3 7 8 13⟩, ⟨1 3 7 10 11⟩, ⟨1 3 8 12 13⟩,
⟨1 3 8 12 14⟩, ⟨1 3 9 12 14⟩, ⟨1 4 5 7 13⟩, ⟨1 4 5 10 12⟩, ⟨1 4 5 12 13⟩,
⟨1 4 8 9 10⟩, ⟨1 4 8 9 15⟩, ⟨1 4 8 15 16⟩, ⟨1 4 9 10 12⟩, ⟨1 4 9 15 16⟩,
⟨1 5 6 11 12⟩, ⟨1 5 6 11 16⟩, ⟨1 5 6 12 15⟩, ⟨1 5 6 15 16⟩, ⟨1 5 9 13 16⟩,
⟨1 5 10 11 12⟩, ⟨1 5 12 13 15⟩, ⟨1 5 13 15 16⟩, ⟨1 6 7 8 10⟩, ⟨1 6 8 12 14⟩,
⟨1 6 11 12 14⟩, ⟨1 6 11 14 16⟩, ⟨1 6 14 15 16⟩, ⟨1 7 8 9 10⟩, ⟨1 7 8 9 15⟩,
⟨1 7 8 12 13⟩, ⟨1 7 8 12 15⟩, ⟨1 7 9 10 11⟩, ⟨1 7 9 13 15⟩, ⟨1 7 12 13 15⟩,
⟨1 9 10 11 12⟩, ⟨1 9 11 12 14⟩, ⟨1 9 13 15 16⟩, ⟨2 3 4 6 13⟩, ⟨2 3 4 6 15⟩,
⟨2 3 4 13 14⟩, ⟨2 3 4 14 15⟩, ⟨2 3 5 7 13⟩, ⟨2 3 5 7 16⟩, ⟨2 3 5 9 13⟩,
⟨2 3 5 10 16⟩, ⟨2 3 6 10 13⟩, ⟨2 3 7 8 13⟩, ⟨2 3 7 8 16⟩, ⟨2 3 8 13 16⟩,
⟨2 3 9 13 14⟩, ⟨2 3 10 13 16⟩, ⟨2 4 5 6 7⟩, ⟨2 4 5 6 15⟩, ⟨2 4 5 10 12⟩,
⟨2 4 5 12 15⟩, ⟨2 4 6 7 11⟩, ⟨2 4 6 11 13⟩, ⟨2 4 8 10 14⟩, ⟨2 4 8 13 14⟩,
⟨2 4 8 13 16⟩, ⟨2 4 10 12 14⟩, ⟨2 4 11 13 16⟩, ⟨2 4 12 14 15⟩, ⟨2 5 6 7 12⟩,
⟨2 5 6 12 15⟩, ⟨2 5 7 10 12⟩, ⟨2 5 7 10 16⟩, ⟨2 6 7 9 11⟩, ⟨2 6 7 9 12⟩,
⟨2 6 8 9 11⟩, ⟨2 6 8 9 12⟩, ⟨2 6 8 10 11⟩, ⟨2 6 10 11 13⟩, ⟨2 7 8 12 13⟩,
⟨2 7 8 12 15⟩, ⟨2 7 8 15 16⟩, ⟨2 7 9 12 13⟩, ⟨2 7 10 12 14⟩, ⟨2 7 10 14 16⟩,
⟨2 7 12 14 15⟩, ⟨2 7 14 15 16⟩, ⟨2 8 9 11 14⟩, ⟨2 8 9 12 13⟩, ⟨2 8 9 13 14⟩,
⟨2 8 10 11 14⟩, ⟨2 10 11 13 16⟩, ⟨2 10 11 14 16⟩, ⟨3 4 6 9 12⟩, ⟨3 4 6 9 15⟩,
⟨3 4 6 12 13⟩, ⟨3 4 7 11 14⟩, ⟨3 4 7 13 14⟩, ⟨3 4 9 11 15⟩, ⟨3 4 9 11 16⟩,
⟨3 4 11 14 15⟩, ⟨3 5 6 11 15⟩, ⟨3 5 6 11 16⟩, ⟨3 5 6 15 16⟩, ⟨3 5 7 9 14⟩,
⟨3 5 7 9 16⟩, ⟨3 5 7 13 14⟩, ⟨3 5 8 10 11⟩, ⟨3 5 8 10 16⟩, ⟨3 5 8 11 15⟩,
⟨3 5 8 15 16⟩, ⟨3 5 9 13 14⟩, ⟨3 6 7 8 16⟩, ⟨3 6 7 9 12⟩, ⟨3 6 7 9 16⟩,
⟨3 6 7 10 12⟩, ⟨3 6 8 14 16⟩, ⟨3 6 9 11 15⟩, ⟨3 6 9 11 16⟩, ⟨3 6 10 12 13⟩,
⟨3 6 14 15 16⟩, ⟨3 7 9 12 14⟩, ⟨3 7 10 11 14⟩, ⟨3 7 10 12 14⟩, ⟨3 8 10 11 14⟩,
⟨3 8 10 12 13⟩, ⟨3 8 10 12 14⟩, ⟨3 8 10 13 16⟩, ⟨3 8 11 14 15⟩, ⟨3 8 14 15 16⟩,
⟨4 5 6 7 14⟩, ⟨4 5 6 8 14⟩, ⟨4 5 6 8 15⟩, ⟨4 5 7 13 14⟩, ⟨4 5 8 13 14⟩,
⟨4 5 8 13 15⟩, ⟨4 5 12 13 15⟩, ⟨4 6 7 11 14⟩, ⟨4 6 8 9 12⟩, ⟨4 6 8 9 15⟩,
⟨4 6 8 12 14⟩, ⟨4 6 11 12 13⟩, ⟨4 6 11 12 14⟩, ⟨4 8 9 10 12⟩, ⟨4 8 10 12 14⟩,
⟨4 8 13 15 16⟩, ⟨4 9 11 13 15⟩, ⟨4 9 11 13 16⟩, ⟨4 9 13 15 16⟩, ⟨4 11 12 13 15⟩,
⟨4 11 12 14 15⟩, ⟨5 6 7 8 10⟩, ⟨5 6 7 8 14⟩, ⟨5 6 7 10 12⟩, ⟨5 6 8 10 11⟩,

Continuation on next page –

Table B.5 – Continuation from last page

⟨5 6 8 11 15⟩,	⟨5 6 10 11 12⟩,	⟨5 7 8 9 10⟩,	⟨5 7 8 9 14⟩,	⟨5 7 9 10 16⟩,
⟨5 8 9 10 16⟩,	⟨5 8 9 13 14⟩,	⟨5 8 9 13 16⟩,	⟨5 8 13 15 16⟩,	⟨6 7 8 14 16⟩,
⟨6 7 9 11 16⟩,	⟨6 7 11 14 16⟩,	⟨6 8 9 11 15⟩,	⟨6 10 11 12 13⟩,	⟨7 8 9 14 15⟩,
⟨7 8 14 15 16⟩,	⟨7 9 10 11 16⟩,	⟨7 9 12 13 15⟩,	⟨7 9 12 14 15⟩,	⟨7 10 11 14 16⟩,
⟨8 9 10 12 13⟩,	⟨8 9 10 13 16⟩,	⟨8 9 11 14 15⟩,	⟨9 10 11 12 13⟩,	⟨9 10 11 13 16⟩,
⟨9 11 12 13 15⟩,	⟨9 11 12 14 15⟩			

f-vector:	$(16, 118, 432, 530, 212)$
# singularities:	12
Simplicial homology:	$(\mathbb{Z}, 0, \mathbb{Z}^{10} \times \mathbb{Z}_2^2, 0, \mathbb{Z})$
Euler characteristic:	12

Table B.6: 5th blowup

$C_{16}^5 =$	⟨1 2 3 5 10⟩,	⟨1 2 3 5 11⟩,	⟨1 2 3 9 14⟩,	⟨1 2 3 9 15⟩,	⟨1 2 3 10 15⟩,
	⟨1 2 3 11 14⟩,	⟨1 2 4 6 11⟩,	⟨1 2 4 6 15⟩,	⟨1 2 4 9 14⟩,	⟨1 2 4 9 15⟩,
	⟨1 2 4 11 14⟩,	⟨1 2 5 6 11⟩,	⟨1 2 5 6 15⟩,	⟨1 2 5 10 15⟩,	⟨1 3 4 6 8⟩,
	⟨1 3 4 6 12⟩,	⟨1 3 4 8 16⟩,	⟨1 3 4 12 16⟩,	⟨1 3 5 7 10⟩,	⟨1 3 5 7 12⟩,
	⟨1 3 5 11 12⟩,	⟨1 3 6 7 12⟩,	⟨1 3 6 7 13⟩,	⟨1 3 6 8 13⟩,	⟨1 3 7 10 13⟩,
	⟨1 3 8 9 15⟩,	⟨1 3 8 9 16⟩,	⟨1 3 8 13 15⟩,	⟨1 3 9 14 16⟩,	⟨1 3 10 13 15⟩,
	⟨1 3 11 12 16⟩,	⟨1 3 11 14 16⟩,	⟨1 4 6 7 12⟩,	⟨1 4 6 7 13⟩,	⟨1 4 6 8 10⟩,
	⟨1 4 6 10 15⟩,	⟨1 4 6 11 13⟩,	⟨1 4 7 11 13⟩,	⟨1 4 7 11 14⟩,	⟨1 4 7 12 14⟩,
	⟨1 4 8 10 16⟩,	⟨1 4 9 14 15⟩,	⟨1 4 10 15 16⟩,	⟨1 4 12 14 15⟩,	⟨1 4 12 15 16⟩,
	⟨1 5 6 8 11⟩,	⟨1 5 6 8 14⟩,	⟨1 5 6 10 14⟩,	⟨1 5 6 10 15⟩,	⟨1 5 7 9 10⟩,
	⟨1 5 7 9 12⟩,	⟨1 5 8 9 12⟩,	⟨1 5 8 9 14⟩,	⟨1 5 8 11 12⟩,	⟨1 5 9 10 14⟩,
	⟨1 6 8 10 14⟩,	⟨1 6 8 11 13⟩,	⟨1 7 9 10 14⟩,	⟨1 7 9 12 14⟩,	⟨1 7 10 13 15⟩,
	⟨1 7 10 14 16⟩,	⟨1 7 10 15 16⟩,	⟨1 7 11 13 15⟩,	⟨1 7 11 14 16⟩,	⟨1 7 11 15 16⟩,
	⟨1 8 9 12 15⟩,	⟨1 8 9 14 16⟩,	⟨1 8 10 14 16⟩,	⟨1 8 11 12 15⟩,	⟨1 8 11 13 15⟩,
	⟨1 9 12 14 15⟩,	⟨1 11 12 15 16⟩,	⟨2 3 4 7 9⟩,	⟨2 3 4 7 14⟩,	⟨2 3 4 9 14⟩,
	⟨2 3 5 8 10⟩,	⟨2 3 5 8 13⟩,	⟨2 3 5 11 13⟩,	⟨2 3 6 7 12⟩,	⟨2 3 6 7 14⟩,
	⟨2 3 6 8 10⟩,	⟨2 3 6 8 13⟩,	⟨2 3 6 10 12⟩,	⟨2 3 6 13 14⟩,	⟨2 3 7 9 15⟩,
	⟨2 3 7 12 15⟩,	⟨2 3 10 12 15⟩,	⟨2 3 11 13 14⟩,	⟨2 4 6 9 10⟩,	⟨2 4 6 9 15⟩,
	⟨2 4 6 10 12⟩,	⟨2 4 6 11 12⟩,	⟨2 4 7 9 16⟩,	⟨2 4 7 14 16⟩,	⟨2 4 8 11 12⟩,
	⟨2 4 8 11 16⟩,	⟨2 4 8 12 16⟩,	⟨2 4 9 10 12⟩,	⟨2 4 9 12 16⟩,	⟨2 4 11 14 16⟩,
	⟨2 5 6 7 11⟩,	⟨2 5 6 7 14⟩,	⟨2 5 6 14 15⟩,	⟨2 5 7 9 13⟩,	⟨2 5 7 9 16⟩,
	⟨2 5 7 11 13⟩,	⟨2 5 7 14 16⟩,	⟨2 5 8 9 12⟩,	⟨2 5 8 9 13⟩,	⟨2 5 8 10 16⟩,

Continuation on next page –

Table B.6 – Continuation from last page

⟨2 5 8 12 16⟩, ⟨2 5 9 12 16⟩, ⟨2 5 10 15 16⟩, ⟨2 5 14 15 16⟩, ⟨2 6 7 11 12⟩,
⟨2 6 8 9 10⟩, ⟨2 6 8 9 13⟩, ⟨2 6 9 13 15⟩, ⟨2 6 13 14 15⟩, ⟨2 7 9 13 15⟩,
⟨2 7 11 12 15⟩, ⟨2 7 11 13 15⟩, ⟨2 8 9 10 12⟩, ⟨2 8 10 12 15⟩, ⟨2 8 10 15 16⟩,
⟨2 8 11 12 15⟩, ⟨2 8 11 14 15⟩, ⟨2 8 11 14 16⟩, ⟨2 8 14 15 16⟩, ⟨2 11 13 14 15⟩,
⟨3 4 5 7 8⟩, ⟨3 4 5 7 10⟩, ⟨3 4 5 8 10⟩, ⟨3 4 6 8 10⟩, ⟨3 4 6 10 12⟩,
⟨3 4 7 8 16⟩, ⟨3 4 7 9 16⟩, ⟨3 4 7 10 13⟩, ⟨3 4 7 13 14⟩, ⟨3 4 9 10 12⟩,
⟨3 4 9 10 13⟩, ⟨3 4 9 12 16⟩, ⟨3 4 9 13 14⟩, ⟨3 5 7 8 15⟩, ⟨3 5 7 12 15⟩,
⟨3 5 8 13 15⟩, ⟨3 5 11 12 13⟩, ⟨3 5 12 13 15⟩, ⟨3 6 7 13 14⟩, ⟨3 7 8 9 15⟩,
⟨3 7 8 9 16⟩, ⟨3 9 10 12 13⟩, ⟨3 9 11 12 13⟩, ⟨3 9 11 12 16⟩, ⟨3 9 11 13 14⟩,
⟨3 9 11 14 16⟩, ⟨3 10 12 13 15⟩, ⟨4 5 6 10 14⟩, ⟨4 5 6 10 15⟩, ⟨4 5 6 14 15⟩,
⟨4 5 7 8 11⟩, ⟨4 5 7 10 13⟩, ⟨4 5 7 11 13⟩, ⟨4 5 8 10 16⟩, ⟨4 5 8 11 12⟩,
⟨4 5 8 12 16⟩, ⟨4 5 9 10 13⟩, ⟨4 5 9 10 14⟩, ⟨4 5 9 13 14⟩, ⟨4 5 10 15 16⟩,
⟨4 5 11 12 13⟩, ⟨4 5 12 13 15⟩, ⟨4 5 12 15 16⟩, ⟨4 5 13 14 15⟩, ⟨4 6 7 12 13⟩,
⟨4 6 9 10 14⟩, ⟨4 6 9 14 15⟩, ⟨4 6 11 12 13⟩, ⟨4 7 8 11 16⟩, ⟨4 7 11 14 16⟩,
⟨4 7 12 13 14⟩, ⟨4 12 13 14 15⟩, ⟨5 6 7 8 11⟩, ⟨5 6 7 8 14⟩, ⟨5 7 8 14 15⟩,
⟨5 7 9 10 13⟩, ⟨5 7 9 12 16⟩, ⟨5 7 12 15 16⟩, ⟨5 7 14 15 16⟩, ⟨5 8 9 13 14⟩,
⟨5 8 13 14 15⟩, ⟨6 7 8 9 10⟩, ⟨6 7 8 9 11⟩, ⟨6 7 8 10 14⟩, ⟨6 7 9 10 14⟩,
⟨6 7 9 11 12⟩, ⟨6 7 9 12 14⟩, ⟨6 7 12 13 14⟩, ⟨6 8 9 11 13⟩, ⟨6 9 11 12 13⟩,
⟨6 9 12 13 15⟩, ⟨6 9 12 14 15⟩, ⟨6 12 13 14 15⟩, ⟨7 8 9 10 15⟩, ⟨7 8 9 11 16⟩,
⟨7 8 10 14 15⟩, ⟨7 9 10 13 15⟩, ⟨7 9 11 12 16⟩, ⟨7 10 14 15 16⟩, ⟨7 11 12 15 16⟩,
⟨8 9 10 12 15⟩, ⟨8 9 11 13 14⟩, ⟨8 9 11 14 16⟩, ⟨8 10 14 15 16⟩, ⟨8 11 13 14 15⟩,
⟨9 10 12 13 15⟩

f-vector:	(16, 117, 438, 540, 216)
# singularities:	11
Simplicial homology:	$(\mathbb{Z}, 0, \mathbb{Z}^{11} \times \mathbb{Z}_2, 0, \mathbb{Z})$
Euler characteristic:	13

Table B.7: 6th blowup : First complex with free second homology group.

C^6_{16} = ⟨1 2 3 5 9⟩, ⟨1 2 3 5 12⟩, ⟨1 2 3 6 8⟩, ⟨1 2 3 6 14⟩, ⟨1 2 3 8 9⟩,
⟨1 2 3 12 14⟩, ⟨1 2 4 5 9⟩, ⟨1 2 4 5 12⟩, ⟨1 2 4 9 15⟩, ⟨1 2 4 11 12⟩,
⟨1 2 4 11 15⟩, ⟨1 2 6 7 8⟩, ⟨1 2 6 7 14⟩, ⟨1 2 7 8 15⟩, ⟨1 2 7 13 14⟩,
⟨1 2 7 13 16⟩, ⟨1 2 7 15 16⟩, ⟨1 2 8 9 15⟩, ⟨1 2 11 12 14⟩, ⟨1 2 11 13 14⟩,
⟨1 2 11 13 15⟩, ⟨1 2 13 15 16⟩, ⟨1 3 5 6 10⟩, ⟨1 3 5 6 11⟩, ⟨1 3 5 7 9⟩,
⟨1 3 5 7 11⟩, ⟨1 3 5 10 12⟩, ⟨1 3 6 8 11⟩, ⟨1 3 6 10 14⟩, ⟨1 3 7 9 16⟩,

Continuation on next page –

Table B.7 – Continuation from last page

⟨1 3 7 11 15⟩,	⟨1 3 7 15 16⟩,	⟨1 3 8 9 16⟩,	⟨1 3 8 11 16⟩,	⟨1 3 10 12 14⟩,
⟨1 3 11 15 16⟩,	⟨1 4 5 7 12⟩,	⟨1 4 5 7 13⟩,	⟨1 4 5 9 16⟩,	⟨1 4 5 13 16⟩,
⟨1 4 6 7 12⟩,	⟨1 4 6 7 13⟩,	⟨1 4 6 10 11⟩,	⟨1 4 6 10 13⟩,	⟨1 4 6 11 12⟩,
⟨1 4 8 9 10⟩,	⟨1 4 8 9 16⟩,	⟨1 4 8 10 16⟩,	⟨1 4 9 10 15⟩,	⟨1 4 10 11 15⟩,
⟨1 4 10 13 16⟩,	⟨1 5 6 10 11⟩,	⟨1 5 7 9 13⟩,	⟨1 5 7 10 11⟩,	⟨1 5 7 10 12⟩,
⟨1 5 9 13 16⟩,	⟨1 6 7 8 12⟩,	⟨1 6 7 13 14⟩,	⟨1 6 8 11 12⟩,	⟨1 6 10 13 14⟩,
⟨1 7 8 10 12⟩,	⟨1 7 8 10 15⟩,	⟨1 7 9 13 16⟩,	⟨1 7 10 11 15⟩,	⟨1 8 9 10 15⟩,
⟨1 8 10 12 16⟩,	⟨1 8 11 12 14⟩,	⟨1 8 11 14 16⟩,	⟨1 8 12 14 16⟩,	⟨1 10 12 14 16⟩,
⟨1 10 13 14 16⟩,	⟨1 11 13 14 15⟩,	⟨1 11 14 15 16⟩,	⟨1 13 14 15 16⟩,	⟨2 3 4 6 13⟩,
⟨2 3 4 6 15⟩,	⟨2 3 4 13 15⟩,	⟨2 3 5 9 13⟩,	⟨2 3 5 10 12⟩,	⟨2 3 5 10 16⟩,
⟨2 3 5 13 16⟩,	⟨2 3 6 8 13⟩,	⟨2 3 6 12 14⟩,	⟨2 3 6 12 15⟩,	⟨2 3 8 9 13⟩,
⟨2 3 10 12 15⟩,	⟨2 3 10 13 15⟩,	⟨2 3 10 13 16⟩,	⟨2 4 5 6 8⟩,	⟨2 4 5 6 16⟩,
⟨2 4 5 8 12⟩,	⟨2 4 5 9 16⟩,	⟨2 4 6 8 13⟩,	⟨2 4 6 9 14⟩,	⟨2 4 6 9 15⟩,
⟨2 4 6 14 16⟩,	⟨2 4 7 9 14⟩,	⟨2 4 7 9 16⟩,	⟨2 4 7 14 16⟩,	⟨2 4 8 12 13⟩,
⟨2 4 11 12 13⟩,	⟨2 4 11 13 15⟩,	⟨2 5 6 7 8⟩,	⟨2 5 6 7 16⟩,	⟨2 5 7 8 15⟩,
⟨2 5 7 15 16⟩,	⟨2 5 8 12 15⟩,	⟨2 5 9 13 16⟩,	⟨2 5 10 12 15⟩,	⟨2 5 10 15 16⟩,
⟨2 6 7 14 16⟩,	⟨2 6 9 12 14⟩,	⟨2 6 9 12 15⟩,	⟨2 7 9 13 14⟩,	⟨2 7 9 13 16⟩,
⟨2 8 9 12 13⟩,	⟨2 8 9 12 15⟩,	⟨2 9 12 13 14⟩,	⟨2 10 13 15 16⟩,	⟨2 11 12 13 14⟩,
⟨3 4 6 7 12⟩,	⟨3 4 6 7 13⟩,	⟨3 4 6 12 15⟩,	⟨3 4 7 9 14⟩,	⟨3 4 7 9 16⟩,
⟨3 4 7 12 14⟩,	⟨3 4 7 13 15⟩,	⟨3 4 7 15 16⟩,	⟨3 4 8 9 10⟩,	⟨3 4 8 9 16⟩,
⟨3 4 8 10 14⟩,	⟨3 4 8 12 14⟩,	⟨3 4 8 12 16⟩,	⟨3 4 9 10 14⟩,	⟨3 4 12 15 16⟩,
⟨3 5 6 7 9⟩,	⟨3 5 6 7 11⟩,	⟨3 5 6 9 10⟩,	⟨3 5 8 9 10⟩,	⟨3 5 8 9 13⟩,
⟨3 5 8 10 14⟩,	⟨3 5 8 13 16⟩,	⟨3 5 8 14 16⟩,	⟨3 5 10 14 16⟩,	⟨3 6 7 9 12⟩,
⟨3 6 7 11 13⟩,	⟨3 6 8 11 13⟩,	⟨3 6 9 10 14⟩,	⟨3 6 9 12 14⟩,	⟨3 7 9 12 14⟩,
⟨3 7 10 11 13⟩,	⟨3 7 10 11 15⟩,	⟨3 7 10 13 15⟩,	⟨3 8 11 13 16⟩,	⟨3 8 12 14 16⟩,
⟨3 10 11 12 15⟩,	⟨3 10 11 12 16⟩,	⟨3 10 11 13 16⟩,	⟨3 10 12 14 16⟩,	⟨3 11 12 15 16⟩,
⟨4 5 6 8 10⟩,	⟨4 5 6 10 11⟩,	⟨4 5 6 11 16⟩,	⟨4 5 7 12 14⟩,	⟨4 5 7 13 15⟩,
⟨4 5 7 14 15⟩,	⟨4 5 8 10 14⟩,	⟨4 5 8 12 14⟩,	⟨4 5 10 11 15⟩,	⟨4 5 10 14 15⟩,
⟨4 5 11 13 15⟩,	⟨4 5 11 13 16⟩,	⟨4 6 8 10 13⟩,	⟨4 6 9 14 15⟩,	⟨4 6 11 12 16⟩,
⟨4 6 12 15 16⟩,	⟨4 6 14 15 16⟩,	⟨4 7 14 15 16⟩,	⟨4 8 10 12 13⟩,	⟨4 8 10 12 16⟩,
⟨4 9 10 14 15⟩,	⟨4 10 12 13 16⟩,	⟨4 11 12 13 16⟩,	⟨5 6 7 8 15⟩,	⟨5 6 7 9 15⟩,
⟨5 6 7 11 16⟩,	⟨5 6 8 10 13⟩,	⟨5 6 8 13 15⟩,	⟨5 6 9 10 13⟩,	⟨5 6 9 13 15⟩,
⟨5 7 9 13 15⟩,	⟨5 7 10 11 12⟩,	⟨5 7 11 12 14⟩,	⟨5 7 11 14 16⟩,	⟨5 7 14 15 16⟩,
⟨5 8 9 10 13⟩,	⟨5 8 11 12 14⟩,	⟨5 8 11 12 15⟩,	⟨5 8 11 13 15⟩,	⟨5 8 11 13 16⟩,
⟨5 8 11 14 16⟩,	⟨5 10 11 12 15⟩,	⟨5 10 14 15 16⟩,	⟨6 7 8 12 15⟩,	⟨6 7 9 12 15⟩,
⟨6 7 11 13 14⟩,	⟨6 7 11 14 16⟩,	⟨6 8 11 12 15⟩,	⟨6 8 11 13 15⟩,	⟨6 9 10 13 15⟩,
⟨6 9 10 14 15⟩,	⟨6 10 13 14 15⟩,	⟨6 11 12 15 16⟩,	⟨6 11 13 14 15⟩,	⟨6 11 14 15 16⟩,
⟨7 8 9 10 13⟩,	⟨7 8 9 10 15⟩,	⟨7 8 9 12 13⟩,	⟨7 8 9 12 15⟩,	⟨7 8 10 12 13⟩,

Continuation on next page –

Table B.7 – Continuation from last page

⟨7 9 10 13 15⟩, ⟨7 9 12 13 14⟩, ⟨7 10 11 12 13⟩, ⟨7 11 12 13 14⟩, ⟨10 11 12 13 16⟩,
⟨10 13 14 15 16⟩

f-vector:	$(16, 119, 456, 565, 226)$
# singularities:	10
Simplicial homology:	$(\mathbb{Z}, 0, \mathbb{Z}^{12}, 0, \mathbb{Z})$
Euler characteristic:	14

Table B.8: 7th blowup

C^7_{16} = ⟨1 2 3 4 13⟩, ⟨1 2 3 4 14⟩, ⟨1 2 3 7 14⟩, ⟨1 2 3 7 16⟩, ⟨1 2 3 8 9⟩,
⟨1 2 3 8 10⟩, ⟨1 2 3 9 13⟩, ⟨1 2 3 10 16⟩, ⟨1 2 4 6 11⟩, ⟨1 2 4 6 14⟩,
⟨1 2 4 9 11⟩, ⟨1 2 4 9 13⟩, ⟨1 2 5 6 12⟩, ⟨1 2 5 6 15⟩, ⟨1 2 5 12 15⟩,
⟨1 2 6 8 9⟩, ⟨1 2 6 8 12⟩, ⟨1 2 6 9 11⟩, ⟨1 2 6 14 15⟩, ⟨1 2 7 14 16⟩,
⟨1 2 8 10 12⟩, ⟨1 2 10 12 15⟩, ⟨1 2 10 14 15⟩, ⟨1 2 10 14 16⟩, ⟨1 3 4 5 7⟩,
⟨1 3 4 5 15⟩, ⟨1 3 4 7 14⟩, ⟨1 3 4 13 15⟩, ⟨1 3 5 7 11⟩, ⟨1 3 5 11 15⟩,
⟨1 3 7 11 15⟩, ⟨1 3 7 13 15⟩, ⟨1 3 7 13 16⟩, ⟨1 3 8 9 13⟩, ⟨1 3 8 10 16⟩,
⟨1 3 8 13 16⟩, ⟨1 4 5 7 14⟩, ⟨1 4 5 12 14⟩, ⟨1 4 5 12 15⟩, ⟨1 4 6 11 14⟩,
⟨1 4 9 10 15⟩, ⟨1 4 9 10 16⟩, ⟨1 4 9 11 16⟩, ⟨1 4 9 13 15⟩, ⟨1 4 10 11 12⟩,
⟨1 4 10 11 16⟩, ⟨1 4 10 12 15⟩, ⟨1 4 11 12 14⟩, ⟨1 5 6 9 13⟩, ⟨1 5 6 9 16⟩,
⟨1 5 6 12 13⟩, ⟨1 5 6 15 16⟩, ⟨1 5 7 8 11⟩, ⟨1 5 7 8 16⟩, ⟨1 5 7 9 14⟩,
⟨1 5 7 9 16⟩, ⟨1 5 8 10 11⟩, ⟨1 5 8 10 16⟩, ⟨1 5 9 13 14⟩, ⟨1 5 10 11 16⟩,
⟨1 5 11 15 16⟩, ⟨1 5 12 13 14⟩, ⟨1 6 8 9 13⟩, ⟨1 6 8 12 13⟩, ⟨1 6 9 11 16⟩,
⟨1 6 11 14 15⟩, ⟨1 6 11 15 16⟩, ⟨1 7 8 11 15⟩, ⟨1 7 8 13 15⟩, ⟨1 7 8 13 16⟩,
⟨1 7 9 14 16⟩, ⟨1 8 10 11 12⟩, ⟨1 8 11 12 14⟩, ⟨1 8 11 14 15⟩, ⟨1 8 12 13 14⟩,
⟨1 8 13 14 15⟩, ⟨1 9 10 14 15⟩, ⟨1 9 10 14 16⟩, ⟨1 9 13 14 15⟩, ⟨2 3 4 6 12⟩,
⟨2 3 4 6 14⟩, ⟨2 3 4 12 13⟩, ⟨2 3 5 6 12⟩, ⟨2 3 5 6 14⟩, ⟨2 3 5 11 12⟩,
⟨2 3 5 11 14⟩, ⟨2 3 7 8 10⟩, ⟨2 3 7 8 14⟩, ⟨2 3 7 10 15⟩, ⟨2 3 7 15 16⟩,
⟨2 3 8 9 14⟩, ⟨2 3 9 13 14⟩, ⟨2 3 10 15 16⟩, ⟨2 3 11 12 13⟩, ⟨2 3 11 13 14⟩,
⟨2 4 6 8 11⟩, ⟨2 4 6 8 12⟩, ⟨2 4 7 8 11⟩, ⟨2 4 7 8 12⟩, ⟨2 4 7 11 16⟩,
⟨2 4 7 12 13⟩, ⟨2 4 7 13 15⟩, ⟨2 4 7 15 16⟩, ⟨2 4 9 11 16⟩, ⟨2 4 9 13 15⟩,
⟨2 4 9 15 16⟩, ⟨2 5 6 14 16⟩, ⟨2 5 6 15 16⟩, ⟨2 5 11 12 15⟩, ⟨2 5 11 14 16⟩,
⟨2 5 11 15 16⟩, ⟨2 6 8 9 11⟩, ⟨2 6 14 15 16⟩, ⟨2 7 8 10 12⟩, ⟨2 7 8 11 14⟩,
⟨2 7 10 12 13⟩, ⟨2 7 10 13 15⟩, ⟨2 7 11 14 16⟩, ⟨2 8 9 11 14⟩, ⟨2 9 11 12 14⟩,
⟨2 9 11 12 15⟩, ⟨2 9 11 15 16⟩, ⟨2 9 12 13 14⟩, ⟨2 9 12 13 15⟩, ⟨2 10 12 13 15⟩,
⟨2 10 14 15 16⟩, ⟨2 11 12 13 14⟩, ⟨3 4 5 7 11⟩, ⟨3 4 5 8 13⟩, ⟨3 4 5 8 15⟩,

Continuation on next page –

Table B.8 – Continuation from last page

⟨3 4 5 11 13⟩, ⟨3 4 6 7 10⟩, ⟨3 4 6 7 14⟩, ⟨3 4 6 10 12⟩, ⟨3 4 7 10 11⟩,
⟨3 4 8 13 16⟩, ⟨3 4 8 15 16⟩, ⟨3 4 10 11 12⟩, ⟨3 4 11 12 13⟩, ⟨3 4 13 15 16⟩,
⟨3 5 6 10 12⟩, ⟨3 5 6 10 14⟩, ⟨3 5 8 9 12⟩, ⟨3 5 8 9 13⟩, ⟨3 5 8 12 15⟩,
⟨3 5 9 10 12⟩, ⟨3 5 9 10 14⟩, ⟨3 5 9 13 14⟩, ⟨3 5 11 12 15⟩, ⟨3 5 11 13 14⟩,
⟨3 6 7 8 10⟩, ⟨3 6 7 8 14⟩, ⟨3 6 8 10 14⟩, ⟨3 7 10 11 15⟩, ⟨3 7 13 15 16⟩,
⟨3 8 9 12 15⟩, ⟨3 8 9 14 15⟩, ⟨3 8 10 14 15⟩, ⟨3 8 10 15 16⟩, ⟨3 9 10 11 12⟩,
⟨3 9 10 11 15⟩, ⟨3 9 10 14 15⟩, ⟨3 9 11 12 15⟩, ⟨4 5 6 8 11⟩, ⟨4 5 6 8 16⟩,
⟨4 5 6 11 13⟩, ⟨4 5 6 13 16⟩, ⟨4 5 7 8 11⟩, ⟨4 5 7 8 12⟩, ⟨4 5 7 12 14⟩,
⟨4 5 8 12 15⟩, ⟨4 5 8 13 16⟩, ⟨4 6 7 10 16⟩, ⟨4 6 7 13 14⟩, ⟨4 6 7 13 16⟩,
⟨4 6 8 12 15⟩, ⟨4 6 8 15 16⟩, ⟨4 6 9 10 15⟩, ⟨4 6 9 10 16⟩, ⟨4 6 9 15 16⟩,
⟨4 6 10 12 15⟩, ⟨4 6 11 13 14⟩, ⟨4 7 10 11 16⟩, ⟨4 7 12 13 14⟩, ⟨4 7 13 15 16⟩,
⟨4 11 12 13 14⟩, ⟨5 6 8 10 11⟩, ⟨5 6 8 10 14⟩, ⟨5 6 8 14 16⟩, ⟨5 6 9 13 16⟩,
⟨5 6 10 11 13⟩, ⟨5 6 10 12 13⟩, ⟨5 7 8 9 12⟩, ⟨5 7 8 9 16⟩, ⟨5 7 9 10 12⟩,
⟨5 7 9 10 14⟩, ⟨5 7 10 12 14⟩, ⟨5 8 9 13 16⟩, ⟨5 8 10 14 16⟩, ⟨5 10 11 13 14⟩,
⟨5 10 11 14 16⟩, ⟨5 10 12 13 14⟩, ⟨6 7 8 9 10⟩, ⟨6 7 8 9 13⟩, ⟨6 7 8 13 15⟩,
⟨6 7 8 14 15⟩, ⟨6 7 9 10 16⟩, ⟨6 7 9 13 16⟩, ⟨6 7 11 13 14⟩, ⟨6 7 11 13 15⟩,
⟨6 7 11 14 15⟩, ⟨6 8 9 10 11⟩, ⟨6 8 12 13 15⟩, ⟨6 8 14 15 16⟩, ⟨6 9 10 11 15⟩,
⟨6 9 11 15 16⟩, ⟨6 10 11 13 15⟩, ⟨6 10 12 13 15⟩, ⟨7 8 9 10 12⟩, ⟨7 8 9 13 16⟩,
⟨7 8 11 14 15⟩, ⟨7 9 10 14 16⟩, ⟨7 10 11 13 14⟩, ⟨7 10 11 13 15⟩, ⟨7 10 11 14 16⟩,
⟨7 10 12 13 14⟩, ⟨8 9 10 11 12⟩, ⟨8 9 11 12 14⟩, ⟨8 9 12 14 15⟩, ⟨8 10 14 15 16⟩,
⟨8 12 13 14 15⟩, ⟨9 12 13 14 15⟩

f-vector:	$(16, 119, 466, 580, 232)$
# singularities:	9
Simplicial homology:	$(\mathbb{Z}, 0, \mathbb{Z}^{13}, 0, \mathbb{Z})$
Euler characteristic:	15

Table B.9: 8th blowup: For the first time a 16 vertex version could not be found.

C^8_{17} = ⟨1 2 3 4 9⟩, ⟨1 2 3 4 17⟩, ⟨1 2 3 6 7⟩, ⟨1 2 3 6 17⟩, ⟨1 2 3 7 9⟩,
⟨1 2 4 5 9⟩, ⟨1 2 4 5 10⟩, ⟨1 2 4 8 10⟩, ⟨1 2 4 8 17⟩, ⟨1 2 5 8 9⟩,
⟨1 2 5 8 11⟩, ⟨1 2 5 10 11⟩, ⟨1 2 6 7 17⟩, ⟨1 2 7 8 16⟩, ⟨1 2 7 8 17⟩,
⟨1 2 7 9 16⟩, ⟨1 2 8 9 16⟩, ⟨1 2 8 10 12⟩, ⟨1 2 8 11 12⟩, ⟨1 2 10 11 12⟩,
⟨1 3 4 7 9⟩, ⟨1 3 4 7 11⟩, ⟨1 3 4 11 17⟩, ⟨1 3 6 7 11⟩, ⟨1 3 6 11 13⟩,
⟨1 3 6 13 16⟩, ⟨1 3 6 16 17⟩, ⟨1 3 8 13 16⟩, ⟨1 3 8 13 17⟩, ⟨1 3 8 16 17⟩,
⟨1 3 10 11 13⟩, ⟨1 3 10 11 17⟩, ⟨1 3 10 13 17⟩, ⟨1 4 5 9 16⟩, ⟨1 4 5 10 15⟩,

Continuation on next page –

Table B.9 – Continuation from last page

⟨1 4 5 15 16⟩, ⟨1 4 6 7 9⟩, ⟨1 4 6 7 11⟩, ⟨1 4 6 9 11⟩, ⟨1 4 8 10 12⟩,
⟨1 4 8 11 12⟩, ⟨1 4 8 11 17⟩, ⟨1 4 9 11 16⟩, ⟨1 4 10 12 15⟩, ⟨1 4 11 12 16⟩,
⟨1 4 12 15 16⟩, ⟨1 5 6 13 15⟩, ⟨1 5 6 13 16⟩, ⟨1 5 6 15 16⟩, ⟨1 5 8 9 16⟩,
⟨1 5 8 11 17⟩, ⟨1 5 8 13 16⟩, ⟨1 5 8 13 17⟩, ⟨1 5 10 11 17⟩, ⟨1 5 10 13 15⟩,
⟨1 5 10 13 17⟩, ⟨1 6 7 9 10⟩, ⟨1 6 7 10 12⟩, ⟨1 6 7 12 15⟩, ⟨1 6 7 15 16⟩,
⟨1 6 7 16 17⟩, ⟨1 6 9 10 11⟩, ⟨1 6 10 11 13⟩, ⟨1 6 10 12 15⟩, ⟨1 6 10 13 15⟩,
⟨1 7 8 16 17⟩, ⟨1 7 9 10 12⟩, ⟨1 7 9 12 15⟩, ⟨1 7 9 15 16⟩, ⟨1 9 10 11 12⟩,
⟨1 9 11 12 16⟩, ⟨1 9 12 15 16⟩, ⟨2 3 4 9 12⟩, ⟨2 3 4 12 15⟩, ⟨2 3 4 15 17⟩,
⟨2 3 6 7 11⟩, ⟨2 3 6 10 15⟩, ⟨2 3 6 10 16⟩, ⟨2 3 6 11 12⟩, ⟨2 3 6 12 15⟩,
⟨2 3 6 16 17⟩, ⟨2 3 7 9 11⟩, ⟨2 3 9 11 12⟩, ⟨2 3 10 15 16⟩, ⟨2 3 15 16 17⟩,
⟨2 4 5 6 7⟩, ⟨2 4 5 6 11⟩, ⟨2 4 5 7 13⟩, ⟨2 4 5 9 13⟩, ⟨2 4 5 10 11⟩,
⟨2 4 6 7 11⟩, ⟨2 4 7 11 16⟩, ⟨2 4 7 12 13⟩, ⟨2 4 7 12 16⟩, ⟨2 4 8 10 17⟩,
⟨2 4 9 12 13⟩, ⟨2 4 10 11 16⟩, ⟨2 4 10 16 17⟩, ⟨2 4 12 15 16⟩, ⟨2 4 15 16 17⟩,
⟨2 5 6 7 13⟩, ⟨2 5 6 8 11⟩, ⟨2 5 6 8 15⟩, ⟨2 5 6 13 15⟩, ⟨2 5 8 9 15⟩,
⟨2 5 9 13 15⟩, ⟨2 6 7 13 17⟩, ⟨2 6 8 11 12⟩, ⟨2 6 8 12 15⟩, ⟨2 6 10 13 15⟩,
⟨2 6 10 13 17⟩, ⟨2 6 10 16 17⟩, ⟨2 7 8 12 16⟩, ⟨2 7 8 12 17⟩, ⟨2 7 9 11 15⟩,
⟨2 7 9 15 16⟩, ⟨2 7 11 15 16⟩, ⟨2 7 12 13 17⟩, ⟨2 8 9 15 16⟩, ⟨2 8 10 12 17⟩,
⟨2 8 12 15 16⟩, ⟨2 9 10 11 12⟩, ⟨2 9 10 11 15⟩, ⟨2 9 10 12 17⟩, ⟨2 9 10 13 15⟩,
⟨2 9 10 13 17⟩, ⟨2 9 12 13 17⟩, ⟨2 10 11 15 16⟩, ⟨3 4 7 9 10⟩, ⟨3 4 7 10 15⟩,
⟨3 4 7 11 17⟩, ⟨3 4 7 15 17⟩, ⟨3 4 8 9 10⟩, ⟨3 4 8 9 12⟩, ⟨3 4 8 10 12⟩,
⟨3 4 10 12 15⟩, ⟨3 5 7 8 9⟩, ⟨3 5 7 8 12⟩, ⟨3 5 7 9 11⟩, ⟨3 5 7 11 17⟩,
⟨3 5 7 12 17⟩, ⟨3 5 8 9 12⟩, ⟨3 5 9 11 16⟩, ⟨3 5 9 12 16⟩, ⟨3 5 10 11 16⟩,
⟨3 5 10 11 17⟩, ⟨3 5 10 12 16⟩, ⟨3 5 10 12 17⟩, ⟨3 6 10 12 15⟩, ⟨3 6 10 12 16⟩,
⟨3 6 11 12 13⟩, ⟨3 6 12 13 16⟩, ⟨3 7 8 9 10⟩, ⟨3 7 8 10 16⟩, ⟨3 7 8 12 17⟩,
⟨3 7 8 16 17⟩, ⟨3 7 10 15 16⟩, ⟨3 7 15 16 17⟩, ⟨3 8 10 12 17⟩, ⟨3 8 10 13 16⟩,
⟨3 8 10 13 17⟩, ⟨3 9 11 12 16⟩, ⟨3 10 11 13 16⟩, ⟨3 11 12 13 16⟩, ⟨4 5 6 7 10⟩,
⟨4 5 6 10 14⟩, ⟨4 5 6 11 14⟩, ⟨4 5 7 8 13⟩, ⟨4 5 7 8 15⟩, ⟨4 5 7 10 15⟩,
⟨4 5 8 13 17⟩, ⟨4 5 8 15 17⟩, ⟨4 5 9 13 17⟩, ⟨4 5 9 16 17⟩, ⟨4 5 10 11 14⟩,
⟨4 5 15 16 17⟩, ⟨4 6 7 9 10⟩, ⟨4 6 8 9 10⟩, ⟨4 6 8 9 12⟩, ⟨4 6 8 10 17⟩,
⟨4 6 8 12 13⟩, ⟨4 6 8 13 17⟩, ⟨4 6 9 11 16⟩, ⟨4 6 9 12 13⟩, ⟨4 6 9 13 17⟩,
⟨4 6 9 16 17⟩, ⟨4 6 10 14 16⟩, ⟨4 6 10 16 17⟩, ⟨4 6 11 14 16⟩, ⟨4 7 8 11 13⟩,
⟨4 7 8 11 15⟩, ⟨4 7 11 12 13⟩, ⟨4 7 11 12 16⟩, ⟨4 7 11 15 17⟩, ⟨4 8 11 12 13⟩,
⟨4 8 11 15 17⟩, ⟨4 10 11 14 16⟩, ⟨5 6 7 10 12⟩, ⟨5 6 7 12 13⟩, ⟨5 6 8 11 15⟩,
⟨5 6 9 11 16⟩, ⟨5 6 9 11 17⟩, ⟨5 6 9 16 17⟩, ⟨5 6 10 12 16⟩, ⟨5 6 10 14 16⟩,
⟨5 6 11 14 16⟩, ⟨5 6 11 15 17⟩, ⟨5 6 12 13 16⟩, ⟨5 6 15 16 17⟩, ⟨5 7 8 9 10⟩,
⟨5 7 8 10 15⟩, ⟨5 7 8 12 16⟩, ⟨5 7 8 13 16⟩, ⟨5 7 9 10 17⟩, ⟨5 7 9 11 17⟩,
⟨5 7 10 12 17⟩, ⟨5 7 12 13 16⟩, ⟨5 8 9 10 15⟩, ⟨5 8 9 12 16⟩, ⟨5 8 11 15 17⟩,
⟨5 9 10 13 15⟩, ⟨5 9 10 13 17⟩, ⟨5 10 11 14 16⟩, ⟨6 7 12 13 17⟩, ⟨6 7 12 15 17⟩,

Continuation on next page –

Table B.9 – Continuation from last page

⟨6 7 15 16 17⟩, ⟨6 8 9 10 11⟩, ⟨6 8 9 11 15⟩, ⟨6 8 9 12 15⟩, ⟨6 8 10 11 13⟩,
⟨6 8 10 13 17⟩, ⟨6 8 11 12 13⟩, ⟨6 9 11 15 17⟩, ⟨6 9 12 13 17⟩, ⟨6 9 12 15 17⟩,
⟨7 8 10 11 15⟩, ⟨7 8 10 11 16⟩, ⟨7 8 11 13 16⟩, ⟨7 9 10 12 17⟩, ⟨7 9 11 15 17⟩,
⟨7 9 12 15 17⟩, ⟨7 10 11 15 16⟩, ⟨7 11 12 13 16⟩, ⟨8 9 10 11 15⟩, ⟨8 9 12 15 16⟩,
⟨8 10 11 13 16⟩

f-vector:	$(17, 126, 494, 615, 246)$
# singularities:	8
Simplicial homology:	$(\mathbb{Z}, 0, \mathbb{Z}^{14}, 0, \mathbb{Z})$
Euler characteristic:	16

Table B.10: 9th blowup

C^9_{17} = ⟨1 2 3 4 11⟩, ⟨1 2 3 4 17⟩, ⟨1 2 3 11 17⟩, ⟨1 2 4 8 9⟩, ⟨1 2 4 8 17⟩,
⟨1 2 4 9 11⟩, ⟨1 2 5 12 13⟩, ⟨1 2 5 12 14⟩, ⟨1 2 5 13 14⟩, ⟨1 2 6 10 12⟩,
⟨1 2 6 10 14⟩, ⟨1 2 6 12 14⟩, ⟨1 2 7 8 14⟩, ⟨1 2 7 8 17⟩, ⟨1 2 7 11 14⟩,
⟨1 2 7 11 17⟩, ⟨1 2 8 9 13⟩, ⟨1 2 8 13 14⟩, ⟨1 2 9 10 13⟩, ⟨1 2 9 10 14⟩,
⟨1 2 9 11 14⟩, ⟨1 2 10 12 13⟩, ⟨1 3 4 5 12⟩, ⟨1 3 4 5 15⟩, ⟨1 3 4 8 11⟩,
⟨1 3 4 8 14⟩, ⟨1 3 4 12 14⟩, ⟨1 3 4 15 17⟩, ⟨1 3 5 8 14⟩, ⟨1 3 5 8 15⟩,
⟨1 3 5 12 14⟩, ⟨1 3 7 13 16⟩, ⟨1 3 7 13 17⟩, ⟨1 3 7 15 16⟩, ⟨1 3 7 15 17⟩,
⟨1 3 8 11 15⟩, ⟨1 3 9 13 15⟩, ⟨1 3 9 13 16⟩, ⟨1 3 9 15 16⟩, ⟨1 3 11 13 15⟩,
⟨1 3 11 13 17⟩, ⟨1 4 5 12 15⟩, ⟨1 4 6 8 14⟩, ⟨1 4 6 8 17⟩, ⟨1 4 6 14 17⟩,
⟨1 4 8 9 11⟩, ⟨1 4 12 14 15⟩, ⟨1 4 14 15 17⟩, ⟨1 5 7 9 10⟩, ⟨1 5 7 9 12⟩,
⟨1 5 7 10 16⟩, ⟨1 5 7 12 13⟩, ⟨1 5 7 13 16⟩, ⟨1 5 8 11 15⟩, ⟨1 5 8 11 16⟩,
⟨1 5 8 14 16⟩, ⟨1 5 9 10 15⟩, ⟨1 5 9 12 15⟩, ⟨1 5 10 11 15⟩, ⟨1 5 10 11 16⟩,
⟨1 5 13 14 16⟩, ⟨1 6 8 10 14⟩, ⟨1 6 8 10 17⟩, ⟨1 6 10 12 17⟩, ⟨1 6 12 14 17⟩,
⟨1 7 8 10 14⟩, ⟨1 7 8 10 17⟩, ⟨1 7 9 10 14⟩, ⟨1 7 9 11 12⟩, ⟨1 7 9 11 14⟩,
⟨1 7 10 16 17⟩, ⟨1 7 11 12 13⟩, ⟨1 7 11 13 17⟩, ⟨1 7 15 16 17⟩, ⟨1 8 9 11 16⟩,
⟨1 8 9 13 16⟩, ⟨1 8 13 14 16⟩, ⟨1 9 10 13 15⟩, ⟨1 9 11 12 16⟩, ⟨1 9 12 15 16⟩,
⟨1 10 11 12 13⟩, ⟨1 10 11 12 16⟩, ⟨1 10 11 13 15⟩, ⟨1 10 12 16 17⟩, ⟨1 12 14 15 17⟩,
⟨1 12 15 16 17⟩, ⟨2 3 4 10 12⟩, ⟨2 3 4 10 17⟩, ⟨2 3 4 11 12⟩, ⟨2 3 5 7 9⟩,
⟨2 3 5 7 15⟩, ⟨2 3 5 8 9⟩, ⟨2 3 5 8 15⟩, ⟨2 3 7 9 12⟩, ⟨2 3 7 10 12⟩,
⟨2 3 7 10 15⟩, ⟨2 3 8 9 10⟩, ⟨2 3 8 10 15⟩, ⟨2 3 9 10 17⟩, ⟨2 3 9 12 17⟩,
⟨2 3 11 12 17⟩, ⟨2 4 6 10 12⟩, ⟨2 4 6 10 14⟩, ⟨2 4 6 12 13⟩, ⟨2 4 6 13 15⟩,
⟨2 4 6 14 15⟩, ⟨2 4 8 9 17⟩, ⟨2 4 9 10 14⟩, ⟨2 4 9 10 17⟩, ⟨2 4 9 11 15⟩,
⟨2 4 9 14 15⟩, ⟨2 4 11 12 13⟩, ⟨2 4 11 13 15⟩, ⟨2 5 6 8 12⟩, ⟨2 5 6 8 15⟩,

Continuation on next page –

Table B.10 – Continuation from last page

⟨2 5 6 12 14⟩,	⟨2 5 6 14 17⟩,	⟨2 5 6 15 17⟩,	⟨2 5 7 8 9⟩,	⟨2 5 7 8 13⟩,
⟨2 5 7 13 17⟩,	⟨2 5 7 15 17⟩,	⟨2 5 8 12 13⟩,	⟨2 5 13 14 17⟩,	⟨2 6 8 12 13⟩,
⟨2 6 8 13 15⟩,	⟨2 6 14 15 17⟩,	⟨2 7 8 9 17⟩,	⟨2 7 8 13 14⟩,	⟨2 7 9 12 17⟩,
⟨2 7 10 12 15⟩,	⟨2 7 11 14 17⟩,	⟨2 7 12 15 17⟩,	⟨2 7 13 14 17⟩,	⟨2 8 9 10 13⟩,
⟨2 8 10 13 15⟩,	⟨2 9 11 14 15⟩,	⟨2 10 12 13 15⟩,	⟨2 11 12 13 15⟩,	⟨2 11 12 15 17⟩,
⟨2 11 14 15 17⟩,	⟨3 4 5 10 12⟩,	⟨3 4 5 10 16⟩,	⟨3 4 5 13 16⟩,	⟨3 4 5 13 17⟩,
⟨3 4 5 15 17⟩,	⟨3 4 8 11 14⟩,	⟨3 4 10 16 17⟩,	⟨3 4 11 12 14⟩,	⟨3 4 13 16 17⟩,
⟨3 5 7 9 14⟩,	⟨3 5 7 10 14⟩,	⟨3 5 7 10 16⟩,	⟨3 5 7 13 16⟩,	⟨3 5 7 13 17⟩,
⟨3 5 7 15 17⟩,	⟨3 5 8 9 14⟩,	⟨3 5 10 12 14⟩,	⟨3 7 9 11 12⟩,	⟨3 7 9 11 14⟩,
⟨3 7 10 12 14⟩,	⟨3 7 10 15 16⟩,	⟨3 7 11 12 14⟩,	⟨3 8 9 10 13⟩,	⟨3 8 9 13 16⟩,
⟨3 8 9 14 15⟩,	⟨3 8 9 15 16⟩,	⟨3 8 10 13 17⟩,	⟨3 8 10 15 16⟩,	⟨3 8 10 16 17⟩,
⟨3 8 11 14 15⟩,	⟨3 8 13 16 17⟩,	⟨3 9 10 13 17⟩,	⟨3 9 11 12 17⟩,	⟨3 9 11 13 15⟩,
⟨3 9 11 13 17⟩,	⟨3 9 11 14 15⟩,	⟨4 5 6 8 10⟩,	⟨4 5 6 8 12⟩,	⟨4 5 6 10 12⟩,
⟨4 5 8 9 11⟩,	⟨4 5 8 9 12⟩,	⟨4 5 8 10 11⟩,	⟨4 5 9 11 13⟩,	⟨4 5 9 12 15⟩,
⟨4 5 9 13 15⟩,	⟨4 5 10 11 16⟩,	⟨4 5 11 13 14⟩,	⟨4 5 11 14 16⟩,	⟨4 5 13 14 16⟩,
⟨4 5 13 15 17⟩,	⟨4 6 8 10 14⟩,	⟨4 6 8 12 13⟩,	⟨4 6 8 13 17⟩,	⟨4 6 13 15 17⟩,
⟨4 6 14 15 17⟩,	⟨4 8 9 12 17⟩,	⟨4 8 10 11 14⟩,	⟨4 8 12 13 17⟩,	⟨4 9 10 14 16⟩,
⟨4 9 10 16 17⟩,	⟨4 9 11 13 15⟩,	⟨4 9 12 15 16⟩,	⟨4 9 12 16 17⟩,	⟨4 9 14 15 16⟩,
⟨4 10 11 14 16⟩,	⟨4 11 12 13 14⟩,	⟨4 12 13 14 16⟩,	⟨4 12 13 16 17⟩,	⟨4 12 14 15 16⟩,
⟨5 6 8 10 15⟩,	⟨5 6 10 12 17⟩,	⟨5 6 10 13 15⟩,	⟨5 6 10 13 17⟩,	⟨5 6 12 14 17⟩,
⟨5 6 13 15 17⟩,	⟨5 7 8 9 12⟩,	⟨5 7 8 12 13⟩,	⟨5 7 9 10 14⟩,	⟨5 8 9 11 16⟩,
⟨5 8 9 14 16⟩,	⟨5 8 10 11 15⟩,	⟨5 9 10 13 15⟩,	⟨5 9 10 13 17⟩,	⟨5 9 10 14 17⟩,
⟨5 9 11 13 17⟩,	⟨5 9 11 16 17⟩,	⟨5 9 14 16 17⟩,	⟨5 10 12 14 17⟩,	⟨5 11 13 14 17⟩,
⟨5 11 14 16 17⟩,	⟨6 8 10 13 15⟩,	⟨6 8 10 13 17⟩,	⟨7 8 9 12 17⟩,	⟨7 8 10 14 15⟩,
⟨7 8 10 15 16⟩,	⟨7 8 10 16 17⟩,	⟨7 8 12 13 14⟩,	⟨7 8 12 14 15⟩,	⟨7 8 12 15 16⟩,
⟨7 8 12 16 17⟩,	⟨7 10 12 14 15⟩,	⟨7 11 12 13 14⟩,	⟨7 11 13 14 17⟩,	⟨7 12 15 16 17⟩,
⟨8 9 14 15 16⟩,	⟨8 10 11 14 15⟩,	⟨8 12 13 14 16⟩,	⟨8 12 13 16 17⟩,	⟨8 12 14 15 16⟩,
⟨9 10 14 16 17⟩,	⟨9 11 12 16 17⟩,	⟨10 11 12 13 15⟩,	⟨10 11 12 15 17⟩,	⟨10 11 12 16 17⟩,
⟨10 11 14 15 17⟩,	⟨10 11 14 16 17⟩,	⟨10 12 14 15 17⟩		

f-vector:	$(17, 129, 516, 645, 258)$
# singularities:	7
Simplicial homology:	$(\mathbb{Z}, 0, \mathbb{Z}^{15}, 0, \mathbb{Z})$
Euler characteristic:	17

Table B.11: 10th blowup

C^{10}_{17} = ⟨1 2 3 5 13⟩, ⟨1 2 3 5 15⟩, ⟨1 2 3 13 15⟩, ⟨1 2 4 7 14⟩, ⟨1 2 4 7 17⟩,
⟨1 2 4 11 12⟩, ⟨1 2 4 11 17⟩, ⟨1 2 4 12 14⟩, ⟨1 2 5 7 13⟩, ⟨1 2 5 7 16⟩,
⟨1 2 5 15 16⟩, ⟨1 2 7 13 17⟩, ⟨1 2 7 14 16⟩, ⟨1 2 8 11 15⟩, ⟨1 2 8 11 17⟩,
⟨1 2 8 13 15⟩, ⟨1 2 8 13 17⟩, ⟨1 2 11 12 15⟩, ⟨1 2 12 14 16⟩, ⟨1 2 12 15 16⟩,
⟨1 3 4 7 8⟩, ⟨1 3 4 7 14⟩, ⟨1 3 4 8 16⟩, ⟨1 3 4 14 16⟩, ⟨1 3 5 6 10⟩,
⟨1 3 5 6 13⟩, ⟨1 3 5 7 9⟩, ⟨1 3 5 7 10⟩, ⟨1 3 5 9 15⟩, ⟨1 3 6 9 10⟩,
⟨1 3 6 9 15⟩, ⟨1 3 6 13 15⟩, ⟨1 3 7 8 10⟩, ⟨1 3 7 9 16⟩, ⟨1 3 7 14 16⟩,
⟨1 3 8 10 16⟩, ⟨1 3 9 10 16⟩, ⟨1 4 6 10 12⟩, ⟨1 4 6 10 13⟩, ⟨1 4 6 11 12⟩,
⟨1 4 6 11 17⟩, ⟨1 4 6 13 15⟩, ⟨1 4 6 15 17⟩, ⟨1 4 7 8 15⟩, ⟨1 4 7 15 17⟩,
⟨1 4 8 12 13⟩, ⟨1 4 8 12 14⟩, ⟨1 4 8 13 15⟩, ⟨1 4 8 14 16⟩, ⟨1 4 10 12 13⟩,
⟨1 5 6 10 13⟩, ⟨1 5 7 8 10⟩, ⟨1 5 7 8 11⟩, ⟨1 5 7 9 16⟩, ⟨1 5 7 11 13⟩,
⟨1 5 8 10 11⟩, ⟨1 5 9 15 16⟩, ⟨1 5 10 11 13⟩, ⟨1 6 9 10 15⟩, ⟨1 6 10 11 12⟩,
⟨1 6 10 11 17⟩, ⟨1 6 10 15 17⟩, ⟨1 7 8 11 15⟩, ⟨1 7 11 12 13⟩, ⟨1 7 11 12 15⟩,
⟨1 7 12 13 17⟩, ⟨1 7 12 15 17⟩, ⟨1 8 9 12 13⟩, ⟨1 8 9 12 14⟩, ⟨1 8 9 13 16⟩,
⟨1 8 9 14 16⟩, ⟨1 8 10 11 16⟩, ⟨1 8 11 16 17⟩, ⟨1 8 13 16 17⟩, ⟨1 9 10 15 17⟩,
⟨1 9 10 16 17⟩, ⟨1 9 12 13 17⟩, ⟨1 9 12 14 16⟩, ⟨1 9 12 15 16⟩, ⟨1 9 12 15 17⟩,
⟨1 9 13 16 17⟩, ⟨1 10 11 12 13⟩, ⟨1 10 11 16 17⟩, ⟨2 3 5 6 10⟩, ⟨2 3 5 6 13⟩,
⟨2 3 5 10 15⟩, ⟨2 3 6 9 10⟩, ⟨2 3 6 9 14⟩, ⟨2 3 6 12 13⟩, ⟨2 3 6 12 14⟩,
⟨2 3 7 9 14⟩, ⟨2 3 7 9 16⟩, ⟨2 3 7 14 16⟩, ⟨2 3 9 10 16⟩, ⟨2 3 10 13 15⟩,
⟨2 3 10 13 16⟩, ⟨2 3 12 13 16⟩, ⟨2 3 12 14 16⟩, ⟨2 4 5 6 10⟩, ⟨2 4 5 6 12⟩,
⟨2 4 5 10 15⟩, ⟨2 4 5 12 16⟩, ⟨2 4 5 15 16⟩, ⟨2 4 6 10 12⟩, ⟨2 4 7 9 14⟩,
⟨2 4 7 9 17⟩, ⟨2 4 8 10 12⟩, ⟨2 4 8 10 14⟩, ⟨2 4 8 12 14⟩, ⟨2 4 9 11 14⟩,
⟨2 4 9 11 17⟩, ⟨2 4 10 11 14⟩, ⟨2 4 10 11 15⟩, ⟨2 4 11 12 15⟩, ⟨2 4 12 15 16⟩,
⟨2 5 6 12 13⟩, ⟨2 5 7 13 16⟩, ⟨2 5 12 13 16⟩, ⟨2 6 8 9 11⟩, ⟨2 6 8 9 14⟩,
⟨2 6 8 11 17⟩, ⟨2 6 8 12 14⟩, ⟨2 6 8 12 17⟩, ⟨2 6 9 10 12⟩, ⟨2 6 9 11 17⟩,
⟨2 6 9 12 17⟩, ⟨2 7 8 10 13⟩, ⟨2 7 8 10 17⟩, ⟨2 7 8 13 17⟩, ⟨2 7 9 10 16⟩,
⟨2 7 9 10 17⟩, ⟨2 7 10 13 16⟩, ⟨2 8 9 11 15⟩, ⟨2 8 9 14 15⟩, ⟨2 8 10 12 17⟩,
⟨2 8 10 13 14⟩, ⟨2 8 13 14 15⟩, ⟨2 9 10 12 17⟩, ⟨2 9 11 14 15⟩, ⟨2 10 11 14 15⟩,
⟨2 10 13 14 15⟩, ⟨3 4 5 8 9⟩, ⟨3 4 5 8 17⟩, ⟨3 4 5 9 17⟩, ⟨3 4 6 7 8⟩,
⟨3 4 6 7 14⟩, ⟨3 4 6 8 17⟩, ⟨3 4 6 14 15⟩, ⟨3 4 6 15 17⟩, ⟨3 4 8 9 13⟩,
⟨3 4 8 10 12⟩, ⟨3 4 8 10 16⟩, ⟨3 4 8 12 13⟩, ⟨3 4 9 13 17⟩, ⟨3 4 10 12 13⟩,
⟨3 4 10 13 17⟩, ⟨3 4 10 16 17⟩, ⟨3 4 14 15 16⟩, ⟨3 4 15 16 17⟩, ⟨3 5 7 9 12⟩,
⟨3 5 7 10 12⟩, ⟨3 5 8 9 15⟩, ⟨3 5 8 15 17⟩, ⟨3 5 9 12 17⟩, ⟨3 5 10 12 15⟩,
⟨3 5 11 12 15⟩, ⟨3 5 11 12 16⟩, ⟨3 5 11 15 17⟩, ⟨3 5 11 16 17⟩, ⟨3 5 12 16 17⟩,
⟨3 6 7 8 9⟩, ⟨3 6 7 9 14⟩, ⟨3 6 8 9 15⟩, ⟨3 6 8 15 17⟩, ⟨3 6 12 13 14⟩,
⟨3 6 13 14 15⟩, ⟨3 7 8 9 12⟩, ⟨3 7 8 10 12⟩, ⟨3 8 9 12 13⟩, ⟨3 9 12 13 17⟩,
⟨3 10 12 13 15⟩, ⟨3 10 13 16 17⟩, ⟨3 11 12 15 16⟩, ⟨3 11 15 16 17⟩, ⟨3 12 13 14 15⟩,

Continuation on next page –

Table B.11 – Continuation from last page

⟨3 12 13 16 17⟩,	⟨3 12 14 15 16⟩,	⟨4 5 6 7 8⟩,	⟨4 5 6 7 10⟩,	⟨4 5 6 8 17⟩,
⟨4 5 6 9 12⟩,	⟨4 5 6 9 17⟩,	⟨4 5 7 8 15⟩,	⟨4 5 7 10 15⟩,	⟨4 5 8 9 15⟩,
⟨4 5 9 11 12⟩,	⟨4 5 9 11 13⟩,	⟨4 5 9 13 15⟩,	⟨4 5 11 12 16⟩,	⟨4 5 11 13 16⟩,
⟨4 5 13 15 16⟩,	⟨4 6 7 10 13⟩,	⟨4 6 7 13 14⟩,	⟨4 6 9 11 12⟩,	⟨4 6 9 11 17⟩,
⟨4 6 13 14 15⟩,	⟨4 7 9 10 13⟩,	⟨4 7 9 10 17⟩,	⟨4 7 9 11 13⟩,	⟨4 7 9 11 14⟩,
⟨4 7 10 15 17⟩,	⟨4 7 11 13 14⟩,	⟨4 8 9 13 15⟩,	⟨4 8 10 11 14⟩,	⟨4 8 10 11 16⟩,
⟨4 8 11 14 16⟩,	⟨4 9 10 13 17⟩,	⟨4 10 11 15 16⟩,	⟨4 10 15 16 17⟩,	⟨4 11 12 15 16⟩,
⟨4 11 13 14 16⟩,	⟨4 13 14 15 16⟩,	⟨5 6 7 8 10⟩,	⟨5 6 8 10 13⟩,	⟨5 6 8 13 17⟩,
⟨5 6 9 12 17⟩,	⟨5 6 12 13 17⟩,	⟨5 7 8 11 15⟩,	⟨5 7 9 11 12⟩,	⟨5 7 9 11 13⟩,
⟨5 7 9 13 16⟩,	⟨5 7 10 12 15⟩,	⟨5 7 11 12 15⟩,	⟨5 8 10 11 13⟩,	⟨5 8 11 13 16⟩,
⟨5 8 11 15 17⟩,	⟨5 8 11 16 17⟩,	⟨5 8 13 16 17⟩,	⟨5 9 13 15 16⟩,	⟨5 12 13 16 17⟩,
⟨6 7 8 9 14⟩,	⟨6 7 8 10 13⟩,	⟨6 7 8 12 14⟩,	⟨6 7 8 12 17⟩,	⟨6 7 8 13 17⟩,
⟨6 7 12 13 14⟩,	⟨6 7 12 13 17⟩,	⟨6 8 9 11 15⟩,	⟨6 8 11 15 17⟩,	⟨6 9 10 11 12⟩,
⟨6 9 10 11 15⟩,	⟨6 10 11 15 17⟩,	⟨7 8 9 12 14⟩,	⟨7 8 10 12 17⟩,	⟨7 9 10 13 16⟩,
⟨7 9 11 12 14⟩,	⟨7 10 12 15 17⟩,	⟨7 11 12 13 14⟩,	⟨8 9 13 15 16⟩,	⟨8 9 14 15 16⟩,
⟨8 10 11 13 14⟩,	⟨8 11 13 14 16⟩,	⟨8 13 14 15 16⟩,	⟨9 10 11 12 14⟩,	⟨9 10 11 14 15⟩,
⟨9 10 12 14 15⟩,	⟨9 10 12 15 17⟩,	⟨9 10 13 16 17⟩,	⟨9 12 14 15 16⟩,	⟨10 11 12 13 14⟩,
⟨10 11 15 16 17⟩,	⟨10 12 13 14 15⟩			

f-vector:	$(17, 133, 542, 680, 272)$
# singularities:	6
Simplicial homology:	$(\mathbb{Z}, 0, \mathbb{Z}^{16}, 0, \mathbb{Z})$
Euler characteristic:	18

Table B.12: 11th blowup

C_{17}^{11} =	⟨2 3 4 9 12⟩,	⟨2 3 4 9 14⟩,	⟨2 3 4 12 15⟩,	⟨2 3 4 14 21⟩,	⟨2 3 4 15 21⟩,
	⟨2 3 6 15 22⟩,	⟨2 3 6 15 23⟩,	⟨2 3 6 17 18⟩,	⟨2 3 6 17 22⟩,	⟨2 3 6 18 23⟩,
	⟨2 3 9 12 16⟩,	⟨2 3 9 14 16⟩,	⟨2 3 10 12 15⟩,	⟨2 3 10 12 17⟩,	⟨2 3 10 15 23⟩,
	⟨2 3 10 17 18⟩,	⟨2 3 10 18 23⟩,	⟨2 3 12 13 16⟩,	⟨2 3 12 13 17⟩,	⟨2 3 13 16 17⟩,
	⟨2 3 14 15 21⟩,	⟨2 3 14 15 22⟩,	⟨2 3 14 16 22⟩,	⟨2 3 16 17 22⟩,	⟨2 4 6 14 18⟩,
	⟨2 4 6 14 23⟩,	⟨2 4 6 18 23⟩,	⟨2 4 7 12 15⟩,	⟨2 4 7 12 22⟩,	⟨2 4 7 15 21⟩,
	⟨2 4 7 21 22⟩,	⟨2 4 9 10 13⟩,	⟨2 4 9 10 14⟩,	⟨2 4 9 12 16⟩,	⟨2 4 9 13 17⟩,
	⟨2 4 9 16 17⟩,	⟨2 4 10 13 22⟩,	⟨2 4 10 14 18⟩,	⟨2 4 10 18 22⟩,	⟨2 4 12 13 16⟩,
	⟨2 4 12 13 22⟩,	⟨2 4 13 16 17⟩,	⟨2 4 14 21 23⟩,	⟨2 4 18 22 23⟩,	⟨2 4 21 22 23⟩,
	⟨2 6 9 17 21⟩,	⟨2 6 9 17 22⟩,	⟨2 6 9 21 23⟩,	⟨2 6 9 22 23⟩,	⟨2 6 10 13 15⟩,

Continuation on next page –

Table B.12 – Continuation from last page

⟨2 6 10 13 22⟩, ⟨2 6 10 15 23⟩, ⟨2 6 10 22 23⟩, ⟨2 6 12 14 18⟩, ⟨2 6 12 14 23⟩,
⟨2 6 12 17 18⟩, ⟨2 6 12 17 21⟩, ⟨2 6 12 21 23⟩, ⟨2 6 13 15 22⟩, ⟨2 7 9 15 16⟩,
⟨2 7 9 15 21⟩, ⟨2 7 9 16 22⟩, ⟨2 7 9 21 23⟩, ⟨2 7 9 22 23⟩, ⟨2 7 10 14 16⟩,
⟨2 7 10 14 18⟩, ⟨2 7 10 16 18⟩, ⟨2 7 12 14 18⟩, ⟨2 7 12 14 22⟩, ⟨2 7 12 15 18⟩,
⟨2 7 14 16 22⟩, ⟨2 7 15 16 18⟩, ⟨2 7 21 22 23⟩, ⟨2 9 10 13 18⟩, ⟨2 9 10 14 16⟩,
⟨2 9 10 16 21⟩, ⟨2 9 10 18 21⟩, ⟨2 9 13 17 21⟩, ⟨2 9 13 18 21⟩, ⟨2 9 15 16 21⟩,
⟨2 9 16 17 22⟩, ⟨2 10 12 15 18⟩, ⟨2 10 12 17 18⟩, ⟨2 10 13 15 18⟩, ⟨2 10 16 18 21⟩,
⟨2 10 18 22 23⟩, ⟨2 12 13 14 22⟩, ⟨2 12 13 14 23⟩, ⟨2 12 13 17 21⟩, ⟨2 12 13 21 23⟩,
⟨2 13 14 15 21⟩, ⟨2 13 14 15 22⟩, ⟨2 13 14 21 23⟩, ⟨2 13 15 18 21⟩, ⟨2 15 16 18 21⟩,
⟨3 4 6 15 16⟩, ⟨3 4 6 15 23⟩, ⟨3 4 6 16 17⟩, ⟨3 4 6 17 18⟩, ⟨3 4 6 18 23⟩,
⟨3 4 9 10 12⟩, ⟨3 4 9 10 14⟩, ⟨3 4 10 12 15⟩, ⟨3 4 10 13 14⟩, ⟨3 4 10 13 23⟩,
⟨3 4 10 15 23⟩, ⟨3 4 13 14 21⟩, ⟨3 4 13 21 23⟩, ⟨3 4 15 16 17⟩, ⟨3 4 15 17 18⟩,
⟨3 4 15 18 22⟩, ⟨3 4 15 21 22⟩, ⟨3 4 18 22 23⟩, ⟨3 4 21 22 23⟩, ⟨3 6 9 13 14⟩,
⟨3 6 9 13 15⟩, ⟨3 6 9 14 16⟩, ⟨3 6 9 15 16⟩, ⟨3 6 10 13 21⟩, ⟨3 6 10 13 22⟩,
⟨3 6 10 21 22⟩, ⟨3 6 13 14 21⟩, ⟨3 6 13 15 22⟩, ⟨3 6 14 16 21⟩, ⟨3 6 16 17 22⟩,
⟨3 6 16 21 22⟩, ⟨3 7 9 13 14⟩, ⟨3 7 9 13 15⟩, ⟨3 7 9 14 18⟩, ⟨3 7 9 15 16⟩,
⟨3 7 9 16 18⟩, ⟨3 7 10 12 14⟩, ⟨3 7 10 12 21⟩, ⟨3 7 10 13 14⟩, ⟨3 7 10 13 22⟩,
⟨3 7 10 21 22⟩, ⟨3 7 12 14 18⟩, ⟨3 7 12 18 23⟩, ⟨3 7 12 21 23⟩, ⟨3 7 13 15 22⟩,
⟨3 7 15 16 18⟩, ⟨3 7 15 18 22⟩, ⟨3 7 18 22 23⟩, ⟨3 7 21 22 23⟩, ⟨3 9 10 12 14⟩,
⟨3 9 12 14 18⟩, ⟨3 9 12 16 18⟩, ⟨3 10 12 17 21⟩, ⟨3 10 13 17 18⟩, ⟨3 10 13 17 21⟩,
⟨3 10 13 18 23⟩, ⟨3 12 13 16 18⟩, ⟨3 12 13 17 21⟩, ⟨3 12 13 18 23⟩, ⟨3 12 13 21 23⟩,
⟨3 13 16 17 18⟩, ⟨3 14 15 21 22⟩, ⟨3 14 16 21 22⟩, ⟨3 15 16 17 18⟩, ⟨4 6 9 12 18⟩,
⟨4 6 9 12 21⟩, ⟨4 6 9 18 21⟩, ⟨4 6 12 16 17⟩, ⟨4 6 12 16 21⟩, ⟨4 6 12 17 18⟩,
⟨4 6 14 15 16⟩, ⟨4 6 14 15 23⟩, ⟨4 6 14 16 18⟩, ⟨4 6 16 18 21⟩, ⟨4 7 9 12 15⟩,
⟨4 7 9 12 21⟩, ⟨4 7 9 15 21⟩, ⟨4 7 10 13 14⟩, ⟨4 7 10 13 22⟩, ⟨4 7 10 14 18⟩,
⟨4 7 10 16 18⟩, ⟨4 7 10 16 21⟩, ⟨4 7 10 21 22⟩, ⟨4 7 12 13 16⟩, ⟨4 7 12 13 22⟩,
⟨4 7 12 16 21⟩, ⟨4 7 13 14 18⟩, ⟨4 7 13 16 18⟩, ⟨4 9 10 12 23⟩, ⟨4 9 10 13 23⟩,
⟨4 9 12 15 23⟩, ⟨4 9 12 16 17⟩, ⟨4 9 12 17 18⟩, ⟨4 9 13 15 17⟩, ⟨4 9 13 15 23⟩,
⟨4 9 15 17 22⟩, ⟨4 9 15 21 22⟩, ⟨4 9 17 18 22⟩, ⟨4 9 18 21 22⟩, ⟨4 10 12 15 23⟩,
⟨4 10 16 18 21⟩, ⟨4 10 18 21 22⟩, ⟨4 13 14 15 16⟩, ⟨4 13 14 15 23⟩, ⟨4 13 14 16 18⟩,
⟨4 13 14 21 23⟩, ⟨4 13 15 16 17⟩, ⟨4 15 17 18 22⟩, ⟨6 9 10 12 14⟩, ⟨6 9 10 12 23⟩,
⟨6 9 10 14 15⟩, ⟨6 9 10 15 17⟩, ⟨6 9 10 17 22⟩, ⟨6 9 10 22 23⟩, ⟨6 9 12 14 18⟩,
⟨6 9 12 21 23⟩, ⟨6 9 13 14 18⟩, ⟨6 9 13 15 17⟩, ⟨6 9 13 17 21⟩, ⟨6 9 13 18 21⟩,
⟨6 9 14 15 16⟩, ⟨6 10 12 14 15⟩, ⟨6 10 12 15 23⟩, ⟨6 10 13 15 17⟩, ⟨6 10 13 17 21⟩,
⟨6 10 17 21 22⟩, ⟨6 12 14 15 23⟩, ⟨6 12 16 17 21⟩, ⟨6 13 14 16 18⟩, ⟨6 13 14 16 21⟩,
⟨6 13 16 18 21⟩, ⟨6 16 17 21 22⟩, ⟨7 9 12 15 23⟩, ⟨7 9 12 21 23⟩, ⟨7 9 13 14 18⟩,
⟨7 9 13 15 23⟩, ⟨7 9 13 16 18⟩, ⟨7 9 13 16 22⟩, ⟨7 9 13 22 23⟩, ⟨7 10 12 14 21⟩,
⟨7 10 14 16 21⟩, ⟨7 12 13 16 22⟩, ⟨7 12 14 16 21⟩, ⟨7 12 14 16 22⟩, ⟨7 12 15 18 23⟩,

Continuation on next page –

Table B.12 – Continuation from last page

⟨7 13 15 22 23⟩,	⟨7 15 18 22 23⟩,	⟨9 10 13 18 23⟩,	⟨9 10 14 15 21⟩,	⟨9 10 14 16 21⟩,
⟨9 10 15 17 22⟩,	⟨9 10 15 21 22⟩,	⟨9 10 18 21 22⟩,	⟨9 10 18 22 23⟩,	⟨9 12 13 16 18⟩,
⟨9 12 13 16 22⟩,	⟨9 12 13 18 22⟩,	⟨9 12 16 17 22⟩,	⟨9 12 17 18 22⟩,	⟨9 13 18 22 23⟩,
⟨9 14 15 16 21⟩,	⟨10 12 14 15 21⟩,	⟨10 12 15 17 18⟩,	⟨10 12 15 17 21⟩,	⟨10 13 15 17 18⟩,
⟨10 15 17 21 22⟩,	⟨12 13 14 22 23⟩,	⟨12 13 18 22 23⟩,	⟨12 14 15 21 22⟩,	⟨12 14 15 22 23⟩,
⟨12 14 16 21 22⟩,	⟨12 15 17 18 22⟩,	⟨12 15 17 21 22⟩,	⟨12 15 18 22 23⟩,	⟨12 16 17 21 22⟩,
⟨13 14 15 16 21⟩,	⟨13 14 15 22 23⟩,	⟨13 15 16 17 18⟩,	⟨13 15 16 18 21⟩	

f-vector:	(17, 131, 544, 685, 274)
# singularities:	5
Simplicial homology:	$(\mathbb{Z}, 0, \mathbb{Z}^{17}, 0, \mathbb{Z})$
Euler characteristic:	19

Table B.13: 12th blowup

C_{17}^{12} =	⟨3 4 6 9 18⟩,	⟨3 4 6 9 48⟩,	⟨3 4 6 18 27⟩,	⟨3 4 6 27 48⟩,	⟨3 4 9 18 42⟩,
	⟨3 4 9 21 37⟩,	⟨3 4 9 21 42⟩,	⟨3 4 9 37 48⟩,	⟨3 4 12 21 42⟩,	⟨3 4 12 21 50⟩,
	⟨3 4 12 42 50⟩,	⟨3 4 18 27 42⟩,	⟨3 4 21 24 36⟩,	⟨3 4 21 24 50⟩,	⟨3 4 21 36 37⟩,
	⟨3 4 24 27 36⟩,	⟨3 4 24 27 45⟩,	⟨3 4 24 45 50⟩,	⟨3 4 27 30 36⟩,	⟨3 4 27 30 48⟩,
	⟨3 4 27 42 45⟩,	⟨3 4 30 36 37⟩,	⟨3 4 30 37 48⟩,	⟨3 4 42 45 50⟩,	⟨3 6 9 18 48⟩,
	⟨3 6 12 33 36⟩,	⟨3 6 12 33 37⟩,	⟨3 6 12 36 48⟩,	⟨3 6 12 37 50⟩,	⟨3 6 12 48 50⟩,
	⟨3 6 18 27 37⟩,	⟨3 6 18 37 50⟩,	⟨3 6 18 48 50⟩,	⟨3 6 21 33 36⟩,	⟨3 6 21 33 48⟩,
	⟨3 6 21 36 48⟩,	⟨3 6 27 33 37⟩,	⟨3 6 27 33 48⟩,	⟨3 9 12 18 33⟩,	⟨3 9 12 18 42⟩,
	⟨3 9 12 21 37⟩,	⟨3 9 12 21 42⟩,	⟨3 9 12 24 33⟩,	⟨3 9 12 24 45⟩,	⟨3 9 12 37 45⟩,
	⟨3 9 18 24 33⟩,	⟨3 9 18 24 48⟩,	⟨3 9 24 45 48⟩,	⟨3 9 30 36 37⟩,	⟨3 9 30 36 50⟩,
	⟨3 9 30 37 48⟩,	⟨3 9 30 45 48⟩,	⟨3 9 30 45 50⟩,	⟨3 9 36 37 45⟩,	⟨3 9 36 45 50⟩,
	⟨3 12 18 33 36⟩,	⟨3 12 18 36 42⟩,	⟨3 12 21 37 50⟩,	⟨3 12 24 27 33⟩,	⟨3 12 24 27 45⟩,
	⟨3 12 27 33 45⟩,	⟨3 12 33 37 45⟩,	⟨3 12 36 42 50⟩,	⟨3 12 36 48 50⟩,	⟨3 18 21 24 33⟩,
	⟨3 18 21 24 37⟩,	⟨3 18 21 33 36⟩,	⟨3 18 21 36 37⟩,	⟨3 18 24 37 50⟩,	⟨3 18 24 48 50⟩,
	⟨3 18 27 37 45⟩,	⟨3 18 27 42 45⟩,	⟨3 18 36 37 45⟩,	⟨3 18 36 42 45⟩,	⟨3 21 24 30 33⟩,
	⟨3 21 24 30 36⟩,	⟨3 21 24 37 50⟩,	⟨3 21 30 33 48⟩,	⟨3 21 30 36 48⟩,	⟨3 24 27 30 33⟩,
	⟨3 24 27 30 36⟩,	⟨3 24 45 48 50⟩,	⟨3 27 30 33 48⟩,	⟨3 27 33 37 45⟩,	⟨3 30 36 48 50⟩,
	⟨3 30 45 48 50⟩,	⟨3 36 42 45 50⟩,	⟨4 6 9 18 45⟩,	⟨4 6 9 45 48⟩,	⟨4 6 18 24 27⟩,
	⟨4 6 18 24 33⟩,	⟨4 6 18 33 45⟩,	⟨4 6 21 24 36⟩,	⟨4 6 21 24 50⟩,	⟨4 6 21 30 42⟩,
	⟨4 6 21 30 50⟩,	⟨4 6 21 36 42⟩,	⟨4 6 24 27 36⟩,	⟨4 6 24 30 33⟩,	⟨4 6 24 30 50⟩,
	⟨4 6 27 30 36⟩,	⟨4 6 27 30 48⟩,	⟨4 6 30 33 37⟩,	⟨4 6 30 36 42⟩,	⟨4 6 30 37 48⟩,

Continuation on next page –

Table B.13 – Continuation from last page

⟨4 6 33 37 45⟩,	⟨4 6 37 45 48⟩,	⟨4 9 12 18 27⟩,	⟨4 9 12 18 50⟩,	⟨4 9 12 27 50⟩,
⟨4 9 18 21 36⟩,	⟨4 9 18 21 45⟩,	⟨4 9 18 27 42⟩,	⟨4 9 18 36 50⟩,	⟨4 9 21 36 42⟩,
⟨4 9 21 37 48⟩,	⟨4 9 21 45 48⟩,	⟨4 9 27 42 50⟩,	⟨4 9 36 42 50⟩,	⟨4 12 18 24 27⟩,
⟨4 12 18 24 45⟩,	⟨4 12 18 45 48⟩,	⟨4 12 18 48 50⟩,	⟨4 12 21 33 42⟩,	⟨4 12 21 33 50⟩,
⟨4 12 24 27 45⟩,	⟨4 12 27 45 50⟩,	⟨4 12 33 36 37⟩,	⟨4 12 33 36 42⟩,	⟨4 12 33 37 45⟩,
⟨4 12 33 45 50⟩,	⟨4 12 36 37 45⟩,	⟨4 12 36 42 50⟩,	⟨4 12 36 45 48⟩,	⟨4 12 36 48 50⟩,
⟨4 18 21 36 37⟩,	⟨4 18 21 37 48⟩,	⟨4 18 21 45 48⟩,	⟨4 18 24 33 45⟩,	⟨4 18 36 37 48⟩,
⟨4 18 36 48 50⟩,	⟨4 21 30 33 42⟩,	⟨4 21 30 33 50⟩,	⟨4 24 30 33 45⟩,	⟨4 24 30 45 50⟩,
⟨4 27 42 45 50⟩,	⟨4 30 33 36 37⟩,	⟨4 30 33 36 42⟩,	⟨4 30 33 45 50⟩,	⟨4 36 37 45 48⟩,
⟨6 9 12 24 37⟩,	⟨6 9 12 24 45⟩,	⟨6 9 12 36 37⟩,	⟨6 9 12 36 45⟩,	⟨6 9 18 21 36⟩,
⟨6 9 18 21 45⟩,	⟨6 9 18 36 50⟩,	⟨6 9 18 48 50⟩,	⟨6 9 21 36 42⟩,	⟨6 9 21 42 45⟩,
⟨6 9 24 33 37⟩,	⟨6 9 24 33 48⟩,	⟨6 9 24 45 48⟩,	⟨6 9 33 36 37⟩,	⟨6 9 33 36 50⟩,
⟨6 9 33 48 50⟩,	⟨6 9 36 42 45⟩,	⟨6 12 21 24 36⟩,	⟨6 12 21 24 37⟩,	⟨6 12 21 36 48⟩,
⟨6 12 21 37 50⟩,	⟨6 12 21 48 50⟩,	⟨6 12 24 36 45⟩,	⟨6 12 33 36 37⟩,	⟨6 18 21 36 50⟩,
⟨6 18 21 42 45⟩,	⟨6 18 21 42 50⟩,	⟨6 18 24 27 42⟩,	⟨6 18 24 33 42⟩,	⟨6 18 27 37 50⟩,
⟨6 18 27 42 50⟩,	⟨6 18 33 42 45⟩,	⟨6 21 24 37 50⟩,	⟨6 21 30 42 50⟩,	⟨6 21 33 36 50⟩,
⟨6 21 33 48 50⟩,	⟨6 24 27 36 42⟩,	⟨6 24 30 33 37⟩,	⟨6 24 30 37 50⟩,	⟨6 24 33 42 48⟩,
⟨6 24 36 42 45⟩,	⟨6 24 42 45 48⟩,	⟨6 27 30 36 42⟩,	⟨6 27 30 37 48⟩,	⟨6 27 30 37 50⟩,
⟨6 27 30 42 50⟩,	⟨6 27 33 37 45⟩,	⟨6 27 33 45 48⟩,	⟨6 27 37 45 48⟩,	⟨6 33 42 45 48⟩,
⟨9 12 18 27 42⟩,	⟨9 12 18 33 48⟩,	⟨9 12 18 48 50⟩,	⟨9 12 21 37 42⟩,	⟨9 12 24 27 37⟩,
⟨9 12 24 27 50⟩,	⟨9 12 24 33 50⟩,	⟨9 12 27 37 42⟩,	⟨9 12 33 48 50⟩,	⟨9 12 36 37 45⟩,
⟨9 18 24 33 48⟩,	⟨9 21 30 42 48⟩,	⟨9 21 30 42 50⟩,	⟨9 21 30 45 48⟩,	⟨9 21 30 45 50⟩,
⟨9 21 37 42 48⟩,	⟨9 21 42 45 50⟩,	⟨9 24 27 37 50⟩,	⟨9 24 30 33 36⟩,	⟨9 24 30 33 37⟩,
⟨9 24 30 36 50⟩,	⟨9 24 30 37 50⟩,	⟨9 24 33 36 50⟩,	⟨9 27 37 42 50⟩,	⟨9 30 33 36 37⟩,
⟨9 30 37 42 48⟩,	⟨9 30 37 42 50⟩,	⟨9 36 42 45 50⟩,	⟨12 18 21 24 37⟩,	⟨12 18 21 24 45⟩,
⟨12 18 21 37 48⟩,	⟨12 18 21 45 48⟩,	⟨12 18 24 27 42⟩,	⟨12 18 24 37 42⟩,	⟨12 18 33 36 42⟩,
⟨12 18 33 42 48⟩,	⟨12 18 37 42 48⟩,	⟨12 21 24 36 48⟩,	⟨12 21 24 45 48⟩,	⟨12 21 33 42 48⟩,
⟨12 21 33 48 50⟩,	⟨12 21 37 42 48⟩,	⟨12 24 27 33 50⟩,	⟨12 24 27 37 42⟩,	⟨12 24 36 45 48⟩,
⟨12 27 33 45 50⟩,	⟨18 21 24 33 45⟩,	⟨18 21 33 36 50⟩,	⟨18 21 33 45 50⟩,	⟨18 21 42 45 50⟩,
⟨18 24 33 42 48⟩,	⟨18 24 37 42 48⟩,	⟨18 24 37 48 50⟩,	⟨18 27 33 36 45⟩,	⟨18 27 33 36 50⟩,
⟨18 27 33 45 50⟩,	⟨18 27 36 37 45⟩,	⟨18 27 36 37 50⟩,	⟨18 27 42 45 50⟩,	⟨18 33 36 42 45⟩,
⟨18 36 37 48 50⟩,	⟨21 24 30 33 45⟩,	⟨21 24 30 36 48⟩,	⟨21 24 30 45 48⟩,	⟨21 30 33 42 48⟩,
⟨21 30 33 45 50⟩,	⟨24 27 30 33 36⟩,	⟨24 27 33 36 50⟩,	⟨24 27 36 37 48⟩,	⟨24 27 36 37 50⟩,
⟨24 27 36 42 48⟩,	⟨24 27 37 42 48⟩,	⟨24 30 36 48 50⟩,	⟨24 30 45 48 50⟩,	⟨24 36 37 48 50⟩,
⟨24 36 42 45 48⟩,	⟨27 30 33 36 42⟩,	⟨27 30 33 42 48⟩,	⟨27 30 37 42 48⟩,	⟨27 30 37 42 50⟩,
⟨27 33 36 42 48⟩,	⟨27 33 36 45 48⟩,	⟨27 36 37 45 48⟩,	⟨33 36 42 45 48⟩	

f-vector: $(17, 133, 562, 710, 284)$

Continuation on next page –

Table B.13 – Continuation from last page

# singularities:	4
Simplicial homology:	$(\mathbb{Z}, 0, \mathbb{Z}^{18}, 0, \mathbb{Z})$
Euler characteristic:	20

Table B.14: 13th blowup

$C^{13}_{18} =$	$\langle 1\,2\,3\,9\,12\rangle$,	$\langle 1\,2\,3\,9\,14\rangle$,	$\langle 1\,2\,3\,10\,11\rangle$,	$\langle 1\,2\,3\,10\,15\rangle$,	$\langle 1\,2\,3\,11\,12\rangle$,
	$\langle 1\,2\,3\,14\,15\rangle$,	$\langle 1\,2\,4\,6\,11\rangle$,	$\langle 1\,2\,4\,6\,12\rangle$,	$\langle 1\,2\,4\,11\,12\rangle$,	$\langle 1\,2\,5\,7\,9\rangle$,
	$\langle 1\,2\,5\,7\,13\rangle$,	$\langle 1\,2\,5\,9\,14\rangle$,	$\langle 1\,2\,5\,13\,14\rangle$,	$\langle 1\,2\,6\,9\,13\rangle$,	$\langle 1\,2\,6\,9\,16\rangle$,
	$\langle 1\,2\,6\,11\,13\rangle$,	$\langle 1\,2\,6\,12\,17\rangle$,	$\langle 1\,2\,6\,16\,17\rangle$,	$\langle 1\,2\,7\,9\,13\rangle$,	$\langle 1\,2\,9\,12\,16\rangle$,
	$\langle 1\,2\,10\,11\,13\rangle$,	$\langle 1\,2\,10\,13\,14\rangle$,	$\langle 1\,2\,10\,14\,15\rangle$,	$\langle 1\,2\,12\,16\,17\rangle$,	$\langle 1\,3\,5\,10\,17\rangle$,
	$\langle 1\,3\,5\,10\,18\rangle$,	$\langle 1\,3\,5\,14\,15\rangle$,	$\langle 1\,3\,5\,14\,18\rangle$,	$\langle 1\,3\,5\,15\,17\rangle$,	$\langle 1\,3\,7\,12\,13\rangle$,
	$\langle 1\,3\,7\,12\,14\rangle$,	$\langle 1\,3\,7\,13\,16\rangle$,	$\langle 1\,3\,7\,14\,16\rangle$,	$\langle 1\,3\,9\,12\,16\rangle$,	$\langle 1\,3\,9\,14\,16\rangle$,
	$\langle 1\,3\,10\,11\,12\rangle$,	$\langle 1\,3\,10\,12\,18\rangle$,	$\langle 1\,3\,10\,15\,17\rangle$,	$\langle 1\,3\,12\,13\,16\rangle$,	$\langle 1\,3\,12\,14\,18\rangle$,
	$\langle 1\,4\,5\,6\,12\rangle$,	$\langle 1\,4\,5\,6\,17\rangle$,	$\langle 1\,4\,5\,7\,9\rangle$,	$\langle 1\,4\,5\,7\,13\rangle$,	$\langle 1\,4\,5\,9\,17\rangle$,
	$\langle 1\,4\,5\,12\,13\rangle$,	$\langle 1\,4\,6\,11\,17\rangle$,	$\langle 1\,4\,7\,9\,10\rangle$,	$\langle 1\,4\,7\,10\,12\rangle$,	$\langle 1\,4\,7\,12\,13\rangle$,
	$\langle 1\,4\,8\,9\,11\rangle$,	$\langle 1\,4\,8\,9\,17\rangle$,	$\langle 1\,4\,8\,11\,17\rangle$,	$\langle 1\,4\,9\,10\,16\rangle$,	$\langle 1\,4\,9\,11\,16\rangle$,
	$\langle 1\,4\,10\,11\,12\rangle$,	$\langle 1\,4\,10\,11\,15\rangle$,	$\langle 1\,4\,10\,15\,16\rangle$,	$\langle 1\,4\,11\,15\,16\rangle$,	$\langle 1\,5\,6\,12\,17\rangle$,
	$\langle 1\,5\,9\,14\,17\rangle$,	$\langle 1\,5\,10\,17\,18\rangle$,	$\langle 1\,5\,12\,13\,15\rangle$,	$\langle 1\,5\,12\,15\,16\rangle$,	$\langle 1\,5\,12\,16\,17\rangle$,
	$\langle 1\,5\,13\,14\,15\rangle$,	$\langle 1\,5\,14\,17\,18\rangle$,	$\langle 1\,5\,15\,16\,17\rangle$,	$\langle 1\,6\,8\,11\,17\rangle$,	$\langle 1\,6\,8\,11\,18\rangle$,
	$\langle 1\,6\,8\,17\,18\rangle$,	$\langle 1\,6\,9\,10\,16\rangle$,	$\langle 1\,6\,9\,10\,18\rangle$,	$\langle 1\,6\,9\,13\,18\rangle$,	$\langle 1\,6\,10\,16\,18\rangle$,
	$\langle 1\,6\,11\,13\,18\rangle$,	$\langle 1\,6\,16\,17\,18\rangle$,	$\langle 1\,7\,8\,10\,12\rangle$,	$\langle 1\,7\,8\,10\,18\rangle$,	$\langle 1\,7\,8\,11\,13\rangle$,
	$\langle 1\,7\,8\,11\,18\rangle$,	$\langle 1\,7\,8\,12\,14\rangle$,	$\langle 1\,7\,8\,13\,16\rangle$,	$\langle 1\,7\,8\,14\,16\rangle$,	$\langle 1\,7\,9\,10\,18\rangle$,
	$\langle 1\,7\,9\,13\,18\rangle$,	$\langle 1\,7\,11\,13\,18\rangle$,	$\langle 1\,8\,9\,11\,14\rangle$,	$\langle 1\,8\,9\,14\,17\rangle$,	$\langle 1\,8\,10\,12\,18\rangle$,
	$\langle 1\,8\,11\,13\,15\rangle$,	$\langle 1\,8\,11\,14\,16\rangle$,	$\langle 1\,8\,11\,15\,16\rangle$,	$\langle 1\,8\,12\,14\,18\rangle$,	$\langle 1\,8\,13\,15\,16\rangle$,
	$\langle 1\,8\,14\,17\,18\rangle$,	$\langle 1\,9\,11\,14\,16\rangle$,	$\langle 1\,10\,11\,13\,15\rangle$,	$\langle 1\,10\,13\,14\,15\rangle$,	$\langle 1\,10\,15\,16\,17\rangle$,
	$\langle 1\,10\,16\,17\,18\rangle$,	$\langle 1\,12\,13\,15\,16\rangle$,	$\langle 2\,3\,5\,6\,7\rangle$,	$\langle 2\,3\,5\,6\,16\rangle$,	$\langle 2\,3\,5\,7\,17\rangle$,
	$\langle 2\,3\,5\,15\,16\rangle$,	$\langle 2\,3\,5\,15\,17\rangle$,	$\langle 2\,3\,6\,7\,10\rangle$,	$\langle 2\,3\,6\,10\,11\rangle$,	$\langle 2\,3\,6\,11\,14\rangle$,
	$\langle 2\,3\,6\,14\,16\rangle$,	$\langle 2\,3\,7\,10\,17\rangle$,	$\langle 2\,3\,9\,11\,12\rangle$,	$\langle 2\,3\,9\,11\,14\rangle$,	$\langle 2\,3\,10\,15\,17\rangle$,
	$\langle 2\,3\,14\,15\,16\rangle$,	$\langle 2\,4\,6\,11\,14\rangle$,	$\langle 2\,4\,6\,12\,18\rangle$,	$\langle 2\,4\,6\,14\,16\rangle$,	$\langle 2\,4\,6\,16\,18\rangle$,
	$\langle 2\,4\,11\,12\,18\rangle$,	$\langle 2\,4\,11\,14\,17\rangle$,	$\langle 2\,4\,11\,17\,18\rangle$,	$\langle 2\,4\,14\,16\,17\rangle$,	$\langle 2\,4\,16\,17\,18\rangle$,
	$\langle 2\,5\,6\,7\,10\rangle$,	$\langle 2\,5\,6\,10\,13\rangle$,	$\langle 2\,5\,6\,13\,16\rangle$,	$\langle 2\,5\,7\,9\,10\rangle$,	$\langle 2\,5\,7\,13\,17\rangle$,
	$\langle 2\,5\,9\,10\,14\rangle$,	$\langle 2\,5\,10\,13\,14\rangle$,	$\langle 2\,5\,13\,16\,17\rangle$,	$\langle 2\,5\,15\,16\,17\rangle$,	$\langle 2\,6\,9\,13\,16\rangle$,
	$\langle 2\,6\,10\,11\,13\rangle$,	$\langle 2\,6\,12\,17\,18\rangle$,	$\langle 2\,6\,16\,17\,18\rangle$,	$\langle 2\,7\,8\,10\,12\rangle$,	$\langle 2\,7\,8\,10\,14\rangle$,
	$\langle 2\,7\,8\,12\,14\rangle$,	$\langle 2\,7\,9\,10\,14\rangle$,	$\langle 2\,7\,9\,11\,12\rangle$,	$\langle 2\,7\,9\,11\,14\rangle$,	$\langle 2\,7\,9\,12\,15\rangle$,
	$\langle 2\,7\,9\,13\,17\rangle$,	$\langle 2\,7\,9\,15\,17\rangle$,	$\langle 2\,7\,10\,12\,15\rangle$,	$\langle 2\,7\,10\,15\,17\rangle$,	$\langle 2\,7\,11\,12\,18\rangle$,

Continuation on next page –

Table B.14 – Continuation from last page

⟨2 7 11 14 18⟩,	⟨2 7 12 14 18⟩,	⟨2 8 9 13 15⟩,	⟨2 8 9 13 17⟩,	⟨2 8 9 15 17⟩,
⟨2 8 10 12 15⟩,	⟨2 8 10 14 15⟩,	⟨2 8 12 13 15⟩,	⟨2 8 12 13 17⟩,	⟨2 8 12 14 18⟩,
⟨2 8 12 17 18⟩,	⟨2 8 14 15 17⟩,	⟨2 8 14 17 18⟩,	⟨2 9 12 13 15⟩,	⟨2 9 12 13 16⟩,
⟨2 11 14 17 18⟩,	⟨2 12 13 16 17⟩,	⟨2 14 15 16 17⟩,	⟨3 4 5 7 13⟩,	⟨3 4 5 7 15⟩,
⟨3 4 5 13 15⟩,	⟨3 4 6 12 13⟩,	⟨3 4 6 12 18⟩,	⟨3 4 6 13 18⟩,	⟨3 4 7 12 13⟩,
⟨3 4 7 12 14⟩,	⟨3 4 7 14 16⟩,	⟨3 4 7 15 16⟩,	⟨3 4 12 14 18⟩,	⟨3 4 13 15 18⟩,
⟨3 4 14 15 16⟩,	⟨3 4 14 15 18⟩,	⟨3 5 6 7 15⟩,	⟨3 5 6 15 16⟩,	⟨3 5 7 8 13⟩,
⟨3 5 7 8 17⟩,	⟨3 5 8 10 17⟩,	⟨3 5 8 10 18⟩,	⟨3 5 8 13 18⟩,	⟨3 5 13 15 18⟩,
⟨3 5 14 15 18⟩,	⟨3 6 7 10 17⟩,	⟨3 6 7 11 16⟩,	⟨3 6 7 11 18⟩,	⟨3 6 7 15 16⟩,
⟨3 6 7 17 18⟩,	⟨3 6 10 11 13⟩,	⟨3 6 10 12 13⟩,	⟨3 6 10 12 17⟩,	⟨3 6 11 13 18⟩,
⟨3 6 11 14 16⟩,	⟨3 6 12 17 18⟩,	⟨3 7 8 13 16⟩,	⟨3 7 8 16 18⟩,	⟨3 7 8 17 18⟩,
⟨3 7 11 16 18⟩,	⟨3 8 10 12 17⟩,	⟨3 8 10 12 18⟩,	⟨3 8 12 17 18⟩,	⟨3 8 13 16 18⟩,
⟨3 9 10 11 12⟩,	⟨3 9 10 11 16⟩,	⟨3 9 10 12 13⟩,	⟨3 9 10 13 16⟩,	⟨3 9 11 14 16⟩,
⟨3 9 12 13 16⟩,	⟨3 10 11 13 16⟩,	⟨3 11 13 16 18⟩,	⟨4 5 6 10 14⟩,	⟨4 5 6 10 15⟩,
⟨4 5 6 11 14⟩,	⟨4 5 6 11 17⟩,	⟨4 5 6 12 15⟩,	⟨4 5 7 9 10⟩,	⟨4 5 7 10 15⟩,
⟨4 5 9 10 14⟩,	⟨4 5 9 14 17⟩,	⟨4 5 11 14 17⟩,	⟨4 5 12 13 15⟩,	⟨4 6 9 10 14⟩,
⟨4 6 9 10 15⟩,	⟨4 6 9 14 15⟩,	⟨4 6 12 13 15⟩,	⟨4 6 13 14 15⟩,	⟨4 6 13 14 18⟩,
⟨4 6 14 16 18⟩,	⟨4 7 10 11 12⟩,	⟨4 7 10 11 15⟩,	⟨4 7 11 12 18⟩,	⟨4 7 11 15 16⟩,
⟨4 7 11 16 18⟩,	⟨4 7 12 14 18⟩,	⟨4 7 14 16 18⟩,	⟨4 8 9 11 17⟩,	⟨4 9 10 15 17⟩,
⟨4 9 10 16 17⟩,	⟨4 9 11 16 17⟩,	⟨4 9 14 15 17⟩,	⟨4 10 15 16 17⟩,	⟨4 11 16 17 18⟩,
⟨4 13 14 15 18⟩,	⟨4 14 15 16 17⟩,	⟨5 6 7 10 15⟩,	⟨5 6 10 13 16⟩,	⟨5 6 10 14 16⟩,
⟨5 6 11 12 16⟩,	⟨5 6 11 12 17⟩,	⟨5 6 11 14 16⟩,	⟨5 6 12 15 16⟩,	⟨5 7 8 13 17⟩,
⟨5 8 10 11 12⟩,	⟨5 8 10 11 14⟩,	⟨5 8 10 12 17⟩,	⟨5 8 10 14 16⟩,	⟨5 8 10 16 18⟩,
⟨5 8 11 12 16⟩,	⟨5 8 11 14 16⟩,	⟨5 8 12 16 17⟩,	⟨5 8 13 16 17⟩,	⟨5 8 13 16 18⟩,
⟨5 10 11 12 17⟩,	⟨5 10 11 13 14⟩,	⟨5 10 11 13 18⟩,	⟨5 10 11 17 18⟩,	⟨5 10 13 16 18⟩,
⟨5 11 13 14 18⟩,	⟨5 11 14 17 18⟩,	⟨5 13 14 15 18⟩,	⟨6 7 8 11 17⟩,	⟨6 7 8 11 18⟩,
⟨6 7 8 17 18⟩,	⟨6 7 10 15 17⟩,	⟨6 7 11 12 16⟩,	⟨6 7 11 12 17⟩,	⟨6 7 12 15 16⟩,
⟨6 7 12 15 17⟩,	⟨6 9 10 12 13⟩,	⟨6 9 10 12 17⟩,	⟨6 9 10 13 16⟩,	⟨6 9 10 14 18⟩,
⟨6 9 10 15 17⟩,	⟨6 9 12 13 15⟩,	⟨6 9 12 15 17⟩,	⟨6 9 13 14 15⟩,	⟨6 9 13 14 18⟩,
⟨6 10 14 16 18⟩,	⟨7 8 10 14 16⟩,	⟨7 8 10 16 18⟩,	⟨7 8 11 13 17⟩,	⟨7 9 10 14 18⟩,
⟨7 9 11 12 17⟩,	⟨7 9 11 13 14⟩,	⟨7 9 11 13 17⟩,	⟨7 9 12 15 17⟩,	⟨7 9 13 14 18⟩,
⟨7 10 11 12 15⟩,	⟨7 10 14 16 18⟩,	⟨7 11 12 15 16⟩,	⟨7 11 13 14 18⟩,	⟨8 9 11 13 15⟩,
⟨8 9 11 13 17⟩,	⟨8 9 11 14 15⟩,	⟨8 9 14 15 17⟩,	⟨8 10 11 12 15⟩,	⟨8 10 11 14 15⟩,
⟨8 11 12 15 16⟩,	⟨8 12 13 15 16⟩,	⟨8 12 13 16 17⟩,	⟨9 10 11 12 17⟩,	⟨9 10 11 16 17⟩,
⟨9 11 13 14 15⟩,	⟨10 11 13 14 15⟩,	⟨10 11 13 16 18⟩,	⟨10 11 16 17 18⟩	

f-vector:	$(18, 153, 642, 810, 324)$
# singularities:	3

Continuation on next page –

Table B.14 – Continuation from last page

Simplicial homology:	$(\mathbb{Z}, 0, \mathbb{Z}^{19}, 0, \mathbb{Z})$
Euler characteristic:	21

Table B.15: 14th blowup

C^{14}_{18} =	⟨1 2 3 6 11⟩,	⟨1 2 3 6 13⟩,	⟨1 2 3 11 15⟩,	⟨1 2 3 13 15⟩,	⟨1 2 6 11 16⟩,
	⟨1 2 6 13 16⟩,	⟨1 2 8 11 14⟩,	⟨1 2 8 11 16⟩,	⟨1 2 8 14 16⟩,	⟨1 2 9 12 14⟩,
	⟨1 2 9 12 15⟩,	⟨1 2 9 14 15⟩,	⟨1 2 11 14 15⟩,	⟨1 2 12 13 14⟩,	⟨1 2 12 13 18⟩,
	⟨1 2 12 15 18⟩,	⟨1 2 13 14 16⟩,	⟨1 2 13 15 18⟩,	⟨1 3 4 6 13⟩,	⟨1 3 4 6 17⟩,
	⟨1 3 4 7 13⟩,	⟨1 3 4 7 14⟩,	⟨1 3 4 8 10⟩,	⟨1 3 4 8 16⟩,	⟨1 3 4 10 14⟩,
	⟨1 3 4 12 16⟩,	⟨1 3 4 12 17⟩,	⟨1 3 6 11 17⟩,	⟨1 3 7 9 10⟩,	⟨1 3 7 9 14⟩,
	⟨1 3 7 10 13⟩,	⟨1 3 8 10 18⟩,	⟨1 3 8 12 15⟩,	⟨1 3 8 12 16⟩,	⟨1 3 8 15 18⟩,
	⟨1 3 9 10 14⟩,	⟨1 3 10 13 18⟩,	⟨1 3 11 15 17⟩,	⟨1 3 12 15 17⟩,	⟨1 3 13 15 18⟩,
	⟨1 4 6 7 13⟩,	⟨1 4 6 7 14⟩,	⟨1 4 6 14 18⟩,	⟨1 4 6 17 18⟩,	⟨1 4 8 10 14⟩,
	⟨1 4 8 14 16⟩,	⟨1 4 12 15 16⟩,	⟨1 4 12 15 17⟩,	⟨1 4 14 16 18⟩,	⟨1 4 15 16 17⟩,
	⟨1 4 16 17 18⟩,	⟨1 6 7 12 14⟩,	⟨1 6 7 12 16⟩,	⟨1 6 7 13 16⟩,	⟨1 6 11 12 16⟩,
	⟨1 6 11 12 18⟩,	⟨1 6 11 17 18⟩,	⟨1 6 12 14 18⟩,	⟨1 7 9 10 12⟩,	⟨1 7 9 12 14⟩,
	⟨1 7 10 12 16⟩,	⟨1 7 10 13 16⟩,	⟨1 8 10 11 14⟩,	⟨1 8 10 11 18⟩,	⟨1 8 11 12 16⟩,
	⟨1 8 11 12 18⟩,	⟨1 8 12 15 18⟩,	⟨1 9 10 12 15⟩,	⟨1 9 10 14 15⟩,	⟨1 10 11 13 17⟩,
	⟨1 10 11 13 18⟩,	⟨1 10 11 14 17⟩,	⟨1 10 12 15 16⟩,	⟨1 10 13 16 17⟩,	⟨1 10 14 15 17⟩,
	⟨1 10 15 16 17⟩,	⟨1 11 13 17 18⟩,	⟨1 11 14 15 17⟩,	⟨1 12 13 14 18⟩,	⟨1 13 14 16 18⟩,
	⟨1 13 16 17 18⟩,	⟨2 3 4 6 12⟩,	⟨2 3 4 6 15⟩,	⟨2 3 4 10 11⟩,	⟨2 3 4 10 12⟩,
	⟨2 3 4 11 15⟩,	⟨2 3 5 10 12⟩,	⟨2 3 5 10 14⟩,	⟨2 3 5 11 12⟩,	⟨2 3 5 11 14⟩,
	⟨2 3 6 7 16⟩,	⟨2 3 6 7 17⟩,	⟨2 3 6 11 16⟩,	⟨2 3 6 12 17⟩,	⟨2 3 6 13 15⟩,
	⟨2 3 7 8 9⟩,	⟨2 3 7 8 16⟩,	⟨2 3 7 9 17⟩,	⟨2 3 8 9 12⟩,	⟨2 3 8 12 16⟩,
	⟨2 3 9 12 17⟩,	⟨2 3 10 11 14⟩,	⟨2 3 11 12 16⟩,	⟨2 4 6 10 12⟩,	⟨2 4 6 10 15⟩,
	⟨2 4 7 8 10⟩,	⟨2 4 7 8 18⟩,	⟨2 4 7 10 17⟩,	⟨2 4 7 17 18⟩,	⟨2 4 8 10 18⟩,
	⟨2 4 10 11 17⟩,	⟨2 4 10 15 18⟩,	⟨2 4 11 15 18⟩,	⟨2 4 11 17 18⟩,	⟨2 5 6 8 10⟩,
	⟨2 5 6 8 17⟩,	⟨2 5 6 10 12⟩,	⟨2 5 6 12 14⟩,	⟨2 5 6 14 17⟩,	⟨2 5 7 8 10⟩,
	⟨2 5 7 8 16⟩,	⟨2 5 7 10 16⟩,	⟨2 5 8 16 17⟩,	⟨2 5 10 14 17⟩,	⟨2 5 10 16 17⟩,
	⟨2 5 11 12 13⟩,	⟨2 5 11 13 14⟩,	⟨2 5 12 13 14⟩,	⟨2 6 7 13 16⟩,	⟨2 6 7 13 17⟩,
	⟨2 6 8 10 18⟩,	⟨2 6 8 13 17⟩,	⟨2 6 8 13 18⟩,	⟨2 6 10 15 18⟩,	⟨2 6 12 14 17⟩,
	⟨2 6 13 15 18⟩,	⟨2 7 8 9 18⟩,	⟨2 7 9 12 14⟩,	⟨2 7 9 12 17⟩,	⟨2 7 9 14 15⟩,
	⟨2 7 9 15 18⟩,	⟨2 7 10 16 17⟩,	⟨2 7 11 14 15⟩,	⟨2 7 11 14 18⟩,	⟨2 7 11 15 18⟩,
	⟨2 7 12 14 17⟩,	⟨2 7 13 16 17⟩,	⟨2 7 14 17 18⟩,	⟨2 8 9 11 12⟩,	⟨2 8 9 11 13⟩,
	⟨2 8 9 13 18⟩,	⟨2 8 11 12 16⟩,	⟨2 8 11 13 14⟩,	⟨2 8 13 14 16⟩,	⟨2 8 13 16 17⟩,

Continuation on next page –

Table B.15 – Continuation from last page

⟨2 9 11 12 13⟩, ⟨2 9 12 13 18⟩, ⟨2 9 12 15 18⟩, ⟨2 10 11 14 17⟩, ⟨2 11 14 17 18⟩,
⟨3 4 6 12 17⟩, ⟨3 4 6 13 15⟩, ⟨3 4 7 9 11⟩, ⟨3 4 7 9 14⟩, ⟨3 4 7 11 15⟩,
⟨3 4 7 13 15⟩, ⟨3 4 8 10 16⟩, ⟨3 4 9 11 14⟩, ⟨3 4 10 11 14⟩, ⟨3 4 10 12 16⟩,
⟨3 5 8 11 13⟩, ⟨3 5 8 11 14⟩, ⟨3 5 8 13 17⟩, ⟨3 5 8 14 17⟩, ⟨3 5 10 12 16⟩,
⟨3 5 10 14 17⟩, ⟨3 5 10 16 17⟩, ⟨3 5 11 12 16⟩, ⟨3 5 11 13 15⟩, ⟨3 5 11 15 17⟩,
⟨3 5 11 16 17⟩, ⟨3 5 13 15 17⟩, ⟨3 6 7 9 10⟩, ⟨3 6 7 9 17⟩, ⟨3 6 7 10 18⟩,
⟨3 6 7 16 18⟩, ⟨3 6 8 9 14⟩, ⟨3 6 8 9 17⟩, ⟨3 6 8 14 17⟩, ⟨3 6 9 10 14⟩,
⟨3 6 10 14 16⟩, ⟨3 6 10 16 18⟩, ⟨3 6 11 16 17⟩, ⟨3 6 14 16 17⟩, ⟨3 7 8 9 18⟩,
⟨3 7 8 16 18⟩, ⟨3 7 9 11 13⟩, ⟨3 7 9 13 18⟩, ⟨3 7 10 13 18⟩, ⟨3 7 11 13 15⟩,
⟨3 8 9 11 13⟩, ⟨3 8 9 11 14⟩, ⟨3 8 9 12 17⟩, ⟨3 8 9 13 18⟩, ⟨3 8 10 16 18⟩,
⟨3 8 12 15 17⟩, ⟨3 8 13 15 17⟩, ⟨3 8 13 15 18⟩, ⟨3 10 14 16 17⟩, ⟨4 6 7 12 13⟩,
⟨4 6 7 12 14⟩, ⟨4 6 9 10 12⟩, ⟨4 6 9 10 15⟩, ⟨4 6 9 12 13⟩, ⟨4 6 9 13 15⟩,
⟨4 6 12 14 17⟩, ⟨4 6 14 17 18⟩, ⟨4 7 8 10 17⟩, ⟨4 7 8 12 15⟩, ⟨4 7 8 12 17⟩,
⟨4 7 8 14 15⟩, ⟨4 7 8 14 16⟩, ⟨4 7 8 16 18⟩, ⟨4 7 9 11 15⟩, ⟨4 7 9 14 15⟩,
⟨4 7 12 13 15⟩, ⟨4 7 12 14 17⟩, ⟨4 7 14 16 18⟩, ⟨4 7 14 17 18⟩, ⟨4 8 9 11 14⟩,
⟨4 8 9 11 17⟩, ⟨4 8 9 14 15⟩, ⟨4 8 9 15 17⟩, ⟨4 8 10 11 14⟩, ⟨4 8 10 11 17⟩,
⟨4 8 10 16 18⟩, ⟨4 8 12 15 17⟩, ⟨4 9 10 12 18⟩, ⟨4 9 10 15 18⟩, ⟨4 9 11 13 17⟩,
⟨4 9 11 13 18⟩, ⟨4 9 11 15 18⟩, ⟨4 9 12 13 18⟩, ⟨4 9 13 15 17⟩, ⟨4 10 12 16 18⟩,
⟨4 11 13 17 18⟩, ⟨4 12 13 15 16⟩, ⟨4 12 13 16 18⟩, ⟨4 13 15 16 17⟩, ⟨4 13 16 17 18⟩,
⟨5 6 8 10 12⟩, ⟨5 6 8 12 14⟩, ⟨5 6 8 14 17⟩, ⟨5 7 8 10 12⟩, ⟨5 7 8 12 15⟩,
⟨5 7 8 14 15⟩, ⟨5 7 8 14 16⟩, ⟨5 7 10 12 16⟩, ⟨5 7 11 12 13⟩, ⟨5 7 11 12 16⟩,
⟨5 7 11 13 15⟩, ⟨5 7 11 15 16⟩, ⟨5 7 12 13 15⟩, ⟨5 7 14 15 16⟩, ⟨5 8 11 13 14⟩,
⟨5 8 12 14 15⟩, ⟨5 8 13 14 16⟩, ⟨5 8 13 16 17⟩, ⟨5 11 15 16 17⟩, ⟨5 12 13 14 15⟩,
⟨5 13 14 15 16⟩, ⟨5 13 15 16 17⟩, ⟨6 7 9 10 17⟩, ⟨6 7 10 11 13⟩, ⟨6 7 10 11 18⟩,
⟨6 7 10 13 17⟩, ⟨6 7 11 12 13⟩, ⟨6 7 11 12 16⟩, ⟨6 7 11 16 18⟩, ⟨6 8 9 14 15⟩,
⟨6 8 9 15 17⟩, ⟨6 8 10 11 12⟩, ⟨6 8 10 11 18⟩, ⟨6 8 11 12 18⟩, ⟨6 8 12 14 15⟩,
⟨6 8 12 15 18⟩, ⟨6 8 13 15 17⟩, ⟨6 8 13 15 18⟩, ⟨6 9 10 12 13⟩, ⟨6 9 10 13 17⟩,
⟨6 9 10 14 15⟩, ⟨6 9 13 15 17⟩, ⟨6 10 11 12 13⟩, ⟨6 10 14 15 16⟩, ⟨6 10 15 16 18⟩,
⟨6 11 14 16 17⟩, ⟨6 11 14 16 18⟩, ⟨6 11 14 17 18⟩, ⟨6 12 14 15 18⟩, ⟨6 14 15 16 18⟩,
⟨7 8 10 12 17⟩, ⟨7 9 10 12 17⟩, ⟨7 9 11 13 18⟩, ⟨7 9 11 15 18⟩, ⟨7 10 11 13 18⟩,
⟨7 10 13 16 17⟩, ⟨7 11 14 15 16⟩, ⟨7 11 14 16 18⟩, ⟨8 9 11 12 17⟩, ⟨8 10 11 12 17⟩,
⟨9 10 12 13 17⟩, ⟨9 10 12 15 18⟩, ⟨9 11 12 13 17⟩, ⟨10 11 12 13 17⟩, ⟨10 12 15 16 18⟩,
⟨10 14 15 16 17⟩, ⟨11 14 15 16 17⟩, ⟨12 13 14 15 16⟩, ⟨12 13 14 16 18⟩, ⟨12 14 15 16 18⟩

f-vector:	$(18, 148, 632, 800, 320)$
# singularities:	2
Simplicial homology:	$(\mathbb{Z}, 0, \mathbb{Z}^{20}, 0, \mathbb{Z})$
Euler characteristic:	22

Continuation on next page –

Table B.16: 15th blowup

$C^{15}_{19} =$	⟨1 12 15 16 17⟩,	⟨1 2 12 13 15⟩,	⟨1 2 12 14 17⟩,	⟨1 2 12 15 17⟩,	⟨1 2 3 14 17⟩,
	⟨1 2 3 17 18⟩,	⟨1 2 3 4 14⟩,	⟨2 3 4 5 8⟩,	⟨2 3 4 5 9⟩,	⟨1 2 3 4 6⟩,
	⟨2 3 4 6 8⟩,	⟨2 3 4 9 15⟩,	⟨2 3 4 14 15⟩,	⟨2 3 5 6 8⟩,	⟨2 3 5 6 18⟩,
	⟨2 3 5 9 10⟩,	⟨2 3 5 10 17⟩,	⟨2 3 5 17 18⟩,	⟨1 2 3 6 18⟩,	⟨2 3 9 10 15⟩,
	⟨2 3 10 12 13⟩,	⟨2 3 10 12 14⟩,	⟨2 3 10 13 16⟩,	⟨2 3 10 14 17⟩,	⟨2 3 10 15 16⟩,
	⟨2 3 12 13 19⟩,	⟨2 3 12 14 19⟩,	⟨2 3 13 16 19⟩,	⟨1 2 4 12 13⟩,	⟨1 2 4 12 14⟩,
	⟨2 4 5 8 12⟩,	⟨2 4 5 9 16⟩,	⟨2 4 5 12 16⟩,	⟨1 2 4 6 18⟩,	⟨2 4 6 8 18⟩,
	⟨1 2 4 8 18⟩,	⟨2 4 8 10 19⟩,	⟨2 4 8 12 19⟩,	⟨2 4 9 12 14⟩,	⟨2 4 9 12 16⟩,
	⟨2 4 9 14 15⟩,	⟨1 2 4 10 13⟩,	⟨2 4 10 13 19⟩,	⟨2 4 12 13 19⟩,	⟨2 5 6 8 12⟩,
	⟨2 5 6 12 16⟩,	⟨2 5 6 14 16⟩,	⟨2 5 6 14 18⟩,	⟨2 5 9 10 13⟩,	⟨2 5 9 13 18⟩,
	⟨2 5 9 14 16⟩,	⟨2 5 9 14 18⟩,	⟨2 5 10 13 18⟩,	⟨2 5 10 17 18⟩,	⟨2 6 8 9 17⟩,
	⟨2 6 8 12 16⟩,	⟨2 6 8 17 18⟩,	⟨2 6 13 14 15⟩,	⟨2 6 13 14 18⟩,	⟨2 6 13 15 17⟩,
	⟨2 6 13 17 18⟩,	⟨1 2 8 17 18⟩,	⟨1 2 8 9 17⟩,	⟨2 8 10 16 19⟩,	⟨2 8 12 16 19⟩,
	⟨1 2 9 13 15⟩,	⟨1 2 9 15 17⟩,	⟨2 9 12 14 16⟩,	⟨2 9 13 14 15⟩,	⟨2 9 13 14 18⟩,
	⟨2 10 12 13 17⟩,	⟨2 10 12 14 17⟩,	⟨2 10 13 16 19⟩,	⟨2 10 13 17 18⟩,	⟨2 12 13 15 17⟩,
	⟨2 12 14 16 19⟩,	⟨1 3 15 16 17⟩,	⟨1 3 15 17 19⟩,	⟨1 3 17 18 19⟩,	⟨1 3 4 14 17⟩,
	⟨3 4 5 8 13⟩,	⟨3 4 5 9 13⟩,	⟨1 3 4 6 17⟩,	⟨3 4 6 8 17⟩,	⟨3 4 7 8 10⟩,
	⟨3 4 7 8 17⟩,	⟨3 4 7 9 10⟩,	⟨3 4 7 9 15⟩,	⟨3 4 7 15 16⟩,	⟨3 4 7 16 17⟩,
	⟨3 4 8 10 19⟩,	⟨3 4 8 13 19⟩,	⟨3 4 9 10 18⟩,	⟨3 4 9 13 18⟩,	⟨3 4 14 15 16⟩,
	⟨3 4 14 16 17⟩,	⟨3 5 6 8 9⟩,	⟨3 5 6 9 10⟩,	⟨3 5 8 9 13⟩,	⟨3 5 10 12 14⟩,
	⟨3 5 10 14 17⟩,	⟨3 5 12 14 19⟩,	⟨3 5 12 18 19⟩,	⟨3 5 14 15 17⟩,	⟨3 5 14 15 19⟩,
	⟨3 5 15 17 19⟩,	⟨3 5 17 18 19⟩,	⟨1 3 6 7 16⟩,	⟨1 3 6 7 17⟩,	⟨3 6 7 13 16⟩,
	⟨3 6 7 13 17⟩,	⟨3 6 8 9 17⟩,	⟨1 3 6 9 16⟩,	⟨1 3 6 9 18⟩,	⟨3 6 9 10 18⟩,
	⟨3 6 9 13 16⟩,	⟨3 6 9 13 17⟩,	⟨1 3 7 16 17⟩,	⟨3 7 8 10 12⟩,	⟨3 7 8 12 13⟩,
	⟨3 7 8 13 17⟩,	⟨3 7 9 10 15⟩,	⟨3 7 10 12 15⟩,	⟨3 7 12 13 15⟩,	⟨3 7 13 15 16⟩,
	⟨3 8 9 13 17⟩,	⟨3 8 10 12 19⟩,	⟨3 8 12 13 19⟩,	⟨3 10 12 13 16⟩,	⟨3 10 12 15 16⟩,
	⟨3 12 13 15 16⟩,	⟨3 14 15 16 17⟩,	⟨1 4 12 13 14⟩,	⟨1 4 5 15 19⟩,	⟨4 5 6 7 13⟩,
	⟨4 5 6 7 16⟩,	⟨4 5 6 12 13⟩,	⟨4 5 6 12 16⟩,	⟨1 4 5 7 13⟩,	⟨1 4 5 7 15⟩,
	⟨4 5 7 15 16⟩,	⟨4 5 8 12 13⟩,	⟨4 5 9 13 18⟩,	⟨4 5 9 16 18⟩,	⟨1 4 5 10 13⟩,
	⟨1 4 5 10 19⟩,	⟨4 5 10 13 19⟩,	⟨4 5 13 16 18⟩,	⟨4 5 13 16 19⟩,	⟨4 5 14 15 16⟩,
	⟨4 5 14 15 19⟩,	⟨4 5 14 16 19⟩,	⟨1 4 6 13 14⟩,	⟨1 4 6 14 17⟩,	⟨1 4 6 15 18⟩,
	⟨1 4 6 7 13⟩,	⟨1 4 6 7 16⟩,	⟨4 6 8 17 18⟩,	⟨1 4 6 9 15⟩,	⟨1 4 6 9 16⟩,
	⟨4 6 9 12 15⟩,	⟨4 6 9 12 16⟩,	⟨4 6 12 13 15⟩,	⟨4 6 13 14 15⟩,	⟨4 6 14 17 18⟩,

Continuation on next page –

Table B.16 – Continuation from last page

⟨4 7 8 9 10⟩, ⟨4 7 8 9 17⟩, ⟨1 4 7 9 15⟩, ⟨1 4 7 9 16⟩, ⟨4 7 9 16 17⟩,
⟨1 4 8 15 18⟩, ⟨4 8 9 10 18⟩, ⟨4 8 9 17 18⟩, ⟨4 8 12 13 19⟩, ⟨4 9 12 14 15⟩,
⟨4 9 16 17 18⟩, ⟨4 12 13 14 15⟩, ⟨4 13 16 17 18⟩, ⟨1 5 15 17 19⟩, ⟨5 6 7 12 13⟩,
⟨5 6 7 12 15⟩, ⟨5 6 7 15 16⟩, ⟨5 6 8 9 12⟩, ⟨5 6 9 10 12⟩, ⟨5 6 15 16 17⟩,
⟨5 7 8 12 13⟩, ⟨5 7 8 12 17⟩, ⟨5 7 8 13 17⟩, ⟨1 5 7 9 15⟩, ⟨1 5 7 9 10⟩,
⟨5 7 9 10 13⟩, ⟨1 5 7 10 13⟩, ⟨5 8 9 12 17⟩, ⟨5 8 9 13 17⟩, ⟨1 5 9 15 17⟩,
⟨1 5 9 10 17⟩, ⟨5 9 10 12 17⟩, ⟨5 9 14 16 18⟩, ⟨1 5 10 17 19⟩, ⟨5 10 12 14 17⟩,
⟨5 10 13 18 19⟩, ⟨5 10 17 18 19⟩, ⟨5 12 14 18 19⟩, ⟨5 13 16 18 19⟩, ⟨5 14 15 16 17⟩,
⟨5 14 16 18 19⟩, ⟨1 6 13 14 17⟩, ⟨1 6 7 13 17⟩, ⟨6 7 12 13 15⟩, ⟨6 7 13 15 16⟩,
⟨6 8 9 12 16⟩, ⟨1 6 9 15 18⟩, ⟨6 9 10 12 18⟩, ⟨6 9 12 15 18⟩, ⟨6 10 12 15 18⟩,
⟨6 13 14 17 18⟩, ⟨6 13 15 16 17⟩, ⟨1 7 12 16 17⟩, ⟨1 7 8 12 16⟩, ⟨1 7 8 9 16⟩,
⟨1 7 8 9 10⟩, ⟨7 8 9 16 17⟩, ⟨1 7 8 10 12⟩, ⟨7 8 12 16 17⟩, ⟨1 8 12 15 16⟩,
⟨1 8 12 15 18⟩, ⟨1 8 12 18 19⟩, ⟨1 8 17 18 19⟩, ⟨1 8 9 10 17⟩, ⟨8 9 10 17 18⟩,
⟨8 9 12 16 17⟩, ⟨1 8 10 12 19⟩, ⟨1 8 10 17 19⟩, ⟨8 10 16 18 19⟩, ⟨8 10 17 18 19⟩,
⟨8 12 15 16 18⟩, ⟨8 12 16 18 19⟩, ⟨1 9 12 13 15⟩, ⟨1 9 12 15 18⟩, ⟨9 10 12 17 18⟩,
⟨9 12 13 14 15⟩, ⟨9 12 13 14 18⟩, ⟨9 12 14 16 18⟩, ⟨9 12 16 17 18⟩, ⟨10 12 13 16 17⟩,
⟨10 12 15 16 18⟩, ⟨10 12 16 17 18⟩, ⟨10 13 16 17 18⟩, ⟨10 13 16 18 19⟩, ⟨1 11 12 13 14⟩,
⟨1 11 12 14 17⟩, ⟨1 11 12 18 19⟩, ⟨1 11 13 14 17⟩, ⟨2 3 11 14 15⟩, ⟨2 3 11 14 19⟩,
⟨2 3 11 15 16⟩, ⟨2 3 11 16 19⟩, ⟨1 2 4 8 11⟩, ⟨2 4 8 10 11⟩, ⟨1 2 4 10 11⟩,
⟨2 6 8 9 11⟩, ⟨2 6 8 11 16⟩, ⟨2 6 9 11 17⟩, ⟨2 6 11 14 15⟩, ⟨2 6 11 14 16⟩,
⟨2 6 11 15 17⟩, ⟨1 2 8 9 11⟩, ⟨2 8 10 11 16⟩, ⟨1 2 9 11 13⟩, ⟨2 9 10 11 13⟩,
⟨2 9 10 11 15⟩, ⟨2 9 11 15 17⟩, ⟨1 2 10 11 13⟩, ⟨2 10 11 15 16⟩, ⟨2 11 14 16 19⟩,
⟨1 3 11 15 16⟩, ⟨1 3 11 15 19⟩, ⟨1 3 11 18 19⟩, ⟨3 4 10 11 18⟩, ⟨3 4 10 11 19⟩,
⟨3 4 11 13 18⟩, ⟨3 4 11 13 19⟩, ⟨3 5 6 10 11⟩, ⟨3 5 6 11 18⟩, ⟨3 5 10 11 12⟩,
⟨3 5 11 12 18⟩, ⟨3 6 10 11 18⟩, ⟨1 3 9 11 16⟩, ⟨1 3 9 11 18⟩, ⟨3 9 11 13 16⟩,
⟨3 9 11 13 18⟩, ⟨3 10 11 12 19⟩, ⟨3 11 12 18 19⟩, ⟨3 11 13 16 19⟩, ⟨3 11 14 15 19⟩,
⟨1 4 11 15 19⟩, ⟨4 6 11 14 15⟩, ⟨4 6 11 14 18⟩, ⟨4 6 11 15 18⟩, ⟨1 4 8 11 15⟩,
⟨4 8 10 11 18⟩, ⟨4 8 11 15 18⟩, ⟨1 4 10 11 19⟩, ⟨4 11 13 16 17⟩, ⟨4 11 13 16 19⟩,
⟨4 11 13 17 18⟩, ⟨4 11 14 15 19⟩, ⟨4 11 14 16 17⟩, ⟨4 11 14 16 19⟩, ⟨4 11 14 17 18⟩,
⟨5 6 10 11 12⟩, ⟨5 6 11 12 15⟩, ⟨5 6 11 14 16⟩, ⟨5 6 11 14 18⟩, ⟨5 6 11 15 17⟩,
⟨5 6 11 16 17⟩, ⟨5 7 9 11 13⟩, ⟨5 7 9 11 15⟩, ⟨5 7 11 12 15⟩, ⟨5 7 11 12 17⟩,
⟨5 7 11 13 17⟩, ⟨5 9 11 13 17⟩, ⟨5 9 11 15 17⟩, ⟨5 11 12 14 17⟩, ⟨5 11 12 14 18⟩,
⟨5 11 14 16 17⟩, ⟨6 8 9 11 16⟩, ⟨6 9 11 13 16⟩, ⟨6 9 11 13 17⟩, ⟨6 10 11 12 15⟩,
⟨6 10 11 15 18⟩, ⟨6 11 13 16 17⟩, ⟨1 7 11 12 17⟩, ⟨1 7 11 13 17⟩, ⟨7 9 10 11 13⟩,
⟨7 9 10 11 15⟩, ⟨1 7 10 11 12⟩, ⟨1 7 10 11 13⟩, ⟨7 10 11 12 15⟩, ⟨1 8 11 15 16⟩,
⟨1 8 9 11 16⟩, ⟨8 10 11 16 18⟩, ⟨8 11 15 16 18⟩, ⟨1 9 11 12 13⟩, ⟨1 9 11 12 18⟩,
⟨9 11 12 13 18⟩, ⟨1 10 11 12 19⟩, ⟨10 11 15 16 18⟩, ⟨11 12 13 14 18⟩, ⟨11 13 14 17 18⟩,
⟨12 13 15 16 17⟩, ⟨12 14 16 18 19⟩

Continuation on next page –

f-vector:	$(19, 164, 696, 880, 352)$
# singularities:	1
Simplicial homology:	$(\mathbb{Z}, 0, \mathbb{Z}^{21}, 0, \mathbb{Z})$
Euler characteristic:	23

Table B.17: 16th blowup: 17-vertex triangulation of the $K3$ surface $(K3)_{17}$ with the standard PL structure. Note that the complex is not 2-neighborly.

$(K3)_{17}$ =	⟨1 2 3 8 13⟩,	⟨1 2 3 8 14⟩,	⟨1 2 3 12 13⟩,	⟨1 2 3 12 15⟩,	⟨1 2 3 14 15⟩,
	⟨1 2 4 7 13⟩,	⟨1 2 4 7 15⟩,	⟨1 2 4 13 15⟩,	⟨1 2 5 6 9⟩,	⟨1 2 5 6 14⟩,
	⟨1 2 5 9 17⟩,	⟨1 2 5 14 15⟩,	⟨1 2 5 15 17⟩,	⟨1 2 6 8 14⟩,	⟨1 2 6 8 15⟩,
	⟨1 2 6 9 16⟩,	⟨1 2 6 15 17⟩,	⟨1 2 6 16 17⟩,	⟨1 2 7 8 11⟩,	⟨1 2 7 8 15⟩,
	⟨1 2 7 11 13⟩,	⟨1 2 8 10 11⟩,	⟨1 2 8 10 13⟩,	⟨1 2 9 16 17⟩,	⟨1 2 10 11 13⟩,
	⟨1 2 12 13 15⟩,	⟨1 3 4 5 8⟩,	⟨1 3 4 5 17⟩,	⟨1 3 4 6 10⟩,	⟨1 3 4 6 12⟩,
	⟨1 3 4 8 9⟩,	⟨1 3 4 9 10⟩,	⟨1 3 4 12 17⟩,	⟨1 3 5 7 11⟩,	⟨1 3 5 7 14⟩,
	⟨1 3 5 8 16⟩,	⟨1 3 5 11 16⟩,	⟨1 3 5 14 15⟩,	⟨1 3 5 15 17⟩,	⟨1 3 6 10 12⟩,
	⟨1 3 7 8 11⟩,	⟨1 3 7 8 14⟩,	⟨1 3 8 9 12⟩,	⟨1 3 8 11 16⟩,	⟨1 3 8 12 13⟩,
	⟨1 3 9 10 12⟩,	⟨1 3 12 15 17⟩,	⟨1 4 5 8 16⟩,	⟨1 4 5 11 16⟩,	⟨1 4 5 11 17⟩,
	⟨1 4 6 7 12⟩,	⟨1 4 6 7 15⟩,	⟨1 4 6 10 15⟩,	⟨1 4 7 12 13⟩,	⟨1 4 8 9 16⟩,
	⟨1 4 9 10 14⟩,	⟨1 4 9 14 16⟩,	⟨1 4 10 14 16⟩,	⟨1 4 10 15 16⟩,	⟨1 4 11 16 17⟩,
	⟨1 4 12 13 17⟩,	⟨1 4 13 15 16⟩,	⟨1 4 13 16 17⟩,	⟨1 5 6 9 13⟩,	⟨1 5 6 13 14⟩,
	⟨1 5 7 10 12⟩,	⟨1 5 7 10 14⟩,	⟨1 5 7 11 12⟩,	⟨1 5 9 11 13⟩,	⟨1 5 9 11 17⟩,
	⟨1 5 10 12 14⟩,	⟨1 5 11 12 13⟩,	⟨1 5 12 13 14⟩,	⟨1 6 7 8 14⟩,	⟨1 6 7 8 15⟩,
	⟨1 6 7 10 12⟩,	⟨1 6 7 10 16⟩,	⟨1 6 7 14 16⟩,	⟨1 6 9 11 13⟩,	⟨1 6 9 11 14⟩,
	⟨1 6 9 14 16⟩,	⟨1 6 10 15 16⟩,	⟨1 6 11 13 14⟩,	⟨1 6 15 16 17⟩,	⟨1 7 10 14 16⟩,
	⟨1 7 11 12 13⟩,	⟨1 8 9 10 11⟩,	⟨1 8 9 10 12⟩,	⟨1 8 9 11 16⟩,	⟨1 8 10 12 13⟩,
	⟨1 9 10 11 14⟩,	⟨1 9 11 16 17⟩,	⟨1 10 11 13 14⟩,	⟨1 10 12 13 14⟩,	⟨1 12 13 15 16⟩,
	⟨1 12 13 16 17⟩,	⟨1 12 15 16 17⟩,	⟨2 3 4 6 12⟩,	⟨2 3 4 6 16⟩,	⟨2 3 4 7 14⟩,
	⟨2 3 4 7 16⟩,	⟨2 3 4 12 14⟩,	⟨2 3 5 6 12⟩,	⟨2 3 5 6 16⟩,	⟨2 3 5 12 16⟩,
	⟨2 3 7 14 16⟩,	⟨2 3 8 13 14⟩,	⟨2 3 9 10 13⟩,	⟨2 3 9 10 14⟩,	⟨2 3 9 11 14⟩,
	⟨2 3 9 11 16⟩,	⟨2 3 9 12 13⟩,	⟨2 3 9 12 16⟩,	⟨2 3 10 13 14⟩,	⟨2 3 11 14 16⟩,
	⟨2 3 12 14 15⟩,	⟨2 4 5 7 10⟩,	⟨2 4 5 7 11⟩,	⟨2 4 5 8 10⟩,	⟨2 4 5 8 12⟩,
	⟨2 4 5 11 12⟩,	⟨2 4 6 11 12⟩,	⟨2 4 6 11 16⟩,	⟨2 4 7 9 14⟩,	⟨2 4 7 9 15⟩,
	⟨2 4 7 10 13⟩,	⟨2 4 7 11 16⟩,	⟨2 4 8 10 12⟩,	⟨2 4 9 10 13⟩,	⟨2 4 9 10 14⟩,
	⟨2 4 9 13 15⟩,	⟨2 4 10 12 14⟩,	⟨2 5 6 7 10⟩,	⟨2 5 6 7 12⟩,	⟨2 5 6 8 10⟩,

Continuation on next page –

Table B.17 – Continuation from last page

⟨2 5 6 8 14⟩,	⟨2 5 6 9 16⟩,	⟨2 5 7 11 12⟩,	⟨2 5 8 12 14⟩,	⟨2 5 9 15 16⟩,
⟨2 5 9 15 17⟩,	⟨2 5 12 14 15⟩,	⟨2 5 12 15 16⟩,	⟨2 6 7 10 13⟩,	⟨2 6 7 11 12⟩,
⟨2 6 7 11 13⟩,	⟨2 6 8 10 15⟩,	⟨2 6 10 11 13⟩,	⟨2 6 10 11 17⟩,	⟨2 6 10 15 17⟩,
⟨2 6 11 16 17⟩,	⟨2 7 8 11 15⟩,	⟨2 7 9 11 14⟩,	⟨2 7 9 11 15⟩,	⟨2 7 11 14 16⟩,
⟨2 8 10 11 15⟩,	⟨2 8 10 12 14⟩,	⟨2 8 10 13 14⟩,	⟨2 9 11 15 17⟩,	⟨2 9 11 16 17⟩,
⟨2 9 12 13 16⟩,	⟨2 9 13 15 16⟩,	⟨2 10 11 15 17⟩,	⟨2 12 13 15 16⟩,	⟨3 4 5 8 17⟩,
⟨3 4 6 8 9⟩,	⟨3 4 6 8 11⟩,	⟨3 4 6 9 15⟩,	⟨3 4 6 10 15⟩,	⟨3 4 6 11 13⟩,
⟨3 4 6 13 16⟩,	⟨3 4 7 12 14⟩,	⟨3 4 7 12 17⟩,	⟨3 4 7 13 16⟩,	⟨3 4 7 13 17⟩,
⟨3 4 8 11 17⟩,	⟨3 4 9 10 15⟩,	⟨3 4 11 13 17⟩,	⟨3 5 6 10 12⟩,	⟨3 5 6 10 15⟩,
⟨3 5 6 13 15⟩,	⟨3 5 6 13 16⟩,	⟨3 5 7 11 15⟩,	⟨3 5 7 14 15⟩,	⟨3 5 8 16 17⟩,
⟨3 5 9 10 15⟩,	⟨3 5 9 10 17⟩,	⟨3 5 9 15 17⟩,	⟨3 5 10 12 16⟩,	⟨3 5 10 16 17⟩,
⟨3 5 11 15 16⟩,	⟨3 5 13 15 16⟩,	⟨3 6 7 8 9⟩,	⟨3 6 7 8 15⟩,	⟨3 6 7 9 15⟩,
⟨3 6 8 11 15⟩,	⟨3 6 11 13 15⟩,	⟨3 7 8 9 13⟩,	⟨3 7 8 11 15⟩,	⟨3 7 8 13 14⟩,
⟨3 7 9 13 17⟩,	⟨3 7 9 15 17⟩,	⟨3 7 12 14 15⟩,	⟨3 7 12 15 17⟩,	⟨3 7 13 14 16⟩,
⟨3 8 9 12 13⟩,	⟨3 8 10 11 16⟩,	⟨3 8 10 11 17⟩,	⟨3 8 10 16 17⟩,	⟨3 9 10 11 14⟩,
⟨3 9 10 11 16⟩,	⟨3 9 10 12 16⟩,	⟨3 9 10 13 17⟩,	⟨3 10 11 13 14⟩,	⟨3 10 11 13 17⟩,
⟨3 11 13 14 16⟩,	⟨3 11 13 15 16⟩,	⟨4 5 7 8 13⟩,	⟨4 5 7 8 16⟩,	⟨4 5 7 10 13⟩,
⟨4 5 7 11 16⟩,	⟨4 5 8 10 15⟩,	⟨4 5 8 11 12⟩,	⟨4 5 8 11 17⟩,	⟨4 5 8 13 15⟩,
⟨4 5 9 10 13⟩,	⟨4 5 9 10 15⟩,	⟨4 5 9 13 15⟩,	⟨4 6 7 9 12⟩,	⟨4 6 7 9 15⟩,
⟨4 6 8 9 14⟩,	⟨4 6 8 11 14⟩,	⟨4 6 9 12 14⟩,	⟨4 6 11 12 14⟩,	⟨4 6 11 13 17⟩,
⟨4 6 11 16 17⟩,	⟨4 6 13 16 17⟩,	⟨4 7 8 13 16⟩,	⟨4 7 9 12 14⟩,	⟨4 7 12 13 17⟩,
⟨4 8 9 14 16⟩,	⟨4 8 10 11 12⟩,	⟨4 8 10 11 15⟩,	⟨4 8 11 14 15⟩,	⟨4 8 13 15 16⟩,
⟨4 8 14 15 16⟩,	⟨4 10 11 12 15⟩,	⟨4 10 12 14 16⟩,	⟨4 10 12 15 16⟩,	⟨4 11 12 14 15⟩,
⟨4 12 14 15 16⟩,	⟨5 6 7 10 12⟩,	⟨5 6 8 10 15⟩,	⟨5 6 8 13 14⟩,	⟨5 6 8 13 15⟩,
⟨5 6 9 13 16⟩,	⟨5 7 8 9 13⟩,	⟨5 7 8 9 17⟩,	⟨5 7 8 16 17⟩,	⟨5 7 9 10 13⟩,
⟨5 7 9 10 17⟩,	⟨5 7 10 14 16⟩,	⟨5 7 10 16 17⟩,	⟨5 7 11 15 16⟩,	⟨5 7 14 15 16⟩,
⟨5 8 9 12 13⟩,	⟨5 8 9 12 17⟩,	⟨5 8 11 12 17⟩,	⟨5 8 12 13 14⟩,	⟨5 9 11 12 13⟩,
⟨5 9 11 12 17⟩,	⟨5 9 13 15 16⟩,	⟨5 10 12 14 16⟩,	⟨5 12 14 15 16⟩,	⟨6 7 8 9 16⟩,
⟨6 7 8 14 16⟩,	⟨6 7 9 12 16⟩,	⟨6 7 10 13 17⟩,	⟨6 7 10 16 17⟩,	⟨6 7 11 12 13⟩,
⟨6 7 12 13 17⟩,	⟨6 7 12 16 17⟩,	⟨6 8 9 14 16⟩,	⟨6 8 11 13 14⟩,	⟨6 8 11 13 15⟩,
⟨6 9 11 12 13⟩,	⟨6 9 11 12 14⟩,	⟨6 9 12 13 16⟩,	⟨6 10 11 13 17⟩,	⟨6 10 15 16 17⟩,
⟨6 12 13 16 17⟩,	⟨7 8 9 12 16⟩,	⟨7 8 9 12 17⟩,	⟨7 8 12 16 17⟩,	⟨7 8 13 14 16⟩,
⟨7 9 10 13 17⟩,	⟨7 9 11 14 15⟩,	⟨7 9 12 14 15⟩,	⟨7 9 12 15 17⟩,	⟨7 11 14 15 16⟩,
⟨8 9 10 11 16⟩,	⟨8 9 10 12 16⟩,	⟨8 10 11 12 17⟩,	⟨8 10 12 13 14⟩,	⟨8 10 12 16 17⟩,
⟨8 11 13 14 15⟩,	⟨8 13 14 15 16⟩,	⟨9 11 12 14 15⟩,	⟨9 11 12 15 17⟩,	⟨10 11 12 15 17⟩,
⟨10 12 15 16 17⟩,	⟨11 13 14 15 16⟩			

f-vector:	(17, 135, 610, 780, 312)

Continuation on next page –

Table B.17 – Continuation from last page

# singularities:	0
Simplicial homology:	$(\mathbb{Z}, 0, \mathbb{Z}^{22}, 0, \mathbb{Z})$
Euler characteristic:	24

Appendix C

A non-simply connected combinatorial 4-manifold with 44 vertices

In the following, the facet list of K_{17} from Section 2.6 is listed.

Table C.1: The combinatorial 4-manifold K_{17}.

$K_{17} =$	⟨1 2 3 7 19⟩,	⟨1 2 3 7 25⟩,	⟨1 2 3 19 40⟩,	⟨1 2 3 25 40⟩,	⟨1 2 7 19 42⟩,
	⟨1 2 7 25 42⟩,	⟨1 2 19 21 40⟩,	⟨1 2 19 21 42⟩,	⟨1 2 21 25 40⟩,	⟨1 2 21 25 42⟩,
	⟨1 3 5 18 26⟩,	⟨1 3 5 18 32⟩,	⟨1 3 5 26 30⟩,	⟨1 3 5 30 32⟩,	⟨1 3 7 19 26⟩,
	⟨1 3 7 25 30⟩,	⟨1 3 7 26 30⟩,	⟨1 3 18 19 26⟩,	⟨1 3 18 19 41⟩,	⟨1 3 18 32 41⟩,
	⟨1 3 19 40 41⟩,	⟨1 3 25 30 32⟩,	⟨1 3 25 32 40⟩,	⟨1 3 32 40 41⟩,	⟨1 5 9 18 26⟩,
	⟨1 5 9 18 32⟩,	⟨1 5 9 26 30⟩,	⟨1 5 9 30 32⟩,	⟨1 7 12 19 26⟩,	⟨1 7 12 19 42⟩,
	⟨1 7 12 26 30⟩,	⟨1 7 12 28 30⟩,	⟨1 7 12 28 42⟩,	⟨1 7 25 28 30⟩,	⟨1 7 25 28 42⟩,
	⟨1 8 9 12 26⟩,	⟨1 8 9 12 37⟩,	⟨1 8 9 18 26⟩,	⟨1 8 9 18 37⟩,	⟨1 8 12 19 26⟩,
	⟨1 8 12 19 37⟩,	⟨1 8 18 19 26⟩,	⟨1 8 18 19 37⟩,	⟨1 9 12 14 35⟩,	⟨1 9 12 14 37⟩,
	⟨1 9 12 26 30⟩,	⟨1 9 12 30 35⟩,	⟨1 9 14 32 35⟩,	⟨1 9 14 32 37⟩,	⟨1 9 18 32 37⟩,
	⟨1 9 30 32 35⟩,	⟨1 10 12 14 28⟩,	⟨1 10 12 14 37⟩,	⟨1 10 12 28 42⟩,	⟨1 10 12 37 42⟩,
	⟨1 10 14 28 36⟩,	⟨1 10 14 36 37⟩,	⟨1 10 28 36 42⟩,	⟨1 10 36 37 42⟩,	⟨1 12 14 28 35⟩,
	⟨1 12 19 37 42⟩,	⟨1 12 28 30 35⟩,	⟨1 14 16 23 32⟩,	⟨1 14 16 23 36⟩,	⟨1 14 16 28 35⟩,
	⟨1 14 16 28 36⟩,	⟨1 14 16 32 35⟩,	⟨1 14 23 32 37⟩,	⟨1 14 23 36 37⟩,	⟨1 16 23 25 32⟩,
	⟨1 16 23 25 36⟩,	⟨1 16 25 28 30⟩,	⟨1 16 25 28 36⟩,	⟨1 16 25 30 32⟩,	⟨1 16 28 30 35⟩,
	⟨1 16 30 32 35⟩,	⟨1 18 19 37 41⟩,	⟨1 18 32 37 41⟩,	⟨1 19 21 37 41⟩,	⟨1 19 21 37 42⟩,
	⟨1 19 21 40 41⟩,	⟨1 21 23 25 36⟩,	⟨1 21 23 25 40⟩,	⟨1 21 23 36 37⟩,	⟨1 21 23 37 41⟩,
	⟨1 21 23 40 41⟩,	⟨1 21 25 36 42⟩,	⟨1 21 36 37 42⟩,	⟨1 23 25 32 40⟩,	⟨1 23 32 37 41⟩,
	⟨1 23 32 40 41⟩,	⟨1 25 28 36 42⟩,	⟨2 3 7 8 23⟩,	⟨2 3 7 8 25⟩,	⟨2 3 7 19 23⟩,

Continuation on next page –

Table C.1 – Continuation from last page

⟨2 3 8 12 23⟩, ⟨2 3 8 12 25⟩, ⟨2 3 12 17 23⟩, ⟨2 3 12 17 40⟩, ⟨2 3 12 25 40⟩,
⟨2 3 16 17 39⟩, ⟨2 3 16 17 40⟩, ⟨2 3 16 19 22⟩, ⟨2 3 16 19 40⟩, ⟨2 3 16 22 39⟩,
⟨2 3 17 23 39⟩, ⟨2 3 19 22 23⟩, ⟨2 3 22 23 39⟩, ⟨2 7 8 23 42⟩, ⟨2 7 8 25 42⟩,
⟨2 7 19 23 24⟩, ⟨2 7 19 24 42⟩, ⟨2 7 23 24 42⟩, ⟨2 8 9 12 13⟩, ⟨2 8 9 12 43⟩,
⟨2 8 9 13 20⟩, ⟨2 8 9 20 43⟩, ⟨2 8 12 13 23⟩, ⟨2 8 12 25 43⟩,
⟨2 8 13 20 23⟩, ⟨2 8 20 23 42⟩, ⟨2 8 20 42 43⟩, ⟨2 8 25 42 43⟩, ⟨2 9 10 13 15⟩,
⟨2 9 10 13 20⟩, ⟨2 9 10 15 27⟩, ⟨2 9 10 20 41⟩, ⟨2 9 10 27 41⟩, ⟨2 9 12 13 15⟩,
⟨2 9 12 15 27⟩, ⟨2 9 12 27 43⟩, ⟨2 9 20 41 43⟩, ⟨2 9 27 41 43⟩, ⟨2 10 13 15 28⟩,
⟨2 10 13 20 28⟩, ⟨2 10 14 15 27⟩, ⟨2 10 14 15 28⟩, ⟨2 10 14 27 41⟩, ⟨2 10 14 28 41⟩,
⟨2 10 20 28 41⟩, ⟨2 12 13 15 17⟩, ⟨2 12 13 17 23⟩, ⟨2 12 15 17 40⟩, ⟨2 12 15 27 40⟩,
⟨2 12 25 27 40⟩, ⟨2 12 25 27 43⟩, ⟨2 13 15 17 28⟩, ⟨2 13 17 23 28⟩, ⟨2 13 20 23 28⟩,
⟨2 14 15 16 21⟩, ⟨2 14 15 16 28⟩, ⟨2 14 15 21 27⟩, ⟨2 14 16 21 35⟩, ⟨2 14 16 28 35⟩,
⟨2 14 21 27 38⟩, ⟨2 14 21 35 38⟩, ⟨2 14 27 38 41⟩, ⟨2 14 28 35 41⟩, ⟨2 14 35 38 41⟩,
⟨2 15 16 17 28⟩, ⟨2 15 16 17 40⟩, ⟨2 15 16 21 40⟩, ⟨2 15 21 27 40⟩, ⟨2 16 17 28 39⟩,
⟨2 16 19 21 35⟩, ⟨2 16 19 21 40⟩, ⟨2 16 19 22 30⟩, ⟨2 16 19 30 35⟩, ⟨2 16 22 28 30⟩,
⟨2 16 22 28 39⟩, ⟨2 16 28 30 35⟩, ⟨2 17 23 28 39⟩, ⟨2 18 20 23 28⟩, ⟨2 18 20 23 42⟩,
⟨2 18 20 28 41⟩, ⟨2 18 20 41 42⟩, ⟨2 18 23 24 30⟩, ⟨2 18 23 24 42⟩, ⟨2 18 23 28 30⟩,
⟨2 18 24 30 35⟩, ⟨2 18 24 35 42⟩, ⟨2 18 28 30 35⟩, ⟨2 18 28 35 41⟩, ⟨2 18 35 41 42⟩,
⟨2 19 21 35 42⟩, ⟨2 19 22 23 30⟩, ⟨2 19 23 24 30⟩, ⟨2 19 24 30 35⟩, ⟨2 19 24 35 42⟩,
⟨2 20 41 42 43⟩, ⟨2 21 25 27 40⟩, ⟨2 21 25 27 42⟩, ⟨2 21 27 38 42⟩, ⟨2 21 35 38 42⟩,
⟨2 22 23 28 30⟩, ⟨2 22 23 28 39⟩, ⟨2 25 27 42 43⟩, ⟨2 27 38 41 42⟩, ⟨2 27 41 42 43⟩,
⟨2 35 38 41 42⟩, ⟨3 4 9 11 17⟩, ⟨3 4 9 11 31⟩, ⟨3 4 9 17 29⟩, ⟨3 4 9 27 31⟩,
⟨3 4 9 27 34⟩, ⟨3 4 9 29 34⟩, ⟨3 4 11 12 17⟩, ⟨3 4 11 12 31⟩, ⟨3 4 12 17 40⟩,
⟨3 4 12 25 31⟩, ⟨3 4 12 25 40⟩, ⟨3 4 17 29 40⟩, ⟨3 4 25 31 32⟩, ⟨3 4 25 32 40⟩,
⟨3 4 27 28 32⟩, ⟨3 4 27 28 34⟩, ⟨3 4 27 31 32⟩, ⟨3 4 28 29 32⟩, ⟨3 4 28 29 34⟩,
⟨3 4 29 32 40⟩, ⟨3 5 14 15 27⟩, ⟨3 5 14 15 36⟩, ⟨3 5 14 27 30⟩, ⟨3 5 14 30 37⟩,
⟨3 5 14 36 37⟩, ⟨3 5 15 20 36⟩, ⟨3 5 15 20 42⟩, ⟨3 5 15 27 34⟩, ⟨3 5 15 34 42⟩,
⟨3 5 18 20 26⟩, ⟨3 5 18 20 42⟩, ⟨3 5 18 32 42⟩, ⟨3 5 20 26 37⟩, ⟨3 5 20 36 37⟩,
⟨3 5 26 30 37⟩, ⟨3 5 27 28 32⟩, ⟨3 5 27 28 34⟩, ⟨3 5 27 30 32⟩, ⟨3 5 28 32 42⟩,
⟨3 5 28 34 42⟩, ⟨3 7 8 23 31⟩, ⟨3 7 8 25 31⟩, ⟨3 7 10 23 31⟩, ⟨3 7 10 23 38⟩,
⟨3 7 10 30 31⟩, ⟨3 7 10 30 37⟩, ⟨3 7 10 37 38⟩, ⟨3 7 19 23 38⟩, ⟨3 7 19 26 38⟩,
⟨3 7 25 30 31⟩, ⟨3 7 26 30 37⟩, ⟨3 7 26 37 38⟩, ⟨3 8 12 23 31⟩, ⟨3 8 12 25 31⟩,
⟨3 9 10 15 27⟩, ⟨3 9 10 15 39⟩, ⟨3 9 10 23 31⟩, ⟨3 9 10 23 39⟩, ⟨3 9 10 27 31⟩,
⟨3 9 11 17 23⟩, ⟨3 9 11 23 31⟩, ⟨3 9 15 27 34⟩, ⟨3 9 15 29 34⟩, ⟨3 9 15 29 39⟩,
⟨3 9 17 23 39⟩, ⟨3 9 17 29 39⟩, ⟨3 10 14 15 27⟩, ⟨3 10 14 15 36⟩, ⟨3 10 14 27 30⟩,
⟨3 10 14 30 37⟩, ⟨3 10 14 36 37⟩, ⟨3 10 15 36 39⟩, ⟨3 10 22 23 38⟩, ⟨3 10 22 23 39⟩,
⟨3 10 22 36 38⟩, ⟨3 10 22 36 39⟩, ⟨3 10 27 30 31⟩, ⟨3 10 36 37 38⟩, ⟨3 11 12 17 23⟩,
⟨3 11 12 23 31⟩, ⟨3 15 20 36 39⟩, ⟨3 15 20 39 43⟩, ⟨3 15 20 42 43⟩, ⟨3 15 29 34 42⟩,

Continuation on next page –

Table C.1 – Continuation from last page

⟨3 15 29 39 43⟩,	⟨3 15 29 42 43⟩,	⟨3 16 17 39 43⟩,	⟨3 16 17 40 43⟩,	⟨3 16 19 20 22⟩,
⟨3 16 19 20 41⟩,	⟨3 16 19 40 41⟩,	⟨3 16 20 22 39⟩,	⟨3 16 20 39 43⟩,	⟨3 16 20 41 43⟩,
⟨3 16 40 41 43⟩,	⟨3 17 29 39 43⟩,	⟨3 17 29 40 43⟩,	⟨3 18 19 20 26⟩,	⟨3 18 19 20 41⟩,
⟨3 18 20 41 42⟩,	⟨3 18 32 41 42⟩,	⟨3 19 20 22 38⟩,	⟨3 19 20 26 38⟩,	⟨3 19 22 23 38⟩,
⟨3 20 22 36 38⟩,	⟨3 20 22 36 39⟩,	⟨3 20 26 37 38⟩,	⟨3 20 36 37 38⟩,	⟨3 20 41 42 43⟩,
⟨3 25 30 31 32⟩,	⟨3 27 30 31 32⟩,	⟨3 28 29 32 42⟩,	⟨3 28 29 34 42⟩,	⟨3 29 32 40 41⟩,
⟨3 29 32 41 42⟩,	⟨3 29 40 41 43⟩,	⟨3 29 41 42 43⟩,	⟨4 5 6 17 30⟩,	⟨4 5 6 17 41⟩,
⟨4 5 6 30 37⟩,	⟨4 5 6 37 41⟩,	⟨4 5 9 11 17⟩,	⟨4 5 9 11 26⟩,	⟨4 5 9 17 30⟩,
⟨4 5 9 26 30⟩,	⟨4 5 11 17 41⟩,	⟨4 5 11 26 41⟩,	⟨4 5 26 30 37⟩,	⟨4 5 26 37 41⟩,
⟨4 6 17 22 29⟩,	⟨4 6 17 22 41⟩,	⟨4 6 17 29 30⟩,	⟨4 6 22 29 37⟩,	⟨4 6 22 37 41⟩,
⟨4 6 29 30 37⟩,	⟨4 8 15 24 28⟩,	⟨4 8 15 24 39⟩,	⟨4 8 15 28 36⟩,	⟨4 8 15 36 39⟩,
⟨4 8 18 22 37⟩,	⟨4 8 18 22 39⟩,	⟨4 8 18 33 37⟩,	⟨4 8 18 33 39⟩,	⟨4 8 22 29 36⟩,
⟨4 8 22 29 37⟩,	⟨4 8 22 36 39⟩,	⟨4 8 24 28 37⟩,	⟨4 8 24 33 37⟩,	⟨4 8 24 33 39⟩,
⟨4 8 28 29 36⟩,	⟨4 8 28 29 37⟩,	⟨4 9 11 26 43⟩,	⟨4 9 11 31 43⟩,	⟨4 9 17 29 30⟩,
⟨4 9 26 27 34⟩,	⟨4 9 26 27 43⟩,	⟨4 9 26 30 34⟩,	⟨4 9 27 31 43⟩,	⟨4 9 29 30 34⟩,
⟨4 10 13 15 28⟩,	⟨4 10 13 15 39⟩,	⟨4 10 13 25 32⟩,	⟨4 10 13 25 39⟩,	⟨4 10 13 28 32⟩,
⟨4 10 15 28 36⟩,	⟨4 10 15 36 39⟩,	⟨4 10 22 29 36⟩,	⟨4 10 22 29 40⟩,	⟨4 10 22 36 39⟩,
⟨4 10 22 39 40⟩,	⟨4 10 25 32 40⟩,	⟨4 10 25 39 40⟩,	⟨4 10 28 29 32⟩,	⟨4 10 28 29 36⟩,
⟨4 10 29 32 40⟩,	⟨4 11 12 17 41⟩,	⟨4 11 12 31 44⟩,	⟨4 11 12 33 41⟩,	⟨4 11 12 33 44⟩,
⟨4 11 26 33 41⟩,	⟨4 11 26 33 44⟩,	⟨4 11 26 43 44⟩,	⟨4 11 31 43 44⟩,	⟨4 12 17 22 40⟩,
⟨4 12 17 22 41⟩,	⟨4 12 18 22 39⟩,	⟨4 12 18 22 41⟩,	⟨4 12 18 33 39⟩,	⟨4 12 18 33 41⟩,
⟨4 12 19 24 31⟩,	⟨4 12 19 24 39⟩,	⟨4 12 19 25 31⟩,	⟨4 12 19 25 39⟩,	⟨4 12 22 39 40⟩,
⟨4 12 24 31 44⟩,	⟨4 12 24 33 39⟩,	⟨4 12 24 33 44⟩,	⟨4 12 25 39 40⟩,	⟨4 13 15 24 28⟩,
⟨4 13 15 24 39⟩,	⟨4 13 19 24 31⟩,	⟨4 13 19 24 39⟩,	⟨4 13 19 25 31⟩,	⟨4 13 19 25 39⟩,
⟨4 13 24 27 28⟩,	⟨4 13 24 27 31⟩,	⟨4 13 25 31 32⟩,	⟨4 13 27 28 32⟩,	⟨4 13 27 31 32⟩,
⟨4 17 22 29 40⟩,	⟨4 18 22 37 41⟩,	⟨4 18 33 37 41⟩,	⟨4 24 26 27 28⟩,	⟨4 24 26 27 44⟩,
⟨4 24 26 28 37⟩,	⟨4 24 26 33 37⟩,	⟨4 24 26 33 44⟩,	⟨4 24 27 31 44⟩,	⟨4 26 27 28 34⟩,
⟨4 26 27 43 44⟩,	⟨4 26 28 34 37⟩,	⟨4 26 30 34 37⟩,	⟨4 26 33 37 41⟩,	⟨4 27 31 43 44⟩,
⟨4 28 29 34 37⟩,	⟨4 29 30 34 37⟩,	⟨5 6 7 14 23⟩,	⟨5 6 7 14 27⟩,	⟨5 6 7 23 24⟩,
⟨5 6 7 24 33⟩,	⟨5 6 7 27 33⟩,	⟨5 6 14 23 37⟩,	⟨5 6 14 27 30⟩,	⟨5 6 14 30 37⟩,
⟨5 6 17 24 33⟩,	⟨5 6 17 24 41⟩,	⟨5 6 17 30 33⟩,	⟨5 6 23 24 41⟩,	⟨5 6 23 37 41⟩,
⟨5 6 27 30 33⟩,	⟨5 7 14 15 23⟩,	⟨5 7 14 15 27⟩,	⟨5 7 15 23 42⟩,	⟨5 7 15 27 34⟩,
⟨5 7 15 34 42⟩,	⟨5 7 23 24 42⟩,	⟨5 7 24 33 42⟩,	⟨5 7 27 28 33⟩,	⟨5 7 27 28 34⟩,
⟨5 7 28 33 42⟩,	⟨5 7 28 34 42⟩,	⟨5 9 11 17 24⟩,	⟨5 9 11 18 24⟩,	⟨5 9 11 18 26⟩,
⟨5 9 17 24 32⟩,	⟨5 9 17 30 32⟩,	⟨5 9 18 24 32⟩,	⟨5 11 17 24 41⟩,	⟨5 11 18 21 24⟩,
⟨5 11 18 21 26⟩,	⟨5 11 21 24 41⟩,	⟨5 11 21 26 41⟩,	⟨5 14 15 23 36⟩,	⟨5 14 23 36 37⟩,
⟨5 15 20 23 36⟩,	⟨5 15 20 23 42⟩,	⟨5 17 24 32 33⟩,	⟨5 17 30 32 33⟩,	⟨5 18 20 21 23⟩,
⟨5 18 20 21 26⟩,	⟨5 18 20 23 42⟩,	⟨5 18 21 23 24⟩,	⟨5 18 23 24 42⟩,	⟨5 18 24 32 42⟩,

Continuation on next page –

Table C.1 – Continuation from last page

⟨5 20 21 23 37⟩, ⟨5 20 21 26 37⟩, ⟨5 20 23 36 37⟩, ⟨5 21 23 24 41⟩, ⟨5 21 23 37 41⟩,
⟨5 21 26 37 41⟩, ⟨5 24 32 33 42⟩, ⟨5 27 28 32 33⟩, ⟨5 27 30 32 33⟩, ⟨5 28 32 33 42⟩,
⟨6 7 11 18 21⟩, ⟨6 7 11 18 39⟩, ⟨6 7 11 21 32⟩, ⟨6 7 11 32 38⟩, ⟨6 7 11 38 39⟩,
⟨6 7 14 21 27⟩, ⟨6 7 14 21 32⟩, ⟨6 7 14 23 32⟩, ⟨6 7 18 21 27⟩, ⟨6 7 18 27 33⟩,
⟨6 7 18 33 39⟩, ⟨6 7 23 24 38⟩, ⟨6 7 23 32 38⟩, ⟨6 7 24 33 39⟩, ⟨6 7 24 38 39⟩,
⟨6 8 11 15 26⟩, ⟨6 8 11 15 39⟩, ⟨6 8 11 18 26⟩, ⟨6 8 11 18 39⟩, ⟨6 8 15 17 24⟩,
⟨6 8 15 17 31⟩, ⟨6 8 15 24 39⟩, ⟨6 8 15 25 26⟩, ⟨6 8 15 25 31⟩, ⟨6 8 17 19 31⟩,
⟨6 8 17 19 33⟩, ⟨6 8 17 24 33⟩, ⟨6 8 18 19 26⟩, ⟨6 8 18 19 33⟩, ⟨6 8 18 33 39⟩,
⟨6 8 19 25 26⟩, ⟨6 8 19 25 31⟩, ⟨6 8 24 33 39⟩, ⟨6 11 15 26 32⟩, ⟨6 11 15 32 38⟩,
⟨6 11 15 38 39⟩, ⟨6 11 18 21 26⟩, ⟨6 11 21 26 32⟩, ⟨6 13 14 19 25⟩, ⟨6 13 14 19 30⟩,
⟨6 13 14 22 25⟩, ⟨6 13 14 22 37⟩, ⟨6 13 14 30 37⟩, ⟨6 13 19 25 31⟩, ⟨6 13 19 29 30⟩,
⟨6 13 19 29 31⟩, ⟨6 13 22 25 31⟩, ⟨6 13 22 29 31⟩, ⟨6 13 22 29 37⟩, ⟨6 13 29 30 37⟩,
⟨6 14 19 20 26⟩, ⟨6 14 19 20 30⟩, ⟨6 14 19 25 26⟩, ⟨6 14 20 21 26⟩, ⟨6 14 20 21 27⟩,
⟨6 14 20 27 30⟩, ⟨6 14 21 26 32⟩, ⟨6 14 22 25 26⟩, ⟨6 14 22 26 32⟩, ⟨6 14 22 32 37⟩,
⟨6 14 23 32 37⟩, ⟨6 15 17 22 31⟩, ⟨6 15 17 22 41⟩, ⟨6 15 17 24 41⟩, ⟨6 15 22 25 26⟩,
⟨6 15 22 25 31⟩, ⟨6 15 22 26 32⟩, ⟨6 15 22 32 41⟩, ⟨6 15 24 38 39⟩, ⟨6 15 24 38 41⟩,
⟨6 15 32 38 41⟩, ⟨6 17 19 29 30⟩, ⟨6 17 19 29 31⟩, ⟨6 17 19 30 33⟩, ⟨6 17 22 29 31⟩,
⟨6 18 19 20 26⟩, ⟨6 18 19 20 33⟩, ⟨6 18 20 21 26⟩, ⟨6 18 20 21 27⟩, ⟨6 18 20 27 33⟩,
⟨6 19 20 30 33⟩, ⟨6 20 27 30 33⟩, ⟨6 22 32 37 41⟩, ⟨6 23 24 38 41⟩, ⟨6 23 32 37 41⟩,
⟨6 23 32 38 41⟩, ⟨7 8 15 16 23⟩, ⟨7 8 15 16 31⟩, ⟨7 8 15 23 42⟩, ⟨7 8 15 25 31⟩,
⟨7 8 15 25 42⟩, ⟨7 8 16 23 31⟩, ⟨7 10 11 21 30⟩, ⟨7 10 11 21 32⟩, ⟨7 10 11 30 37⟩,
⟨7 10 11 32 38⟩, ⟨7 10 11 37 38⟩, ⟨7 10 16 21 31⟩, ⟨7 10 16 21 32⟩, ⟨7 10 16 23 31⟩,
⟨7 10 16 23 32⟩, ⟨7 10 21 30 31⟩, ⟨7 10 23 32 38⟩, ⟨7 11 18 21 30⟩, ⟨7 11 18 30 39⟩,
⟨7 11 30 37 39⟩, ⟨7 11 37 38 39⟩, ⟨7 12 18 28 30⟩, ⟨7 12 18 28 33⟩, ⟨7 12 18 30 39⟩,
⟨7 12 18 33 39⟩, ⟨7 12 19 24 39⟩, ⟨7 12 19 24 42⟩, ⟨7 12 19 26 39⟩, ⟨7 12 24 33 39⟩,
⟨7 12 24 33 42⟩, ⟨7 12 26 30 39⟩, ⟨7 12 28 33 42⟩, ⟨7 14 15 16 21⟩, ⟨7 14 15 16 23⟩,
⟨7 14 15 21 27⟩, ⟨7 14 16 21 32⟩, ⟨7 14 16 23 32⟩, ⟨7 15 16 21 31⟩, ⟨7 15 21 22 27⟩,
⟨7 15 21 22 31⟩, ⟨7 15 22 25 31⟩, ⟨7 15 22 25 34⟩, ⟨7 15 22 27 34⟩, ⟨7 15 25 34 42⟩,
⟨7 18 21 27 28⟩, ⟨7 18 21 28 30⟩, ⟨7 18 27 28 33⟩, ⟨7 19 23 24 38⟩, ⟨7 19 24 38 39⟩,
⟨7 19 26 38 39⟩, ⟨7 21 22 27 28⟩, ⟨7 21 22 28 30⟩, ⟨7 21 22 30 31⟩, ⟨7 22 25 28 30⟩,
⟨7 22 25 28 34⟩, ⟨7 22 25 30 31⟩, ⟨7 22 27 28 34⟩, ⟨7 25 28 34 42⟩, ⟨7 26 30 37 39⟩,
⟨7 26 37 38 39⟩, ⟨8 9 11 18 22⟩, ⟨8 9 11 18 26⟩, ⟨8 9 11 22 43⟩, ⟨8 9 11 26 43⟩,
⟨8 9 12 13 37⟩, ⟨8 9 12 26 43⟩, ⟨8 9 13 20 22⟩, ⟨8 9 13 22 37⟩, ⟨8 9 18 22 37⟩,
⟨8 9 20 22 43⟩, ⟨8 11 15 26 43⟩, ⟨8 11 15 39 43⟩, ⟨8 11 18 22 39⟩, ⟨8 11 22 39 43⟩,
⟨8 12 13 16 23⟩, ⟨8 12 13 16 37⟩, ⟨8 12 16 23 31⟩, ⟨8 12 16 31 37⟩, ⟨8 12 19 25 26⟩,
⟨8 12 19 25 31⟩, ⟨8 12 19 31 37⟩, ⟨8 12 25 26 43⟩, ⟨8 13 16 23 29⟩, ⟨8 13 16 29 37⟩,
⟨8 13 20 22 29⟩, ⟨8 13 20 23 29⟩, ⟨8 13 22 29 37⟩, ⟨8 15 16 17 28⟩, ⟨8 15 16 17 31⟩,
⟨8 15 16 23 36⟩, ⟨8 15 16 28 36⟩, ⟨8 15 17 24 28⟩, ⟨8 15 20 23 36⟩, ⟨8 15 20 23 42⟩,

Continuation on next page –

Table C.1 – Continuation from last page

⟨8 15 20 36 39⟩, ⟨8 15 20 39 43⟩, ⟨8 15 20 42 43⟩, ⟨8 15 25 26 43⟩, ⟨8 15 25 42 43⟩,
⟨8 16 17 28 37⟩, ⟨8 16 17 31 37⟩, ⟨8 16 23 29 36⟩, ⟨8 16 28 29 36⟩, ⟨8 16 28 29 37⟩,
⟨8 17 19 31 37⟩, ⟨8 17 19 33 37⟩, ⟨8 17 24 28 37⟩, ⟨8 17 24 33 37⟩, ⟨8 18 19 33 37⟩,
⟨8 20 22 29 36⟩, ⟨8 20 22 36 39⟩, ⟨8 20 22 39 43⟩, ⟨8 20 23 29 36⟩, ⟨9 10 13 15 39⟩,
⟨9 10 13 20 25⟩, ⟨9 10 13 25 39⟩, ⟨9 10 16 20 25⟩, ⟨9 10 16 20 41⟩, ⟨9 10 16 23 25⟩,
⟨9 10 16 23 31⟩, ⟨9 10 16 31 41⟩, ⟨9 10 23 25 39⟩, ⟨9 10 27 31 41⟩, ⟨9 11 16 22 25⟩,
⟨9 11 16 22 43⟩, ⟨9 11 16 23 25⟩, ⟨9 11 16 23 31⟩, ⟨9 11 16 31 43⟩, ⟨9 11 17 23 25⟩,
⟨9 11 17 24 25⟩, ⟨9 11 18 22 24⟩, ⟨9 11 22 24 25⟩, ⟨9 12 13 14 35⟩, ⟨9 12 13 14 37⟩,
⟨9 12 13 15 35⟩, ⟨9 12 15 27 34⟩, ⟨9 12 15 29 34⟩, ⟨9 12 15 29 35⟩, ⟨9 12 26 27 34⟩,
⟨9 12 26 27 43⟩, ⟨9 12 26 30 34⟩, ⟨9 12 29 30 34⟩, ⟨9 12 29 30 35⟩, ⟨9 13 14 22 25⟩,
⟨9 13 14 22 37⟩, ⟨9 13 14 25 39⟩, ⟨9 13 14 35 39⟩, ⟨9 13 15 35 39⟩, ⟨9 13 20 22 25⟩,
⟨9 14 17 24 25⟩, ⟨9 14 17 24 32⟩, ⟨9 14 17 25 39⟩, ⟨9 14 17 32 35⟩, ⟨9 14 17 35 39⟩,
⟨9 14 22 24 25⟩, ⟨9 14 22 24 32⟩, ⟨9 14 22 32 37⟩, ⟨9 15 29 35 39⟩, ⟨9 16 20 22 25⟩,
⟨9 16 20 22 43⟩, ⟨9 16 20 41 43⟩, ⟨9 16 31 41 43⟩, ⟨9 17 23 25 39⟩, ⟨9 17 29 30 35⟩,
⟨9 17 29 35 39⟩, ⟨9 17 30 32 35⟩, ⟨9 18 22 24 32⟩, ⟨9 18 22 32 37⟩, ⟨9 27 31 41 43⟩,
⟨10 11 12 14 37⟩, ⟨10 11 12 14 41⟩, ⟨10 11 12 33 41⟩, ⟨10 11 12 33 42⟩, ⟨10 11 12 37 42⟩,
⟨10 11 14 30 37⟩, ⟨10 11 14 30 41⟩, ⟨10 11 21 30 41⟩, ⟨10 11 21 32 33⟩, ⟨10 11 21 33 41⟩,
⟨10 11 32 33 42⟩, ⟨10 11 32 38 42⟩, ⟨10 11 37 38 42⟩, ⟨10 12 14 28 41⟩, ⟨10 12 28 33 41⟩,
⟨10 12 28 33 42⟩, ⟨10 13 20 25 32⟩, ⟨10 13 20 28 32⟩, ⟨10 14 15 28 36⟩, ⟨10 14 27 30 41⟩,
⟨10 16 20 25 32⟩, ⟨10 16 20 32 33⟩, ⟨10 16 20 33 41⟩, ⟨10 16 21 31 41⟩, ⟨10 16 21 32 33⟩,
⟨10 16 21 33 41⟩, ⟨10 16 23 25 32⟩, ⟨10 20 28 32 33⟩, ⟨10 20 28 33 41⟩, ⟨10 21 30 31 41⟩,
⟨10 22 23 38 40⟩, ⟨10 22 23 39 40⟩, ⟨10 22 29 36 38⟩, ⟨10 22 29 38 40⟩, ⟨10 23 25 32 40⟩,
⟨10 23 25 39 40⟩, ⟨10 23 32 38 40⟩, ⟨10 27 30 31 41⟩, ⟨10 28 29 32 42⟩, ⟨10 28 29 36 42⟩,
⟨10 28 32 33 42⟩, ⟨10 29 32 38 40⟩, ⟨10 29 32 38 42⟩, ⟨10 29 36 38 42⟩, ⟨10 36 37 38 42⟩,
⟨11 12 13 14 37⟩, ⟨11 12 13 14 41⟩, ⟨11 12 13 16 23⟩, ⟨11 12 13 16 37⟩, ⟨11 12 13 17 23⟩,
⟨11 12 13 17 41⟩, ⟨11 12 16 23 31⟩, ⟨11 12 16 31 37⟩, ⟨11 12 31 37 44⟩, ⟨11 12 33 42 44⟩,
⟨11 12 37 42 44⟩, ⟨11 13 14 30 37⟩, ⟨11 13 14 30 41⟩, ⟨11 13 16 23 29⟩, ⟨11 13 16 29 37⟩,
⟨11 13 17 23 27⟩, ⟨11 13 17 24 27⟩, ⟨11 13 17 24 41⟩, ⟨11 13 23 27 29⟩, ⟨11 13 24 27 29⟩,
⟨11 13 24 29 30⟩, ⟨11 13 24 30 41⟩, ⟨11 13 29 30 37⟩, ⟨11 15 26 32 44⟩, ⟨11 15 26 43 44⟩,
⟨11 15 32 38 44⟩, ⟨11 15 38 39 44⟩, ⟨11 15 39 43 44⟩, ⟨11 16 22 25 34⟩, ⟨11 16 22 34 39⟩,
⟨11 16 22 39 43⟩, ⟨11 16 23 25 29⟩, ⟨11 16 25 29 34⟩, ⟨11 16 29 34 37⟩, ⟨11 16 31 37 44⟩,
⟨11 16 31 43 44⟩, ⟨11 16 34 37 39⟩, ⟨11 16 37 39 44⟩, ⟨11 16 39 43 44⟩, ⟨11 17 23 25 27⟩,
⟨11 17 24 25 27⟩, ⟨11 18 21 24 30⟩, ⟨11 18 22 24 25⟩, ⟨11 18 22 25 34⟩, ⟨11 18 22 34 39⟩,
⟨11 18 24 25 29⟩, ⟨11 18 24 29 30⟩, ⟨11 18 25 29 34⟩, ⟨11 18 29 30 34⟩, ⟨11 18 30 34 39⟩,
⟨11 21 24 30 41⟩, ⟨11 21 26 32 33⟩, ⟨11 21 26 33 41⟩, ⟨11 23 25 27 29⟩, ⟨11 24 25 27 29⟩,
⟨11 26 32 33 44⟩, ⟨11 29 30 34 37⟩, ⟨11 30 34 37 39⟩, ⟨11 32 33 42 44⟩, ⟨11 32 38 42 44⟩,
⟨11 37 38 39 44⟩, ⟨11 37 38 42 44⟩, ⟨12 13 14 35 41⟩, ⟨12 13 15 17 41⟩, ⟨12 13 15 35 41⟩,
⟨12 14 28 35 41⟩, ⟨12 15 17 22 40⟩, ⟨12 15 17 22 41⟩, ⟨12 15 18 22 34⟩, ⟨12 15 18 22 41⟩,

Continuation on next page –

Table C.1 – Continuation from last page

⟨12 15 18 29 34⟩, ⟨12 15 18 29 35⟩, ⟨12 15 18 35 41⟩, ⟨12 15 22 27 34⟩, ⟨12 15 22 27 40⟩,
⟨12 18 22 34 39⟩, ⟨12 18 28 30 35⟩, ⟨12 18 28 33 41⟩, ⟨12 18 28 35 41⟩, ⟨12 18 29 30 34⟩,
⟨12 18 29 30 35⟩, ⟨12 18 30 34 39⟩, ⟨12 19 24 31 44⟩, ⟨12 19 24 42 44⟩, ⟨12 19 25 26 39⟩,
⟨12 19 31 37 44⟩, ⟨12 19 37 42 44⟩, ⟨12 22 27 34 39⟩, ⟨12 22 27 39 40⟩, ⟨12 24 33 42 44⟩,
⟨12 25 26 27 39⟩, ⟨12 25 26 27 43⟩, ⟨12 25 27 39 40⟩, ⟨12 26 27 34 39⟩, ⟨12 26 30 34 39⟩,
⟨13 14 19 25 39⟩, ⟨13 14 19 30 38⟩, ⟨13 14 19 38 39⟩, ⟨13 14 30 38 41⟩, ⟨13 14 35 38 39⟩,
⟨13 14 35 38 41⟩, ⟨13 15 17 24 28⟩, ⟨13 15 17 24 41⟩, ⟨13 15 24 38 39⟩, ⟨13 15 24 38 41⟩,
⟨13 15 35 38 39⟩, ⟨13 15 35 38 41⟩, ⟨13 17 23 27 28⟩, ⟨13 17 24 27 28⟩, ⟨13 19 24 29 30⟩,
⟨13 19 24 29 31⟩, ⟨13 19 24 30 38⟩, ⟨13 19 24 38 39⟩, ⟨13 20 22 25 31⟩, ⟨13 20 22 29 31⟩,
⟨13 20 23 27 28⟩, ⟨13 20 23 27 29⟩, ⟨13 20 25 31 32⟩, ⟨13 20 27 28 32⟩, ⟨13 20 27 29 31⟩,
⟨13 20 27 31 32⟩, ⟨13 24 27 29 31⟩, ⟨13 24 30 38 41⟩, ⟨14 15 16 23 36⟩, ⟨14 15 16 28 36⟩,
⟨14 16 21 32 35⟩, ⟨14 17 24 25 26⟩, ⟨14 17 24 26 32⟩, ⟨14 17 25 26 39⟩, ⟨14 17 26 32 35⟩,
⟨14 17 26 35 39⟩, ⟨14 19 20 26 38⟩, ⟨14 19 20 30 38⟩, ⟨14 19 25 26 39⟩, ⟨14 19 26 38 39⟩,
⟨14 20 21 26 38⟩, ⟨14 20 21 27 38⟩, ⟨14 20 27 30 38⟩, ⟨14 21 26 32 35⟩, ⟨14 21 26 35 38⟩,
⟨14 22 24 25 26⟩, ⟨14 22 24 26 32⟩, ⟨14 26 35 38 39⟩, ⟨14 27 30 38 41⟩, ⟨15 16 17 31 40⟩,
⟨15 16 21 31 40⟩, ⟨15 17 22 31 40⟩, ⟨15 18 22 25 34⟩, ⟨15 18 22 25 44⟩, ⟨15 18 22 32 41⟩,
⟨15 18 22 32 44⟩, ⟨15 18 25 29 34⟩, ⟨15 18 25 29 44⟩, ⟨15 18 29 35 44⟩, ⟨15 18 32 38 41⟩,
⟨15 18 32 38 44⟩, ⟨15 18 35 38 41⟩, ⟨15 18 35 38 44⟩, ⟨15 21 22 27 40⟩, ⟨15 21 22 31 40⟩,
⟨15 22 25 26 44⟩, ⟨15 22 26 32 44⟩, ⟨15 25 26 43 44⟩, ⟨15 25 29 34 42⟩, ⟨15 25 29 42 43⟩,
⟨15 25 29 43 44⟩, ⟨15 29 35 39 44⟩, ⟨15 29 39 43 44⟩, ⟨15 35 38 39 44⟩, ⟨16 17 28 37 39⟩,
⟨16 17 31 37 44⟩, ⟨16 17 31 40 43⟩, ⟨16 17 31 43 44⟩, ⟨16 17 37 39 44⟩, ⟨16 17 39 43 44⟩,
⟨16 19 20 22 30⟩, ⟨16 19 20 30 33⟩, ⟨16 19 20 33 41⟩, ⟨16 19 21 33 35⟩, ⟨16 19 21 33 41⟩,
⟨16 19 21 40 41⟩, ⟨16 19 30 33 35⟩, ⟨16 20 22 25 30⟩, ⟨16 20 22 39 43⟩, ⟨16 20 25 30 32⟩,
⟨16 20 30 32 33⟩, ⟨16 21 31 40 41⟩, ⟨16 21 32 33 35⟩, ⟨16 22 25 28 30⟩, ⟨16 22 25 28 34⟩,
⟨16 22 28 34 39⟩, ⟨16 23 25 29 36⟩, ⟨16 25 28 29 34⟩, ⟨16 25 28 29 36⟩, ⟨16 28 29 34 37⟩,
⟨16 28 34 37 39⟩, ⟨16 30 32 33 35⟩, ⟨16 31 40 41 43⟩, ⟨17 19 29 30 35⟩, ⟨17 19 29 31 44⟩,
⟨17 19 29 35 44⟩, ⟨17 19 30 33 35⟩, ⟨17 19 31 37 44⟩, ⟨17 19 33 35 37⟩, ⟨17 19 35 37 44⟩,
⟨17 22 29 31 40⟩, ⟨17 23 25 27 39⟩, ⟨17 23 27 28 39⟩, ⟨17 24 25 26 27⟩, ⟨17 24 26 27 28⟩,
⟨17 24 26 28 37⟩, ⟨17 24 26 32 33⟩, ⟨17 24 26 33 37⟩, ⟨17 25 26 27 39⟩, ⟨17 26 27 28 39⟩,
⟨17 26 28 37 39⟩, ⟨17 26 32 33 35⟩, ⟨17 26 33 35 37⟩, ⟨17 26 35 37 39⟩, ⟨17 29 31 40 43⟩,
⟨17 29 31 43 44⟩, ⟨17 29 35 39 44⟩, ⟨17 29 39 43 44⟩, ⟨17 30 32 33 35⟩, ⟨17 35 37 39 44⟩,
⟨18 19 20 33 41⟩, ⟨18 19 33 37 41⟩, ⟨18 20 21 23 28⟩, ⟨18 20 21 27 28⟩, ⟨18 20 27 28 33⟩,
⟨18 20 28 33 41⟩, ⟨18 21 23 24 30⟩, ⟨18 21 23 28 30⟩, ⟨18 22 24 25 44⟩, ⟨18 22 24 32 44⟩,
⟨18 22 32 37 41⟩, ⟨18 24 25 29 44⟩, ⟨18 24 29 30 35⟩, ⟨18 24 29 35 44⟩, ⟨18 24 32 42 44⟩,
⟨18 24 35 42 44⟩, ⟨18 32 38 41 42⟩, ⟨18 32 38 42 44⟩, ⟨18 35 38 41 42⟩, ⟨18 35 38 42 44⟩,
⟨19 20 22 30 38⟩, ⟨19 21 33 35 37⟩, ⟨19 21 33 37 41⟩, ⟨19 21 35 37 42⟩, ⟨19 22 23 30 38⟩,
⟨19 23 24 30 38⟩, ⟨19 24 29 30 35⟩, ⟨19 24 29 31 44⟩, ⟨19 24 29 35 44⟩, ⟨19 24 35 42 44⟩,
⟨19 35 37 42 44⟩, ⟨20 21 23 27 28⟩, ⟨20 21 23 27 36⟩, ⟨20 21 23 36 37⟩, ⟨20 21 26 37 38⟩,

Continuation on next page –

Table C.1 – Continuation from last page

$\langle 20\,21\,27\,36\,38\rangle$,	$\langle 20\,21\,36\,37\,38\rangle$,	$\langle 20\,22\,25\,30\,31\rangle$,	$\langle 20\,22\,29\,31\,38\rangle$,	$\langle 20\,22\,29\,36\,38\rangle$,
$\langle 20\,22\,30\,31\,38\rangle$,	$\langle 20\,23\,27\,29\,36\rangle$,	$\langle 20\,25\,30\,31\,32\rangle$,	$\langle 20\,27\,28\,32\,33\rangle$,	$\langle 20\,27\,29\,31\,38\rangle$,
$\langle 20\,27\,29\,36\,38\rangle$,	$\langle 20\,27\,30\,31\,32\rangle$,	$\langle 20\,27\,30\,31\,38\rangle$,	$\langle 20\,27\,30\,32\,33\rangle$,	$\langle 21\,22\,23\,27\,28\rangle$,
$\langle 21\,22\,23\,27\,40\rangle$,	$\langle 21\,22\,23\,28\,30\rangle$,	$\langle 21\,22\,23\,30\,40\rangle$,	$\langle 21\,22\,30\,31\,40\rangle$,	$\langle 21\,23\,24\,30\,41\rangle$,
$\langle 21\,23\,25\,27\,36\rangle$,	$\langle 21\,23\,25\,27\,40\rangle$,	$\langle 21\,23\,30\,40\,41\rangle$,	$\langle 21\,25\,27\,36\,42\rangle$,	$\langle 21\,26\,32\,33\,35\rangle$,
$\langle 21\,26\,33\,35\,37\rangle$,	$\langle 21\,26\,33\,37\,41\rangle$,	$\langle 21\,26\,35\,37\,38\rangle$,	$\langle 21\,27\,36\,38\,42\rangle$,	$\langle 21\,30\,31\,40\,41\rangle$,
$\langle 21\,35\,37\,38\,42\rangle$,	$\langle 21\,36\,37\,38\,42\rangle$,	$\langle 22\,23\,27\,28\,39\rangle$,	$\langle 22\,23\,27\,39\,40\rangle$,	$\langle 22\,23\,30\,38\,40\rangle$,
$\langle 22\,24\,25\,26\,44\rangle$,	$\langle 22\,24\,26\,32\,44\rangle$,	$\langle 22\,27\,28\,34\,39\rangle$,	$\langle 22\,29\,31\,38\,40\rangle$,	$\langle 22\,30\,31\,38\,40\rangle$,
$\langle 23\,24\,30\,38\,41\rangle$,	$\langle 23\,25\,27\,29\,36\rangle$,	$\langle 23\,25\,27\,39\,40\rangle$,	$\langle 23\,30\,38\,40\,41\rangle$,	$\langle 23\,32\,38\,40\,41\rangle$,
$\langle 24\,25\,26\,27\,44\rangle$,	$\langle 24\,25\,27\,29\,44\rangle$,	$\langle 24\,26\,32\,33\,44\rangle$,	$\langle 24\,27\,29\,31\,44\rangle$,	$\langle 24\,32\,33\,42\,44\rangle$,
$\langle 25\,26\,27\,43\,44\rangle$,	$\langle 25\,27\,29\,36\,42\rangle$,	$\langle 25\,27\,29\,42\,43\rangle$,	$\langle 25\,27\,29\,43\,44\rangle$,	$\langle 25\,28\,29\,34\,42\rangle$,
$\langle 25\,28\,29\,36\,42\rangle$,	$\langle 26\,27\,28\,34\,39\rangle$,	$\langle 26\,28\,34\,37\,39\rangle$,	$\langle 26\,30\,34\,37\,39\rangle$,	$\langle 26\,35\,37\,38\,39\rangle$,
$\langle 27\,29\,31\,38\,41\rangle$,	$\langle 27\,29\,31\,41\,43\rangle$,	$\langle 27\,29\,31\,43\,44\rangle$,	$\langle 27\,29\,36\,38\,42\rangle$,	$\langle 27\,29\,38\,41\,42\rangle$,
$\langle 27\,29\,41\,42\,43\rangle$,	$\langle 27\,30\,31\,38\,41\rangle$,	$\langle 29\,31\,38\,40\,41\rangle$,	$\langle 29\,31\,40\,41\,43\rangle$,	$\langle 29\,32\,38\,40\,41\rangle$,
$\langle 29\,32\,38\,41\,42\rangle$,	$\langle 30\,31\,38\,40\,41\rangle$,	$\langle 35\,37\,38\,39\,44\rangle$,	$\langle 35\,37\,38\,42\,44\rangle$.	

Continuation on next page –

Table C.1 – Continuation from last page

f-vector:	$(44, 691, 2324, 2795, 1118)$
# singularities:	0
Simplicial homology:	$(\mathbb{Z}, \mathbb{Z} \times \mathbb{Z}_{17}, \mathbb{Z}_5, \mathbb{Z}_2, 0)$
Euler characteristic:	0

Appendix D

Series of cyclic combinatorial 3-manifolds

In this section, an algorithm to classify infinite series of cyclic 3-manifolds with a constant number of cyclic orbits is presented. It was used to detect the infinite series presented in Table 4.3 and Table 4.4.

Let M be a combinatorial 3-manifold with n vertices and cyclic automorphism group $G = \langle(1,\ldots,n)\rangle$ and let k be the order of the series we are looking for.

First, the algorithm computes M in terms of difference cycles $M = \{d_1,\ldots,d_m\}$, $d_i := (a_{i,1}:a_{i,2}:a_{i,3}:a_{i,4})$. Then the set of all multi-indices $A \subset \mathbb{N}_0^4$ with sum k is generated and the $|D|$-fold Cartesian product $A^{|D|}$ of A is formed. In a loop over all $\mathbf{x} = (x_1,\ldots,x_{|D|}) \in A^{|D|}$ the 4-tuples x_i are added componentwise to d_i, transforming the set of difference cycles M into $M_{\mathbf{x}}$ with $n+k$ vertices. If $M_{\mathbf{x}}$ is a combinatorial manifold, a candidate for a series of order k is found and further members are checked. If $M_{\mathbf{x}}$ is not a manifold for all $|D|$-tuples $\mathbf{x}$, M is not part of a series of order k.

The following code is written in the GAP-scripting language and makes use of the GAP-package simpcomp [40, 41, 42]. It checks a list of combinatorial 3-manifolds given in the list `complexes` and searches for series of order `order`.

```
# INPUT:
complexes:= []; # ''list of conjectured first complexes of a series with
                # cyclic symmetry <(1, .. , n)>'';
order:=2; # specifies the order of the series the algorithm should search for

# GLOBAL VARIABLES:
candidates:=[]; # array for conjectured series
```

```
ctr:=1; # counter
checkDepth:=10; # number of members for which a conjectured series is tested

for c in complexes do

  # cyclic cyclic group C_n
  n:=SCNumFaces(c,0); # detect number of vertices
  p := PermList(Concatenation([2..n],[1]));
  H := Group(p);

  G:=SCAutomorphismGroup(c);

  if not IsSubgroup(G,H) then
    Error("complex not cyclic.\n");
  fi;

  # replace full automorphism group by cyclic subgroup
  c!.SCAutomorphismGroup:=H;
  # compute cyclic generators of complex c
  gens := List(SCGenerators(c),x->x[1]);
  # compute difference cycles from cyclic orbit representatives
  diffCycles := [];
  for elm in [1..Size(gens)] do
    diffCycles[elm] := [ gens[elm][2]-gens[elm][1],
                         gens[elm][3]-gens[elm][2],
                         gens[elm][4]-gens[elm][3],
                         gens[elm][1]-gens[elm][4]+n ];
  od;

  # compute all possible extensions of a difference cycle with n vertices to
  # a difference cycle with n+order vertices
  if order = 1 then
    baseSet:=[ [ 1 ], [ 2 ], [ 3 ], [ 4 ] ];
  else
    baseSet:=Filtered(Cartesian(
      ListWithIdenticalEntries(order,[1..4])),x->IsSortedList(x)
    );
  fi;

  # loop over all combinations of extensions of the set of orbit
```

```
# representatives
idx := Cartesian(ListWithIdenticalEntries(Size(gens),baseSet));
for j in idx do

  # compute new generators
  newGens:=SCIntFunc.DeepCopy(diffCycles);
  for k in [1..Size(gens)] do
    for l in [1..order] do
      newGens[k][j[k][l]]:=newGens[k][j[k][l]]+1;
    od;
  od;

  # check for double entries
  double:=false;
  for pair in Combinations(newGens,2) do
    perm:=(1,2,3,4);
    for k in [1..4] do
      if pair[1] = Permuted(pair[2],perm^k) then
        double:=true;
        break;
      fi;
    od;
  od;
  if double = true then
    continue;
  fi;

  # check if new generators define a weak pseudomanifold:
  trigs:=[];
  incidence:=[];
  pm:=true;
  for gen in newGens do

    # case 1: difference cycle of length n/4
    if Size(Set(gen)) = 1 then
      newTrigs:=[[gen[1],gen[1],2*gen[1]] ];

    # case 2: difference cycle of length n/2
    elif gen{[1,2]} = gen{[3,4]} then
      newTrigs:=[[gen[1],gen[2],gen[1]+gen[2]],
```

```
                [gen[2],gen[1],gen[1]+gen[2]] ];

    # case 3: difference cycle of length n
    else
      newTrigs:=[[gen[1],gen[2],gen[3]+gen[4]],
                 [gen[1],gen[2]+gen[3],gen[4]],
                 [gen[1]+gen[2],gen[3],gen[4]],
                 [gen[4]+gen[1],gen[2],gen[3]] ];
    fi;

    for t in newTrigs do
      perm:=(1,2,3);
      for k in [1..3] do
        idx:=Position(trigs,Permuted(t,perm^k));
        if idx <> fail then
          break;
        fi;
      od;

      if idx = fail then
        Add(trigs,t);
        incidence[Position(trigs,t)]:=1;
      else
        incidence[idx]:=incidence[idx]+1;
      fi;
      if Maximum(incidence) > 2 then
        pm:=false;
        break;
      fi;
    od;
  od;

  if Minimum(incidence) <> 2 or pm = false then
    continue;
  fi;

  # if new generators define a weak pseudo manifold construct simplicial
  # complex
  tmp:=SCFromDifferenceCycles(newGens);
```

```
    # check if the new complex is a combinatorial manifold
    mfld:=SCIsManifold(tmp);
    if mfld <> true then
      continue;
    fi;

    # if the new complex is a combinatorial manifold check a number of
    # further members
    series:=true;
    for i in [1..checkDepth] do
      for k in [1..Size(gens)] do
        for l in [1..order] do
          newGens[k][j[k][l]]:=newGens[k][j[k][l]]+1;
        od;
      od;

      tmp2:=SCFromDifferenceCycles(newGens);
      pm:=SCIsPseudoManifold(tmp2);
      if pm <> true then
        series:=false;
        break;
      fi;

      mfld:=SCIsManifold(tmp2);
      if mfld <> true then
        series:=false;
        break;
      fi;
    od;

    # add candidate to the list of results
    if series = true then
      candidates[ctr]:=[Position(complexes,c),diffCycles,j];
      ctr:=ctr+1;
    fi;
  od;
od;
```

Bibliography

[1] S. Akbulut. A fake compact contractible 4-manifold. *J. Differential Geom.*, 33(2):335–356, 1991.

[2] S. Akbulut and K. Yasui. Corks, plugs and exotic structures. *J. Gökova Geom. Topol. GGT*, 2:40–82, 2008.

[3] A. Altshuler. Manifolds in stacked 4-polytopes. *J. Combinatorial Theory Ser. A*, 10:198–239, 1971.

[4] A. Altshuler. Polyhedral realization in R^3 of triangulations of the torus and 2-manifolds in cyclic 4-polytopes. *Discrete Math.*, 1(3):211–238, 1971/1972.

[5] A. Altshuler. Neighborly 4-polytopes and neighborly combinatorial 3-manifolds with ten vertices. *Canad. J. Math.*, 29(2):400–420, 1977.

[6] A. Altshuler and U. Brehm. Nonexistence of weakly neighborly polyhedral maps on the orientable 2-manifold of genus 2. *J. Combin. Theory Ser. A*, 42(1):87–103, 1986.

[7] A. Altshuler and L. Steinberg. Neighborly combinatorial 3-manifolds with 9 vertices. *Discrete Math.*, 8:113–137, 1974.

[8] M. A. Armstrong. *Basic topology.* Undergraduate Texts in Mathematics. Springer-Verlag, New York, 1983. Corrected reprint of the 1979 original.

[9] D. Barnette. The minimum number of vertices of a simple polytope. *Israel J. Math.*, 10:121–125, 1971.

[10] D. Barnette. A proof of the lower bound conjecture for convex polytopes. *Pacific J. Math.*, 46:349–354, 1973.

[11] T. Beth, D. Jungnickel, and H. Lenz. *Design theory. Vol. I*, volume 69 of *Encyclopedia of Mathematics and its Applications.* Cambridge University Press, Cambridge, second edition, 1999.

[12] U. Betke, C. Schulz, and J. M. Wills. Zur Zerlegbarkeit von Skeletten. *Geometriae Dedicata*, 5(4):435–451, 1976.

[13] L. J. Billera and C. W. Lee. A proof of the sufficiency of McMullen's conditions for f-vectors of simplicial convex polytopes. *J. Combin. Theory Ser. A*, 31(3):237–255, 1981.

[14] R. H. Bing. An alternative proof that 3-manifolds can be triangulated. *Ann. of Math. (2)*, 69:37–65, 1959.

[15] A. Björner and F. H. Lutz. Simplicial manifolds, bistellar flips and a 16-vertex triangulation of the Poincaré homology 3-sphere. *Experiment. Math.*, 9(2):275–289, 2000.

[16] J. Bokowski. A geometric realization without self-intersections does exist for Dyck's regular map. *Discrete Comput. Geom.*, 4(6):583–589, 1989.

[17] U. Brehm. Maximally symmetric polyhedral realizations of Dyck's regular map. *Mathematika*, 34(2):229–236, 1987.

[18] U. Brehm and A. Altshuler. The weakly neighborly polyhedral maps on the torus. *Geom. Dedicata*, 18(3):227–238, 1985.

[19] U. Brehm and A. Altshuler. On weakly neighborly polyhedral maps of arbitrary genus. *Israel J. Math.*, 53(2):137–157, 1986.

[20] U. Brehm and W. Kühnel. Combinatorial manifolds with few vertices. *Topology*, 26(4):465–473, 1987.

[21] U. Brehm and W. Kühnel. 15-vertex triangulations of an 8-manifold. *Math. Ann.*, 294(1):167–193, 1992.

[22] U. Brehm and W. Kühnel. Equivelar maps on the torus. *European J. Combin.*, 29(8):1843–1861, 2008.

[23] U. Brehm and W. Kühnel. Lattice triangulations of E^3 and of the 3-torus, 2009. To appear in Israel J. Math.

[24] U. Brehm and J. Swiatkowski. Triangulations of lens spaces with few simplices. Berlin, SFB 288, Preprint No. 59, 1993.

[25] U. Brehm and J. M. Wills. Polyhedral manifolds. In P. M. Gruber and J. M. Wills, editors, *Handbook of convex geometry, Vol. A, B*, pages 535–554. North-Holland, Amsterdam, 1993.

[26] B. A. Burton. Regina: normal surface and 3-manifold topology software. `http://regina.sourceforge.net/`, 1999–2009.

[27] C. Carathéodory. Über den Variabilitätsbereich der Koeffizienten von Potenzreihen, die gegebene Werte nicht annehmen. *Math. Ann.*, 64(1):95–115, 1907.

[28] M. Casella and W. Kühnel. A triangulated $K3$ surface with the minimum number of vertices. *Topology*, 40(4):753–772, 2001.

[29] C. J. Colbourn and J. H. Dinitz, editors. *The CRC handbook of combinatorial designs.* CRC Press Series on Discrete Mathematics and its Applications. CRC Press, Boca Raton, FL, 1996.

[30] H. S. M. Coxeter. *Regular polytopes.* Dover Publications Inc., New York, third edition, 1973.

[31] Á. Császár. A polyhedron without diagonals. *Acta Univ. Szeged. Sect. Sci. Math.*, 13:140–142, 1949.

[32] S. K. Donaldson. An application of gauge theory to four-dimensional topology. *J. Differential Geom.*, 18(2):279–315, 1983.

[33] S. K. Donaldson. Connections, cohomology and the intersection forms of 4-manifolds. *J. Differential Geom.*, 24(3):275–341, 1986.

[34] J.-G. Dumas, F. Heckenbach, B. D. Saunders, and V. Welker. Simplicial Homology, a GAP package, Version 1.4.3. `http://www.cis.udel.edu/~dumas/Homology/`, 2009.

[35] W. Dyck. Notiz über eine reguläre Riemann'sche Fläche vom Geschlechte drei und die zugehörige "Normalcurve" vierter Ordnung. *Math. Ann.*, 17(4):510–516, 1880.

[36] J. J. Eells and N. H. Kuiper. Manifolds which are like projective planes. *Inst. Hautes Études Sci. Publ. Math.*, 14:5–46, 1962.

[37] F. Effenberger. Tätigkeitsbericht zum DFG-Projekt Ku 1203/5, 2006.

[38] F. Effenberger. *Hamiltonian submanifolds of regular polytopes.* PhD thesis, University of Stuttgart, 2011.

[39] F. Effenberger and W. Kühnel. Hamiltonian submanifolds of regular polytopes. *Discrete Comput. Geom.*, 43(2):242–262, 2010.

[40] F. Effenberger and J. Spreer. simpcomp - a GAP toolbox for simplicial complexes. *ACM Communications in Computer Algebra*, 44(4):186 – 189, 2010.

[41] F. Effenberger and J. Spreer. simpcomp - a GAP package, Version 1.5.1. `http://www.igt.uni-stuttgart.de/LstDiffgeo/simpcomp`, 2011. Submitted to the *GAP Group.*

[42] F. Effenberger and J. Spreer. Simplicial blowups and discrete normal surfaces in the GAP package simpcomp, 2011. To appear in ACM Communications in Computer Algebra.

[43] A. Emch. Triple and multiple systems, their geometric configurations and groups. *Trans. Amer. Math. Soc.*, 31(1):25–42, 1929.

[44] L. Euler. Solutio problematis ad geometriam situs pertinentis. *Commentarii academiae scientiarum imperialis Petropolitanae*, 8:128–140, 1736.

[45] D. Ewert. Automorphismen von Flächen und deren Wirkung auf der Homologie, 2008. Diploma thesis, University of Stuttgart, 49 pages, 17 figures.

[46] I. Fáry. Absolute subcomplexes. *Ann. Scuola Norm. Sup. Pisa Cl. Sci. (4)*, 4(3):433–471, 1977.

[47] M. Freedman. The topology of four-dimensional manifolds. *J. Differential Geom.*, 17:357–453, 1982.

[48] M. Freedman and R. Kirby. A geometric proof of Rochlin's theorem. *Algebraic and geometric topology (Proc. Sympos. Pure Math., Stanford Univ., Stanford, Calif.)*, 32:85–97, 1976.

[49] GAP – Groups, Algorithms, and Programming, Version 4.4.12. `http://www.gap-system.org`, 2008.

[50] E. Gawrilow and M. Joswig. polymake: a framework for analyzing convex polytopes. In *Polytopes—combinatorics and computation (Oberwolfach, 1997)*, volume 29 of *DMV Sem.*, pages 43–73. Birkhäuser, Basel, 2000.

[51] R. E. Gompf and A. I. Stipsicz. *4-manifolds and Kirby calculus*, volume 20 of *Graduate Studies in Mathematics*. American Mathematical Society, Providence, RI, 1999.

[52] D. R. Grayson and M. E. Stillman. Macaulay2, a software system for research in algebraic geometry, Version 1.3.1. `http://www.math.uiuc.edu/Macaulay2/`, 2009.

[53] B. Grünbaum. *Convex polytopes*, volume 16 of *Pure and Applied Mathematics*. Interscience Publishers John Wiley & Sons, Inc., New York, 1967. With the cooperation of Victor Klee, M. A. Perles and G. C. Shephard.

[54] B. Grünbaum and J. Malkevitch. Pairs of edge-disjoint Hamiltonian circuits. *Aequationes Math.*, 14(1/2):191–196, 1976.

[55] W. Haken. Theorie der Normalflächen. *Acta Math.*, 105:245–375, 1961.

[56] W. Haken. Über das Homöomorphieproblem der 3-Mannigfaltigkeiten. I. *Math. Z.*, 80:89–120, 1962.

[57] H. Hauser. Resolution of singularities 1860–1999. In *Resolution of singularities (Obergurgl, 1997)*, volume 181 of *Progr. Math.*, pages 5–36. Birkhäuser, Basel, 2000.

[58] P. J. Heawood. Map colour theorem. *Quart. J. Math.*, 24:332–338, 1890.

[59] H. Hironaka. Resolution of singularities of an algebraic variety over a field of characteristic zero. I, II. *The Annals of Mathematics (2)*, 79:109–203, 1964.

[60] F. Hirzebruch. Über vierdimensionale Riemannsche Flächen mehrdeutiger analytischer Funktionen von zwei komplexen Veränderlichen. *Math. Ann.*, 126:1–22, 1953.

[61] H. Hopf. Über komplex-analytische Mannigfaltigkeiten. *Univ. Roma. Ist. Naz. Alta Mat. Rend. Mat. e Appl. (5)*, 10:169–182, 1951.

[62] M. Huber. *Flag-transitive Steiner designs*. Frontiers in Mathematics. Birkhäuser Verlag, Basel, 2009.

[63] M. Jungerman and G. Ringel. The genus of the n-octahedron: regular cases. *J. Graph Theory*, 2(1):69–75, 1978.

[64] G. Kalai. Rigidity and the lower bound theorem. I. *Invent. Math.*, 88(1):125–151, 1987.

[65] T. Kalelkar. Euler characteristic and quadrilaterals of normal surfaces. *Proc. Indian Acad. Sci. Math. Sci.*, 118(2):227–233, 2008.

[66] M. A. Kervaire and J. Milnor. Groups of homotopy spheres. I. *Ann. of Math. (2)*, 77:504–537, 1963.

[67] W. Kimmerle and E. Kouzoudi. Doubly transitive automorphism groups of combinatorial surfaces. *Discrete Comput. Geom.*, 29(3):445–457, 2003.

[68] R. C. Kirby. *The topology of 4-manifolds*, volume 1374 of *Lecture Notes in Mathematics*. Springer-Verlag, Berlin, 1989.

[69] V. Klee. A combinatorial analogue of Poincaré's duality theorem. *Canad. J. Math.*, 16:517–531, 1964.

[70] V. Klee. On the number of vertices of a convex polytope. *Canad. J. Math.*, 16:701–720, 1964.

[71] H. Kneser. Geschlossene Flächen in dreidimensionalen Mannigfaltigkeiten. *Jahresbericht der deutschen Mathematiker-Vereinigung*, 38:248–260, 1929.

[72] M. Knödler. Kombinatorische Mannigfaltigkeiten mit transitiver Automorphismengruppe, 2007. In preparation.

[73] M. Kreck. An inverse to the Poincaré conjecture. *Arch. Math. (Basel)*, 77(1):98–106, 2001. Festschrift: Erich Lamprecht.

[74] W. Kühnel. Higher dimensional analogues of Császár's torus. *Results Math.*, 9:95–106, 1986.

[75] W. Kühnel. Minimal triangulations of Kummer varieties. *Abh. Math. Sem. Univ. Hamburg*, 57:7–20, 1986.

[76] W. Kühnel. Triangulations of manifolds with few vertices. In *Advances in differential geometry and topology*, pages 59–114. World Sci. Publ., Teaneck, NJ, 1990.

[77] W. Kühnel. *Tight polyhedral submanifolds and tight triangulations*, volume 1612 of *Lecture Notes in Math.* Springer-Verlag, Berlin, 1995.

[78] W. Kühnel. Centrally-symmetric tight surfaces and graph embeddings. *Beiträge Algebra Geom.*, 37(2):347–354, 1996.

[79] W. Kühnel. Topological aspects of twofold triple systems. *Exposition. Math.*, 16(4):289–332, 1998.

[80] W. Kühnel and T. F. Banchoff. The 9-vertex complex projective plane. *Math. Intelligencer*, 5(3):11–22, 1983.

[81] W. Kühnel and G. Lassmann. The unique 3-neighborly 4-manifold with few vertices. *J. Combin. Theory Ser. A*, 35(2):173–184, 1983.

[82] W. Kühnel and G. Lassmann. The rhombidodecahedral tessellation of 3-space and a particular 15-vertex triangulation of the 3-dimensional torus. *Manuscripta Math.*, 49(1):61–77, 1984.

[83] W. Kühnel and G. Lassmann. Neighborly combinatorial 3-manifolds with dihedral automorphism group. *Israel J. Math.*, 52(1-2):147–166, 1985.

[84] W. Kühnel and G. Lassmann. Combinatorial d-tori with a large symmetry group. *Discrete Comput. Geom.*, 3(2):169–176, 1988.

[85] W. Kühnel and G. Lassmann. Permuted difference cycles and triangulated sphere bundles. *Discrete Math.*, 162(1-3):215–227, 1996.

[86] W. Kühnel and F. H. Lutz. A census of tight triangulations. *Period. Math. Hungar.*, 39(1-3):161–183, 1999. Discrete geometry and rigidity (Budapest, 1999).

[87] N. H. Kuiper. Morse relations for curvature and tightness. In *Proceedings of Liverpool Singularities Symposium, II (1969/1970)*, volume 209 of *Lecture Notes in Math.*, pages 77–89, Berlin, 1971.

[88] G. Lassmann and E. Sparla. A classification of centrally-symmetric and cyclic 12-vertex triangulations of $S^2 \times S^2$. *Discrete Math.*, 223(1-3):175–187, 2000.

[89] S. Lefschetz. Intersections and transformations of complexes and manifolds. *Trans. Amer. Math. Soc.*, 28(1):1–49, 1926.

[90] W. B. R. Lickorish. *An introduction to knot theory*, volume 175 of *Graduate Texts in Mathematics*. Springer-Verlag, New York, 1997.

[91] F. H. Lutz. The Manifold Page. `http://www.math.tu-berlin.de/diskregeom/stellar`.

[92] F. H. Lutz. *Triangulated manifolds with few vertices and vertex-transitive group actions*. PhD thesis, TU Berlin, Aachen, 1999.

[93] F. H. Lutz. Triangulated manifolds with few vertices: Geometric 3-manifolds. `arXiv:math/0311116v1 [math.GT]`, 1999.

[94] F. H. Lutz. Császár's torus. *Electronic Geometry Model*, No. 2001.02.069, 2002. `http://www.eg-models.de/2001.02.069`.

[95] F. H. Lutz. Triangulated manifolds with few vertices: Combinatorial manifolds. `arXiv:math/0506372v1 [math.CO]`, 2005.

[96] F. H. Lutz. Equivelar and d-covered triangulations of surfaces. II. Cyclic triangulations and tessellations. `arXiv:1001.2779v1 [math.CO]`, 2010. To appear in Contrib. Discr. Math.

[97] F. H. Lutz, T. Sulanke, and E. Swartz. f-vectors of 3-manifolds. *Electron. J. Comb.*, 16(2):Research Paper 13, 33, 2009.

[98] A. Macbeath. Action of automorphisms of a compact Riemann surface on the first homology group. *Bull. London Math. Soc.*, 5:103–108, 1973.

[99] K. V. Madahar and K. S. Sarkaria. A minimal triangulation of the Hopf map and its application. *Geom. Dedicata*, 105:105–114, 2000.

[100] P. Martin. Cycles hamiltoniens dans les graphes 4-réguliers 4-connexes. *Aequationes Math.*, 14(1/2):37–40, 1976.

[101] P. McMullen. The maximum numbers of faces of a convex polytope. *Mathematika*, 17:179–184, 1970.

[102] P. McMullen. On simple polytopes. *Invent. Math.*, 113(2):419–444, 1993.

[103] P. McMullen and G. C. Shephard. *Convex polytopes and the upper bound conjecture*, volume 3 of *London Mathematical Society Lecture Note Series*. Cambridge University Press, London, 1971. Prepared in collaboration with J. E. Reeve and A. A. Ball.

[104] P. McMullen and D. W. Walkup. A generalized lower bound conjecture for simplicial polytopes. *Mathematika*, 18:264–273, 1971.

[105] J. Milnor. On manifolds homeomorphic to the 7-sphere. *Ann. of Math. (2)*, 64:399–405, 1956.

[106] J. Milnor. On simply connected 4-manifolds. In *Symposium internacional de topología algebraica International symposi um on algebraic topology*, pages 122–128. Universidad Nacional Autónoma de México and UNESCO, Mexico City, 1958.

[107] J. Milnor. A unique decomposition theorem for 3-manifolds. *Amer. J. Math.*, 84:1–7, 1962.

[108] J. Milnor and D. Husemoller. *Symmetric bilinear forms*, volume 73 of *Ergebnisse der Mathematik und ihrer Grenzgebiete*. Springer-Verlag, New York, 1973.

[109] E. E. Moise. Affine structures in 3-manifolds. V. The triangulation theorem and Hauptvermutung. *Ann. of Math. (2)*, 56:96–114, 1952.

[110] T. Motzkin. Comonotone curves and polyhedra. *Bull. Amer. Math. Soc.*, 63(1):35, 1957.

[111] J. R. Munkres. *Elements of algebraic topology*. Addison-Wesley Publishing Company, Menlo Park, CA, 1984.

[112] I. Novik. Upper bound theorems for homology manifolds. *Israel J. Math.*, 108:45–82, 1998.

[113] I. Novik and E. Swartz. Socles of Buchsbaum modules, complexes and posets. *Advances in Mathematics*, 222(6):2059–2084, 2009.

[114] U. Pachner. Konstruktionsmethoden und das kombinatorische Homöomorphieproblem für Triangulierungen kompakter semilinearer Mannigfaltigkeiten. *Abh. Math. Sem. Uni. Hamburg*, 57:69–86, 1987.

[115] G. Ringel. Triangular embeddings of graphs. In *Graph theory and applications (Proc. Conf., Western Michigan Univ., Kalamazoo, Mich., 1972; dedicated to the memory of J. W. T. Youngs)*, volume 303 of *Lecture Notes in Math.*, pages 269–281. Springer, Berlin, 1972.

[116] G. Ringel. *Map color theorem*, volume 209 of *Die Grundlehren der mathematischen Wissenschaften*. Springer-Verlag, New York, 1974.

[117] J. Riordan. The number of faces of simplicial polytopes. *J. Combinatorial Theory*, 1:82–95, 1966.

[118] V. A. Rohlin. New results in the theory of four-dimensional manifolds. *Doklady Akad. Nauk SSSR (N.S.)*, 84:221–224, 1952.

[119] C. M. Roney-Dougal. The primitive permutation groups of degree less than 2500. *J. Algebra*, 292(1):154–183, 2005.

[120] N. Saveliev. *Lectures on the topology of 3-manifolds: An introduction to the Casson invariant*. de Gruyter Textbook. Walter de Gruyter & Co., Berlin, 1999.

[121] E. Schulte and J. M. Wills. Geometric realizations for Dyck's regular map on a surface of genus 3. *Discrete Comput. Geom.*, 1(2):141–153, 1986.

[122] L. H. Soicher. GRAPE - GRaph Algorithms using PErmutation groups, a GAP package, Version 4.3. `http://www.gap-system.org/Packages/grape.html`, 2006.

[123] E. Spanier. The homology of Kummer manifolds. *Proc. AMS*, 7:155–160, 1956.

[124] E. Sparla. An upper and a lower bound theorem for combinatorial 4-manifolds. *Discrete Comput. Geom.*, 19(4):575–593, 1998.

[125] J. Spreer. Partitions of the triangles of the cross polytope into surfaces. `arXiv:1009.2642v1 [math.CO]`, 2010.

[126] J. Spreer. Supplemental material to the article "Partitions of the triangles of the cross polytope into surfaces". `arXiv:1009.2640v1 [math.CO]`, 2010.

[127] J. Spreer. Normal surfaces as combinatorial slicings. *Discrete Math.*, 311(14):1295–1309, 2011. `doi:10.1016/j.disc.2011.03.013`.

[128] J. Spreer and W. Kühnel. Combinatorial properties of the K3 surface: Simplicial blowups and slicings. *Experiment. Math.*, 20(2):201–216, 2011.

[129] J. Stallings. The piecewise-linear structure of Euclidean space. *Proc. Cambridge Philos. Soc.*, 58:481–488, 1962.

[130] R. P. Stanley. The upper bound conjecture and Cohen-Macaulay rings. *Studies in Appl. Math.*, 54(2):135–142, 1975.

[131] R. P. Stanley. The number of faces of simplicial polytopes and spheres. In *Discrete geometry and convexity (New York, 1982)*, volume 440 of *Ann. New York Acad. Sci.*, pages 212–223. New York Acad. Sci., New York, 1985.

[132] W. Stein et al. *Sage Mathematics Software (Version 4.6.1)*. The Sage Development Team, 2011. `http://www.sagemath.org`.

[133] E. Swartz. Face enumeration - from spheres to manifolds. *J. Europ. Math. Soc.*, 11:449–485, 2009.

[134] C. H. Taubes. Gauge theory on asymptotically periodic 4-manifolds. *J. Differential Geom.*, 25(3):363–430, 1987.

[135] G. Thorbergsson. Tight immersions of highly connected manifolds. *Comment. Math. Helv.*, 61(1):102–121, 1986.

[136] W. P. Thurston. *Three-dimensional geometry and topology. Vol. 1*, volume 35 of *Princeton Mathematical Series*. Princeton University Press, Princeton, NJ, 1997. Edited by Silvio Levy.

[137] S. Tillmann. Normal surfaces in topologically finite 3-manifolds. *Enseign. Math. (2)*, 54(3-4):329–380, 2008.

[138] D. W. Walkup. The lower bound conjecture for 3- and 4-manifolds. *Acta Math.*, 125:75–107, 1970.

[139] J. Weeks. SnapPea (Software for hyperbolic 3-manifolds), 1999. `http://www.geometrygames.org/SnapPea/`.

[140] J. Weeks. Computation of hyperbolic structures in knot theory. In *Handbook of knot theory*, pages 461–480. Elsevier B. V., Amsterdam, 2005.

[141] J. H. C. Whitehead. *Combinatorial homotopy. I*, volume 55. 1949.

[142] G. M. Ziegler. *Lectures on polytopes*, volume 152 of *Graduate Texts in Mathematics*. Springer-Verlag, New York, 1995.

Index

Jonathan Spreer

Schulbildung

1989–2002 *Waldorfschule Uhlandshöhe*, Stuttgart.

06/2002 **Abitur**, Leistungsfächer Mathematik und Deutsch.

Zivildienst

2002–2003 **Ausbildung zum Rettungssanitäter**, *Johanniter-Unfall-Hilfe e. V.*

2002–2006 **Anstellung als studentische Aushilfe**, *Johanniter-Unfall-Hilfe e. V.*

Studium

2002–2008 **Studium der Mathematik**, *Universität Stuttgart.*
Diplomstudiengang mit Nebenfach Informatik

2006–2007 **Studium der Mathematik**, *Université "Pierre et Marie Curie"*, Paris VI.
Masterstudiengang

06/2007 **Maîtrise**, *Université "Pierre et Marie Curie"*, Paris VI.
Vierjähriger französischer Abschluss

08/2008 **Diplom**, *Universität Stuttgart.*
Diplomarbeit: *Über die Topologie von kombinatorischen 4-Mannigfaltigkeiten, insbesondere der $K3$-Fläche*, Betreuer: Prof. Wolfgang Kühnel.

Wissenschaftliche Tätigkeit

2007–2008 **Studentische Hilfskraft am DFG-Projekt Ku 1203/5-2**, *Institut für Geometrie und Topologie*, Universität Stuttgart.

2008–2009 **Wissenschaftlicher Mitarbeiter**, *Institut für Geometrie und Topologie*, Universität Stuttgart.

seit 2010 **Mitarbeiter am DFG-Projekt Ku 1203/5-3**, *Institut für Geometrie und Topologie*, Universität Stuttgart.

Wissenschaftliche Arbeiten

[1] Effenberger, Felix und Jonathan Spreer: *simpcomp - a GAP toolbox for simplicial complexes.* ACM Communications in Computer Algebra, 44(4):186 – 189, 2010.

[2] Effenberger, Felix und Jonathan Spreer: *simpcomp - A GAP package, Version 1.4.* `http://www.igt.uni-stuttgart.de/LstDiffgeo/simpcomp`, 2011. Submitted to the GAP Group.

[3] Effenberger, Felix und Jonathan Spreer: *Simplicial blowups and discrete normal surfaces in the GAP package simpcomp*, 2011. To appear in ACM Communications in Computer Algebra.

[4] Spreer, Jonathan: *Über die Topologie von kombinatorischen 4-Mannigfaltigkeiten, insbesondere der K3-Fläche*, 2008. Diplomarbeit.

[5] Spreer, Jonathan: *Partitioning the triangles of the cross polytope into surfaces.* `arXiv:1009.2642v1[math.CO]`, preprint, 12 pages, 1 figure, 2010.

[6] Spreer, Jonathan: *Combinatorial 3-manifolds with cyclic automorphism group*, 2011. In preparation.

[7] Spreer, Jonathan: *Normal surfaces as combinatorial slicings.* Discrete Math., 311(14):1295–1309, 2011. `doi:10.1016/j.disc.2011.03.013`.

[8] Spreer, Jonathan und Wolfgang Kühnel: *Combinatorial properties of the K3 surface: Simplicial blowups and slicings.* Experiment. Math., 20(2):201–216, 2011. `http://www.igt.uni-stuttgart.de/LstDiffgeo/Spreer/K3`.

Pfaffenwaldring 57 – 70550 Stuttgart

☎ *+49/711/685-65303* • ✉ *jonathan.spreer@mathematik.uni-stuttgart.de*

• *http://www.igt.uni-stuttgart.de/LstDiffgeo/Spreer/*